PROGRESS IN

Nucleic Acid Research and Molecular Biology

Volume 23

PROGRESS IN
Nucleic Acid Research and Molecular Biology

edited by

WALDO E. COHN
Biology Division
Oak Ridge National Laboratory
Oak Ridge, Tennessee

Volume 23

1979

ACADEMIC PRESS
A Subsidiary of Harcourt Brace Jovanovich, Publishers

New York London Toronto Sydney San Francisco

ACADEMIC PRESS, INC.
111 Fifth Avenue, New York, New York 10003

United Kingdom Edition published by
ACADEMIC PRESS, INC. (LONDON) LTD.
24/28 Oval Road, London NW1 7DX

LIBRARY OF CONGRESS CATALOG CARD NUMBER: 63–15847

ISBN 0–12–540023–3

PRINTED IN THE UNITED STATES OF AMERICA

79 80 81 82   9 8 7 6 5 4 3 2 1

Contents

Participation of Modified Nucleosides in Translation and Transcription

B. Singer and M. Kröger

The Accuracy of Translation

Michael Yarus

Structure, Function, and Evolution of Transfer RNAs (with Appendix Giving Complete Sequences of 178 tRNAs)

Ram P. Singhal and Pamela A. M. Fallis

List of Contributors

Numbers in parentheses indicate the pages on which the authors' contributions begin.

PAMELA A. M. FALLIS (227), *Department of Chemistry, Wichita State University, Wichita, Kansas 67208*

DEZIDER GRUNBERGER (105), *Cancer Center/Institute of Cancer Research, Department of Biochemistry and Division of Environmental Sciences, Columbia University College of Physicians and Surgeons, New York, New York 10032*

ALEXANDER A. KRAYEVSKY (1), *Institute of Molecular Biology, The USSR Academy of Sciences, Moscow-334, 117984, USSR*

M. KRÖGER (151), *Institut für Biologie III der Universität, Schänzlestrasse 1, D 7800 Freiburg, West Germany*

MARINA K. KUKHANOVA (1), *Institute of Molecular Biology, The USSR Academy of Sciences, Moscow-334, 117984, USSR*

B. SINGER (151), *Department of Molecular Biology and Virus Laboratory, University of California, Berkeley, California 94720*

RAM P. SINGHAL (227), *Department of Chemistry, Wichita State University, Wichita, Kansas 67208*

GEOFFREY TURNOCK (53), *Department of Biochemistry, University of Leicester, Leicester, England*

I. BERNARD WEINSTEIN (105), *Cancer Center/Institute of Cancer Research, Department of Medicine and Division of Environmental Sciences, Columbia University College of Physicians and Surgeons, New York, New York 10032*

MICHAEL YARUS (195), *Department of Molecular, Cellular, and Developmental Biology, University of Colorado, Boulder, Colorado 80309*

Abbreviations and Symbols

All contributors to this Series are asked to use the terminology (abbreviations and symbols) recommended by the IUPAC-IUB Commission on Biochemical Nomenclature (CBN) and approved by IUPAC and IUB, and the Editor endeavors to assure conformity. These Recommendations have been published in many journals (*1*, *2*) and compendia (*3*) in four languages and are available in reprint form from the Office of Biochemical Nomenclature (OBN), as stated in each publication, and are therefore considered to be generally known. Those used in nucleic acid work, originally set out in section 5 of the first Recommendations (*1*) and subsequently revised and expanded (*2*, *3*), are given in condensed form (I-V) below for the convenience of the reader. Authors may use them without definition, when necessary.

I. Bases, Nucleosides, Mononucleotides

1. *Bases* (in tables, figures, equations, or chromatograms) are symbolized by Ade, Gua, Hyp, Xan, Cyt, Thy, Oro, Ura; Pur = any purine, Pyr = any pyrimidine, Base = any base. The prefixes S−, H_2, F−, Br, Me, etc., may be used for modifications of these.

2. *Ribonucleosides* (in tables, figures, equations, or chromatograms) are symbolized, in the same order, by Ado, Guo, Ino, Xao, Cyd, Thd, Ord, Urd (Ψrd), Puo, Pyd, Nuc. Modifications may be expressed as indicated in (1) above. Sugar residues may be specified by the prefixes r (optional), d (=deoxyribo), a, x, 1, etc., to these, or by two three-letter symbols, as in Ara-Cyt (for aCyd) or dRib-Ade (for dAdo).

3. *Mono-, di-, and triphosphates of nucleosides* (5′) are designated by NMP, NDP, NTP. The N (for "nucleoside") may be replaced by any one of the nucleoside symbols given in II-1 below. 2′-, 3′-, and 5′- are used as prefixes when necessary. The prefix d signifies "deoxy." [Alternatively, nucleotides may be expressed by attaching P to the symbols in (2) above. Thus: P-Ado = AMP; Ado-P = 3′-AMP.] cNMP-cyclic 3′:5′-NMP; Bt_2cAMP = dibutyryl cAMP, etc.

II. Oligonucleotides and Polynucleotides

1. Ribonucleoside Residues

(a) Common: A, G, I, X, C, T, O, U, Ψ, R, Y, N (in the order of I-2 above).

(b) Base-modified: sI or M for thioinosine = 6-mercaptopurine ribonucleoside; sU or S for thiouridine; brU or B for 5-bromouridine; hU or D for 5,6-dihydrouridine; i for isopentenyl; f for formyl. Other modifications are similarly indicated by appropriate *lower-case* prefixes (in contrast to I-1 above) (*2*, *3*).

(c) Sugar-modified: prefixes are d, a, x, or 1 as in I-2 above, alternatively, by *italics* or boldface type (with definition) unless the entire chain is specified by an appropriate prefix. The 2′-*O*-methyl group is indicated by *suffix* m (e.g., -Am- for 2′-*O*-methyladenosine, but -mA- for 6-methyladenosine).

(d) Locants and multipliers, when necessary, are indicated by superscripts and subscripts, respectively, e.g., -m$^6_2$A- = 6-dimethyladenosine; -s^{4}U- or -^{4}S- = 4-thiouridine; -ac^{4}Cm- = 2′-*O*-methyl-4-acetylcytidine.

(e) When space is limited, as in two-dimensional arrays or in aligning homologous sequences, the prefixes may be placed *over the capital letter*, the suffixes *over the phosphodiester symbol*.

2. Phosphoric Acid Residues [left side = 5′, right side = 3′ (or 2′)]

(a) Terminal: p; e.g., pppN . . . is a polynucleotide with a 5′-triphosphate at one end; Ap is adenosine 3′-phosphate; C > p is cytidine 2′ : 3′-cyclic phosphate (*1, 2, 3*); p < A is adenosine 3′ : 5′-cyclic phosphate.

(b) Internal: hyphen (for known sequence), comma (for unknown sequence); unknown sequences are enclosed in parentheses. E.g., pA-G-A-C(C_2,A,U)A-U-G-C > p is a sequence with a (5′) phosphate at one end, a 2′ : 3′-cyclic phosphate at the other, and a tetranucleotide of unknown sequence in the middle. (**Only codon triplets are written without some punctuation separating the residues.**)

3. Polarity, or Direction of Chain

The symbol for the phosphodiester group (whether hyphen or comma or parentheses, as in 2b) represents a 3′-5′ link (i.e., a 5′ . . . 3′ chain) unless otherwise indicated by appropriate numbers. "Reverse polarity" (a chain proceeding from a 3′ terminus at left to a 5′ terminus at right) may be shown by numerals or by right-to-left arrows. Polarity in any direction, as in a two-dimensional array, may be shown by appropriate rotation of the (capital) letters so that 5′ is at left, 3′ at right when the letter is viewed right-side-up.

4. Synthetic Polymers

The complete name or the appropriate group of symbols (see II-1 above) of the repeating unit, **enclosed in parentheses if complex or a symbol,** is either (a) preceded by "poly," or (b) followed by a subscript "n" or appropriate number. **No space follows "poly"** (2, 5).

The conventions of II-2b are used to specify known or unknown (random) sequence, e.g.,

polyadenylate = poly(A) or A_n, a simple homopolymer;

poly(3 adenylate, 2 cytidylate) = poly(A_3C_2) or $(A_3,C_2)_n$, an *irregular* copolymer of A and C in 3 : 2 proportions;

poly(deoxyadenylate-deoxythymidylate) = poly[d(A-T)] or poly(dA-dT) or $(dA\text{-}dT)_n$ or $d(A\text{-}T)_n$, an *alternating* copolymer of dA and dT;

poly(adenylate,guanylate,cytidylate,uridylate) = poly(A,G,C,U) or $(A,G,C,U)_n$, a random assortment of A, G, C, and U residues, proportions unspecified.

The prefix copoly or oligo may replace poly, if desired. The subscript "n" may be replaced by numerals indicating actual size, e.g., $A_n \cdot dT_{12-18}$.

III. Association of Polynucleotide Chains

1. *Associated* (e.g., H-bonded) chains, or bases within chains, are indicated by a *center dot* (not a hyphen or a plus sign) separating the *complete* names or symbols, e.g.:

poly(A) · poly(U) or $A_n \cdot U_m$
poly(A) · 2 poly (U) or $A_n \cdot 2U_m$
poly(dA-dC) · poly(dG-dT) or $(dA\text{-}dC)_n \cdot (dG\text{-}dT)_m$.

2. *Nonassociated* chains are separated by the plus sign, e.g.:

2[poly(A) · poly(U)] → poly(A) · 2 poly(U) + poly(A)
or $2[A_n \cdot U_m] \rightarrow A_n \cdot 2U_m + A_n$.

3. Unspecified or unknown association is expressed by a comma (again meaning "unknown") between the completely specified chains.

Note: In all cases, each chain is completely specified in one or the other of the two systems described in II-4 above.

IV. Natural Nucleic Acids

RNA	ribonucleic acid or ribonucleate
DNA	deoxyribonucleic acid or deoxyribonucleate
mRNA; rRNA; nRNA	messenger RNA; ribosomal RNA; nuclear RNA
hnRNA	heterogeneous nuclear RNA
D-RNA; cRNA	"DNA-like" RNA; complementary RNA
mtDNA	mitochondrial DNA
tRNA	transfer (or acceptor or amino-acid-accepting) RNA; replaces sRNA, which is not to be used for any purpose
aminoacyl-tRNA	"charged" tRNA (i.e., tRNA's carrying aminoacyl residues); may be abbreviated to AA-tRNA
alanine tRNA or $tRNA^{Ala}$, etc.	tRNA normally capable of accepting alanine, to form alanyl-tRNA, etc.
alanyl-tRNA or alanyl-$tRNA^{Ala}$	The same, with alanyl residue covalently attached. [*Note:* fMet = formylmethionyl; hence $tRNA^{fMet}$, identical with $tRNA_f^{Met}$]

Isoacceptors are indicated by appropriate subscripts, i.e., $tRNA_1^{Ala}$, $tRNA_2^{Ala}$, etc.

V. Miscellaneous Abbreviations

P_i, PP_i	inorganic orthophosphate, pyrophosphate
RNase, DNase	ribonuclease, deoxyribonuclease
t_m (not T_m)	melting temperature (°C)

Others listed in Table II of Reference 1 may also be used without definition. No others, with or without definition, are used unless, in the opinion of the editor, they increase the ease of reading.

Enzymes

In naming enzymes, the 1978 recommendations of the IUB Commission on Biochemical Nomenclature (*4*) are followed as far as possible. At first mention, each enzyme is described *either* by its systematic name *or* by the equation for the reaction catalyzed *or* by the recommended trivial name, followed by its EC number in parentheses. Thereafter, a trivial name may be used. Enzyme names are not to be abbreviated except when the substrate has an approved abbreviation (e.g., ATPase, but not LDH, is acceptable).

REFERENCES*

1. *JBC* **241**, 527 (1966); *Bchem* **5**, 1445 (1966); *BJ* **101**, 1 (1966); *ABB* **115**, 1 (1966), **129**, 1 (1969); and elsewhere.†
2. *EJB* **15**, 203 (1970); *JBC* **245**, 5171 (1970); *JMB* **55**, 299 (1971); and elsewhere.†
3. "Handbook of Biochemistry" (G. Fasman, ed.), 3rd ed. Chemical Rubber Co., Cleveland, Ohio, 1970, 1975, Nucleic Acids, Vols. I and II, pp. 3–59.
4. "Enzyme Nomenclature" [recommendations (1978) of the Nomenclature Committee of the IUB], Academic Press, New York, 1979.
5. "Nomenclature of Synthetic Polypeptides," *JBC* **247**, 323 (1972); *Biopolymers* **11**, 321 (1972); and elsewhere.†

Abbreviations of Journal Titles

Journals	*Abbreviations used*
Annu. Rev. Biochem.	ARB
Arch. Biochem. Biophys.	ABB
Biochem. Biophys. Res. Commun.	BBRC
Biochemistry	Bchem
Biochem. J.	BJ
Biochim. Biophys. Acta	BBA
Cold Spring Harbor Symp. Quant. Biol.	CSHSQB
Eur. J. Biochem.	EJB
Fed. Proc.	FP
Hoppe-Seyler's Z. physiol. Chem.	ZpChem
J. Amer. Chem. Soc.	JACS
J. Bacteriol.	J. Bact.
J. Biol. Chem.	JBC
J. Chem. Soc.	JCS
J. Mol. Biol.	JMB
Nature, New Biology	Nature NB
Nucleic Acid Research	NARes
Proc. Nat. Acad. Sci. U.S.	PNAS
Proc. Soc. Exp. Biol. Med.	PSEBM
Progr. Nucl. Acid Res. Mol. Biol.	This Series

* Contractions for names of journals follow.

† Reprints of all CBN Recommendations are available from the Office of Biochemical Nomenclature (W. E. Cohn, Director), Biology Division, Oak Ridge National Laboratory, Box Y, Oak Ridge, Tennessee 37830, USA.

Some Articles Planned for Future Volumes

The Peptidyltransferase Center of Ribosomes

Alexander A. Krayevsky
and
Marina K. Kukhanova

Institute of Molecular Biology
The USSR Academy of Sciences
Moscow, USSR

Progress in Nucleic Acid Research
and Molecular Biology, Vol. 23

ISBN 0-12-540023-3

Introduction

Ribosmes are characterized by two important properties: (*a*) the selection of substrates and their translocation during peptide chain synthesis; (*b*) the presence of several catalytic activities, the most significant of the latter being the peptidyltransferase activity. In such an intricate supramolecular "machine" as the ribosome, studies of the active center (the peptidyltransferase center) and its interrelation with other functions of ribosomes are very complicated. That is why it has been held for many years that the only function of the peptidyltransferase center is *to catalyze peptide bond formation*. However, it has become more and more obvious that the peptidyltransferase center fulfills other functions as well. On the one hand, the peptidyltransferase center is bifunctional: apart from peptide bond formation, it catalyzes hydrolysis of the ester bond in peptidyl-tRNA and hence is involved in the process of *termination*. On the other hand, recently available evidence suggests that processes occurring in this center in the course of peptide bond formation are connected with the act of *translocation*. There is every ground to believe that the interrelation between the peptidyltransferase center and other functional sites will be more apparent when we gain a better insight into the structure and functions of this center, which is a complex enzymic system with a number of peculiarities and differences from simpler enzymes.

This review deals with the description of the peptidyltransferase center: structural organization, substrate specificity, mechanism of peptide bond formation, and possible dynamic aspects of its functioning.

I. Structural Organization of the Peptidyltransferase Center of Ribosomes[1]

The region of the ribosome that catalyzes synthesis of peptide bonds is referred to as the peptidyltransferase center. It is located on the larger subunit and catalyzes peptide bond formation (Eq. 1) as well

[1] Abbreviations: tRNA = tRNA-C-C-A = total tRNA; $tRNA^{Phe}$ = $tRNA^{Phe}$-C-C-A = phenylalanine transfer RNA; tRNA-C-C-A(2′H), tRNA-C-C-A(3′H), tRNA-C-C-A(2′NH_2), tRNA-C-C-A(3′NH_2), tRNA-C-C-A(2′OMe), and tRNA-C-C-A(3′OMe) = tRNAs in which the 3′-terminal adenosine is replaced by 2′(3′)-deoxyadenosine, 2′(3′)-amino-2′(3′)deoxyadenosine, or 2′(3′)-methyladenosine, respectively. tRNA-C-C-A_{ox} and tRNA-C-C-A_{oxred} = tRNA after periodate oxidation and subsequent reduction with $NaBH_4$, respectively. Oligonucleotides, dinucleoside monophosphates, 5′-nucleotides, and nucleosides are designated likewise. For example, C-A-C-C-A = 3′-terminal pentanucleotide fragment of tRNA; C-A(2′H), pA(2′H), and A(2′H) = cytidylyl(3′ → 5′)-2′-

as hydrolysis of the peptidyl-tRNA ester bond at the termination of polypeptide synthesis (Eq. 2).

$$AA_n\text{-NHCHR'CO-tRNA'} + \text{NH}_2\text{CHR''CO-tRNA''} \rightarrow AA_n\text{-NHCHR'CO-NHCHR''CO-tRNA''} + \text{tRNA'} \quad (1)$$

where AA_n = peptide residue.

$$\text{polypeptide-NHCHRCO-tRNA} + H_2O \rightarrow \text{polypeptide-NHCHRCOOH} + \text{tRNA} \quad (2)$$

The peptidyltransferase center can be considered a bisubstrate enzyme. One of its substrates is the 3′ terminus of a peptidyl-tRNA, which serves as a peptide donor; this substrate is attached to the donor (or peptidyl) site. The other substrate is the 3′ terminus of an aminoacyl-tRNA (or water, upon the termination of polypeptide synthesis) acting as a peptide acceptor; it is bound to the acceptor (or amino acid) site of the center.

The peptidyltransferase center, as a catalytic system, has certain peculiarities.

1. It cannot be separated from the large subunit of the ribosome. Its organization involves a number of proteins and rRNA. That is why the study of this center must be carried out in a complex system.

2. The substrates of ribosomes, i.e., aminoacyl-tRNA and peptidyl-tRNA molecules, are macromolecules that possess several regions for binding to the ribosome carrying the template. Therefore, the structure of the peptidyltransferase center can be studied only with systems that allow one to discriminate between the

---

deoxyadenosine, 2′-deoxyadenosine 5′-phosphate, and 2′-deoxyadenosine, respectively. Other abbreviations: C2′ → 5′A = cytidylyl(2′ → 5′)adenosine; C(2′H)-A = 2′-deoxycytidylyl(3′ → 5′)adenosine; p*br*A = 8-bromoadenosine 5′-monophosphate; araA = arabino analog of adenosine; C3′p and C2′p = cytidine 2′-phosphate and cytidine 3′-phosphate; εC and εA = ethenocytidine and ethenoadenosine.

Amino acid derivatives: Phe-tRNA = tRNA-C-C-A-Phe = phenylalanyl-$tRNA^{Phe}$. C-A-C-C-A-Phe, C-A-Phe, pA-Phe, and A-Phe = mixture of the isomeric 2′ and 3′ esters of phenylalanine and C-A-C-C-A, C-A, pA, and A, respectively. pA(2′H)-Phe and pA(3′H)-Phe = 3′-phenylalanine and 2′-phenylalanine esters of pA(2′H) and pA(3′H) correspondingly. pA(3′NH-Phe) and pA(2′NH-Phe) = amides of phenylalanine with pA(3′NH_2) and pA(2′NH_2). pA(2′OMe)-Phe and pA(3′OMe)-Phe = 3′ and 2′ phenylalanine esters of pA(2′OMe) and pA(3′OMe). pA_{oxred}(3′Phe) and pA_{oxred}(2′Phe) = 3′- and 2′-phenylalanine esters of pA_{oxred}. A-(Phe)(Phe) = 2′,3′-di-*O*-phenylalanyladenosine. All amino acids are L unless specified as D.

Acylaminoacid and peptide derivatives are designated similarly. For example: tRNA-C-C-A(3′H)-(AcPhe) and tRNA-C-C-A(3′H)-(Gly_2Phe) are the 2′-esters of *N*-acetylphenylalanine and diglycylphenylalanine with tRNA-C-C-A(3′H); C-A-C-C-A-(fMet) is an *N*-formylmethionyl-pentanucleotide; pA-(fMet) = (2′(3′)-*O*-(*N*-formylmethionyl)adenosine 5′-phosphate; pA-(fDLeu-LLeu) = *N*-formyl-D-leucyl-L-leucyladenosine 5′-phosphate.

interaction of the substrate 3′ termini from the overall process of their binding.

3. The acceptor and donor sites of the peptidyltransferase center bind substrates that comprise elements with both constant (the C-C-A terminus in tRNA) and variable structures (the amino acid in aminoacyl-tRNA and the C-terminal fragment of the peptide in peptidyl-tRNA). Apparently, the center contains zones capable of recognizing substrate elements with a constant structure.

4. The 3′ end of peptidyl-tRNA is transferred from the acceptor site to the donor site, this being part of the overall translocation process.

In this section, we discuss recent publications concerned with the topography of the peptidyltransferase center. Unfortunately, the only direct method for determining the three-dimensional structure of the enzyme active centers, namely, X-ray diffraction analysis, is not yet suitable for studying the peptidyltransferase center of ribosomes. This necessitates a different strategy for elucidating its structure and functioning.

A. Principal Methods Used for the Study of Ribosomal Topography

All the methods used for establishing the structural organization of the peptidyltransferase center and the results gained thereby have been discussed in detail in recent publications (*1*, *2*). Therefore, we give below only a brief outline of the methods, and consider those findings pertinent to its organization. The most efficient of the methods are the following.

1. Partial Reconstitution

Treatment of 50 S subunits with different concentrations of LiCl or NH_4Cl in the presence of ethanol results in a series of protein-deficient cores. The amount of protein removed depends on the salt concentration used. The protein-deficient cores are then tested for binding and catalytic sites. In this way one may elucidate the involvement of individual proteins in various ribosomal activities.

Here, interpretation of the results is complicated. (*a*) The removed protein can affect the conformation of the ribosome as a whole; as a result the function is disturbed. (*b*) There is no standard technique for preparing homogeneous active ribosomes. Some evidence suggests that active ribosomes are more resistant to various agents than are inactive ribosomes. That is why the same treatment can in some cases remove a particular protein entirely, and in others only partially. These two facts seem to account for contradictory data obtained in different laboratories. This method makes it possible to determine precisely only those proteins not directly involved in the manifestation of functional activities. However, it does not furnish reliable information about the role of those components indispensable for the functional activity of ribosomes.

A somewhat different approach used in studying the role of proteins in the functioning of ribosomes is chemical modification of the isolated protein and subsequent assembly of the subunits (*3*). The advantage of such an approach is that the total structure of the ribosome is damaged far less than when the protein is entirely removed.

2. Affinity Labeling

The method of affinity labeling is based on the introduction, into a substrate or a specific inhibitor of ribosomes, of a chemical group capable of binding covalently to proteins or to rRNA. This method was used for the first time in 1971 in studies of the interaction of the chloroambucil derivative of Phe-tRNA with the ribosome (*4*). Analysis of covalently bound complexes allows one to detect proteins and rRNA regions in close proximity to the substrate binding site. The principles of affinity labeling and the experimental criteria of specific site labeling of ribosomes have been discussed (*1, 5*).

Reactive derivatives of four classes of substrate analogs and inhibitors, namely, the derivatives of peptidyl-tRNA, template oligonucleotides, GTP, and antibiotics, have been used for studying the topography of the active sites of ribosomes. Affinity reagents taken for labeling the structural elements of the peptidyltransferase center in *Escherichia coli* ribosomes are shown in Table I (*8–20*). As a rule, an affinity reagent reacts with several components in the peptidyltransferase center. This is because of the complexity of the structure as well as because the reactive group is well removed from bonds undergoing changes in the enzymic reaction. That is why the reagents are covalently bound not only to the components of the peptidyltransferase center but also to other elements of the molecule located close to this center. The affinity labeling of the ribosomal components as a function of reaction media has been analyzed (*1*); the main factor controlling the reaction of some of these reagents with ribosomal components is the concentration of monovalent cations in the labeling medium.

3. Antibodies and Bifunctional Reagents

Antibodies directed against all individual ribosomal proteins have been obtained (see *b*). The ribosomes treated by the antigen-binding fragments (Fab) of antibodies are then tested to see which ribosomal

TABLE I

PEPTIDYL-tRNAs USED FOR LABELING THE STRUCTURAL ELEMENTS OF THE PEPTIDYLTRANSFERASE CENTER IN *Escherichia coli* RIBOSOMES

Reagent	Symbol	Binding site	Labeling proteins: Major	Labeling proteins: Minor	23 S RNA	References
$BrCH_2CO$—Phe—tRNA	I	Donor	L2, L26/27	L6, (L14 + L16), S18	–	*8, 9*
		Donor	–	–	+ +	*10*
		Acceptor	L16	L2, L26/L27	–	*11*
$BrCH_2CO$—Met—$tRNA^{Met}$	II	Donor	L2	L27	–	*12*
		Acceptor	L2	L26/L27	–	
ICH_2CO—Phe—tRNA	III	Donor	–	–	+ +	*14*
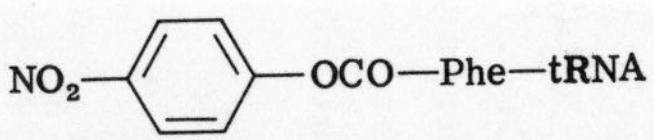	IV	Donor	L26/L27	L2, L16	–	*15, 16*
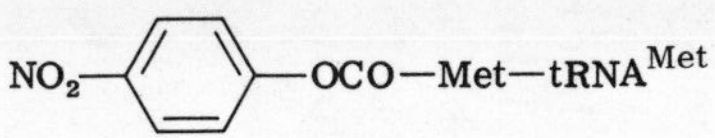	V	Donor	L27	(L14, L15)	–	*13, 17*
$BrCH_2CO$—$(Gly)_n$—Phe—tRNA	VI	Donor				*18*
n = 1–3			L2	L16, L26/L27	–	
= 3–6			L26/L27, L32/L33	L2, L16	–	
= 6–9			L26/L27	L24, L32/L33	–	
= 9–16			L24, L26/L27	L32/L33	–	
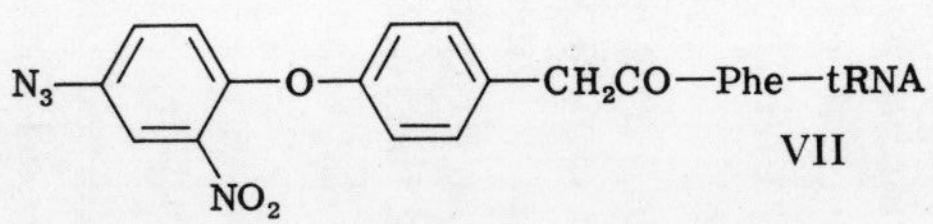	VII	Donor	L11, L18	L2, L5, L27, L32	–	*19*
	VIII	Donor	L11, L18	L1, L5, L27	–	*20*

function is inhibited by a particular Fab. This method revealed which proteins are functionally important. Electron microscopy studies of ribosomes after their reaction with antibodies has also supplied information about proteins residing on the surface of subunits. This method has been used to locate all proteins in the 30 S subunit and 19 proteins in the 50 S subunit of *E. coli* ribosomes (see *7*), as well as to detect proteins in the interface between the subunits.

Application of bifunctional reagents capable of cross-linking protein pairs has yielded the most extensive information on protein–protein neighborhoods. This method was also used to identify rRNA regions in the vicinity of proteins (see *6*).

It should be noted that neither of these methods by itself allows an unambiguous conclusion as to the functional significance of a particular group of ribosomal proteins or rRNA. Indeed, any modification of ribosomes can interfere with the total structure of a particle. Only the totality of data obtained by different methods may provide an answer as to the functional role of a particular component in the ribosomal structure.

B. Proteins of the Peptidyltransferase Center

The evidence from affinity labeling of the peptidyltransferase center has been discussed in detail in Cooperman (*1*). It suffices to note here that the derivatives of *N*-acetylaminoacyl-tRNA that interact with the 70 S ribosome reacted predominantly with the components of the 50 S subunit (Table I). Here, the proteins most modified upon binding of peptidyl-tRNA to the donor site were L2, L11, L18, and L27. Some affinity reagents listed below reacted almost exclusively with the 3′-terminal fragment of 23 S RNA (18 S fragment):

p-$(ClCH_2CH_2)_2NC_6H_4(CH_2)_3CO$-Phe-tRNA (IX) (*4*)

p-N_3-o-$NO_2C_6H_3CO$-Phe-tRNA (X) (*21*)

p-$N_3C_6H_4$-$CH_2CH[NHCOOC(CH_3)_3]CO$-Phe-tRNA (XI) (*22*)

p-$N_3C_6H_4CH_2CH[NHCOOC(CH_3)_3]$-CO-$(Gly)_n$-Phe-tRNA ($n = 2$, 4) (XII) (*23*)

$C_2H_5OOCCN_2CO$-Phe-tRNA (XIII) (*24*)

p-(C_6H_5CO)-$C_6H_4CH_2CH_2CO$-Phe-tRNA (XIV) (*25*)

The location of peptidyl-tRNA derivatives in the donor site has been evidenced, for compounds I, VI–IX, and XI, by their ability to react with puromycin or aminoacyl-tRNA after covalent binding to the ribosome.

Therefore, proteins L2, L11, L18, and L27 as well as a region in the 18 S fragment of 23 S RNA are the principal structural elements that compose the donor site (according to the data of affinity labeling). If a covalent complex with the ribosome is formed after transpeptidation when a peptide residue with the activated group has been transferred to Phe-tRNA in the acceptor site, the same components of the ribosome are modified in the case of compounds I, VI, and IX as when the reagent is located in the donor site. Consequently, the peptide residue does not change its position in the peptidyltransferase center instantaneously after transpeptidation. When I, IX, and XIII are immobilized in the acceptor site (in the presence of deacylated tRNA in the donor site), the same proteins are modified and, in addition, protein L16 (compound I).

It is far more difficult to elucidate the role of rRNA in the operation of ribosomes than that of ribosomal proteins. The method of reconstructing ribosomes from rRNA fragments is of limited application, as such fragments cannot participate in the formation of functionally active ribosomes. The method of studying mutants with a modified rRNA (*26, 27*) is more effective though it also entails a lot of difficulties. That is why the use of substrate analogs with activated groups is nearly the only way for studying the localization of rRNA fragments in the vicinity of the peptidyltransferase center.

The 18 S fragment of 23 S RNA structurally organizes both sites of the peptidyltransferase center (*23*). The interaction of analogs of compound XII having different lengths of the peptide moiety with the donor site has been studied (*23*). Irrespective of the chain length, these compounds reacted mainly with the 50 S subunit and 80% of the label was found in the 18 S fragment of rRNA. Thus, part of the nucleotide sequence in the 18 S fragment is located alongside the peptide chain.

Proteins L2, L11, L18, and L27, which can be labeled with activated analogs of peptidyl-tRNA, reside in the 50 S subunit close to each other, as has been found by electron microscopy using antibodies (Fig. 1) (*28*). Proteins L16, L23, L32/L33, L5, and L30 are located in close proximity to this region, and some of them seem also to be involved in the formation of the peptidyltransferase center. Cross-linking of proteins with bifunctional reagents and a number of functional tests (see *1*) have shown that proteins L1 and L16 as well as L11 and L18 are adjacent.

Affinity reagents also modify some components in the 30 S subunit, which confirms the conclusion that the 3′ terminus of tRNA has binding sites on both the 30 S and 50 S subunits.

Under certain conditions, reagent III alkylates protein S20 (*29*). After the reaction of I or II with 70 S ribosomes, one of the principal labeled proteins is S18 (*8, 18*).

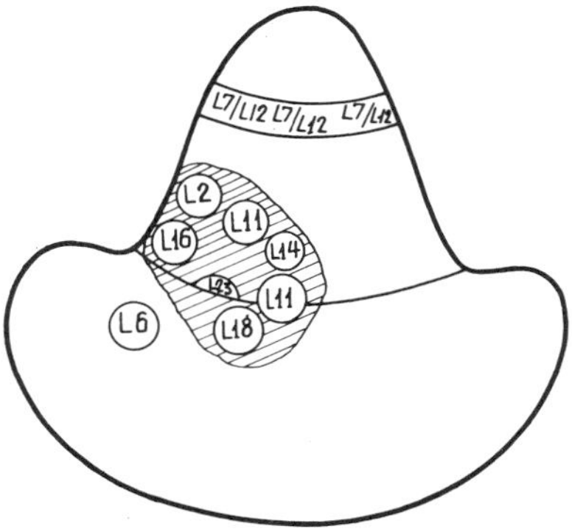

FIG. 1. Location of the proteins organizing the peptidyltransferase center on a three-dimensional model of the 50 S subunits (*7*).

C. Ribosomal Components Composing Sites That Bind Antibiotics

We consider only the antibiotics chloramphenicol and puromycin,[2] which are bound within the peptidyltransferase center. Two chloramphenicol analogs (XV and XVI) and puromycin derivatives and analogs (XVII–XX) were used to study the antibiotic binding site.

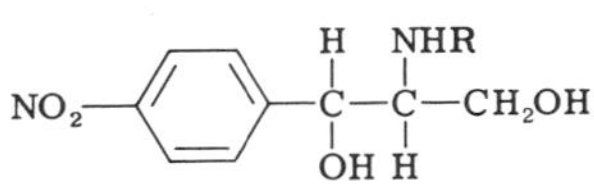

Chloramphenicol:	R = $COCHCl_2$
XV (*30*):	R = $COCH_2Br$
XVI (*31–32*):	R = $COCH_2I$

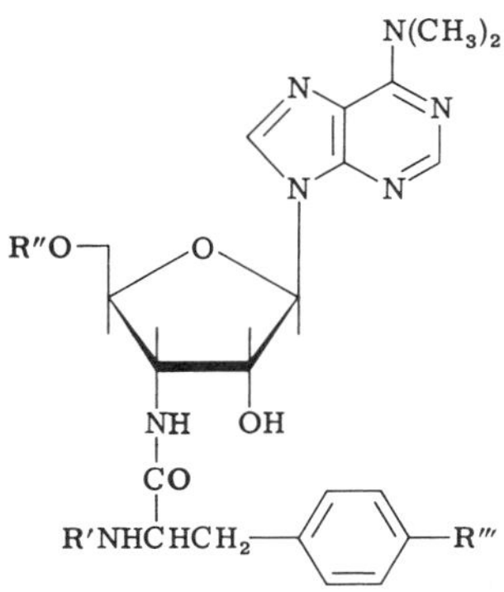

	R′	R″	R‴
Puromycin:	H	H	OCH_3
XVII (*33*):	ICH_2CO	H	OCH_3
XVIII (*34–35*):	H	$C_6H_4PO_3H$— (NH$COCH_2Br$)	OCH_3
XIX (*36*):	H	H	N_3
XX (*37*):	$EtOOCCN_2CO$	H	OCH_3

[2] See article on nucleoside antibiotics by Suhadolnik in Vol. 22 of this series.

Chloramphenicol analog XV reacted mainly with proteins L2 and L26/L27. Apparently, it was bound at both sites but mainly at the donor site. Analog XVI reacted predominantly with L16 and to a lesser extent with L24. The predominant role of L16 in binding chloramphenicol at the acceptor site was also confirmed by the method of partial reconstruction (*38*), and by using antibodies against the complex of chloramphenicol with the 50 S subunit (*39*).

Puromycin analogs were poor substrates for the peptidyltransferase center. All showed pronounced nonspecific binding with the components of ribosomes; up to 21 molecules of compound XIII were incorporated into a ribosome.

Puromycin itself was covalently bound to 70 S ribosomes under the action of UV (*37, 40*). This study yielded the most significant results. The main components of ribosomes specifically attached to puromycin were protein L23 and, presumably, 23 S RNA.

N-(Bromoacetyl)puromycin binds specifically to rat-liver ribosomes, reacting predominantly with proteins L27 and L29 (*41, 42*).

D. Peptidyltransferase Activity of Ribosomes

Although the catalysis of peptide bond formation by the ribosome is the subject of many studies, so far it has not been possible to identify unambiguously a protein (or a group of proteins) responsible for the peptidyltransferase activity, and to understand the mechanism of peptide bond formation.

By the method of partial reconstruction, protein L11 is responsible for the peptidyltransferase activity of ribosomes (*43, 44*). However, other authors, using ribosome cores lacking L11, have suggested that this protein is not essential for the reaction (*45–47*). Their study indicates that L16 is absolutely required for the transferase reaction, while L11 only somewhat stimulates it. Protein L11 also stimulates the activity of other ribosomal reactions: it is definitely involved in the processes of termination (*48*), binding of the antibiotic thiostrepton (*49, 50*), association of subunits (*51*), and formation of the binding sites for chloramphenicol (*38*) and the stringent factor (*52*). However, all of the above processes are also possible in the absence of L11.

The activity of 50 S protein-deficient cores prepared by treating ribosomes with 1 M NH_4Cl in 50% ethanol has been studied (*53*). It has been shown that (*a*) the transferase activity does not depend on the presence of protein L11 in the reactions of C-A-C-C-A-(fMet) with puromycin and AcPhe-tRNA with Phe-tRNA in the presence of a template; (*b*) the subunits are capable of associating in the absence of L11 since the template-dependent formation of AcPhePhe-tRNA takes place; (*c*) the deficient subunits are inactive in the process of translocation, which depends on the presence of the EF-G factor.

The above results seem, at first glance, to be inconsistent with others (*51*) in which protein L11 appeared necessary for the association of subunits. However, the complex [50 S = 30 S] is very unstable and, in the absence of formaldehyde, dissociates upon ultracentrifugation. That is why the complex was not found once L11 had been removed. Apparently, the removal of L11 causes such structural rearrangement of the 50 S subunits that several functions of the ribosome become disordered. The same conclusion has been reached by others (*3, 48*).

Isolated ribosomal proteins were chemically modified and then the subunit was reconstructed using an excess of the modified individual protein (*3*). (The proteins were modified with fluorescein whose fluorescence was measured in order to determine the incorporation of the protein into the subunit.) Modification of L11 did not inactivate entirely either the reaction of C-A-A-C-C-A-fMet with puromycin or the synthesis of polyphenylalanine. Modification of L13 and L17 also had no effect on the polymerase activity of the subunits. In contrast, modification of L 24 and L25 almost completely inactivated all functions of the ribosome that were studied. Presumably, these proteins are among those that determine the structure of the ribosome. Actually, L24 is bound to 23 S RNA in the early stages of reconstruction, and its modification may affect the following steps in this process (*54*). Protein L25 is attached to 5 S RNA and is necessary for the incorporation of 5 S RNA into the subunit during the assembly. Modification of L16 inhibited peptide bond formation, EF-G dependent hydrolysis of GTP, and binding of fragments. Therefore, L16 is required for the operation of the ribosome and, apparently, organizes the structure of the functional sites (*55*).

Similar studies on identifying the components of the catalytic sites are conducted with ribosomes of animal origin. Investigation of the activity of subunits prepared by treating rat-liver ribosomes with 0.6 M KCl has shown that the peptidyltransferase activity may be associated with one of three proteins, L21, L26, or L31 (*56*). In another work (*57*), 80 S ribosomes and 60 S subunits of rat liver were treated with 1 M NH_4Cl in the presence of 50% ethanol. The authors concluded that the peptidyltransferase activity is associated with one (or several) of the proteins L21, L24, L27, L28, and L36.

As follows from the data of chemical modification (*58*), the proteins of both subunits—the large one (L24, L27, L29, L7) and the small one (S4, S13, S18, S20, S21)—are involved in the formation of the peptidyltransferase center of rat liver ribosomes. Therefore, the peptidyltransferase center of animal ribosomes, like that of bacterial ribosomes, resides at the interface between the subunits. However, even more caution must be exercised in interpreting the data obtained with animal ribosomes, as compared to *E. coli* ribosomes, since the former are a more complex object comprising many proteins and undergoing intricate rearrangements.

E. The Role of 5 S RNA in the Functioning of *Escherichia coli* Ribosomes

The 5 S RNA plays a significant role in maintaining the total active conformation of ribosomes and in the manifestation of peptidyltransferase activity. When a 50 S subunit is reconstructed from proteins, 23 S RNA, and 5 S RNA (the process consisting of several steps), 5 S RNA can be incorporated into the subparticle at any step of the assembly (*54*). The particles are more active in a number of tests in the presence of 5 S RNA. Particles lacking 5 S RNA possess only 8–10% of the activity in the reaction of AcPhe-tRNA or C-A-C-C-A-(AcLeu) with puromycin[2]; they bind virtually no chloramphenicol, C-A-C-C-A-Leu←Ac, and the complex [aminoacyl-tRNA + EF-Tu +

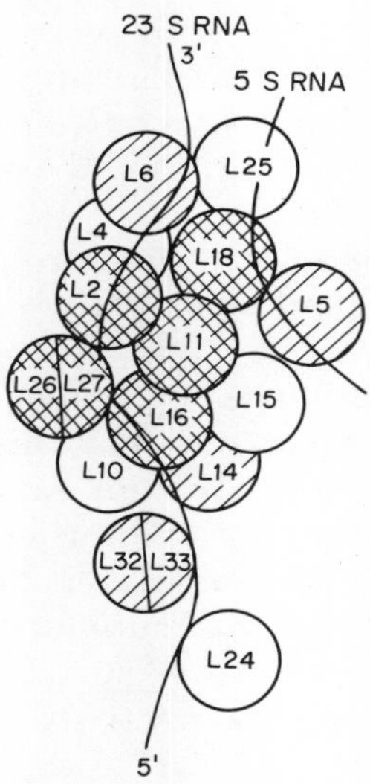

FIG. 2. Schematic illustration of the 50 S subunit proteins of the peptidyltransferase center. L2, L11, L18, L26/L27, L16, Proteins determined by several methods; L6, L5, L14, L32/L33, proteins found by affinity labeling or partial reconstitution.

GTP]. At the same time, they fix AcPhe-tRNA, the process depending on the initiation factor. It is obvious therefore that 5 S RNA is essential for the activity of the 50 S subunit.

By summarizing the observations concerning the structure of the peptidyltransferase center obtained using the above techniques, one can visualize the arrangement of proteins and rRNA that compose the peptidyltransferase center of *E. coli* ribosomes (Fig. 2).

F. Binding Sites for Protein Transfer Factors and Stringent Factor

The presence of transfer factors in the protein-synthesizing system is necessary for the ribosome to function. When these factors form a complex with ribosomes in the presence of GTP, the complex displays GTPase activity and GTP is hydrolyzed to GDP and orthophosphate. The functional sites of the ribosome that mediate, together with translation factors, GTP hydrolysis are located on the 50 S subunit in close proximity to the peptidyltransferase center. Moreover, some of the proteins in this center are involved in organizing the factor binding sites and in GTP hydrolysis.

Proteins L7/L12 play the key role in binding all transfer factors and in GTP hydrolysis (*59*, *60*). Besides L7/L12, protein BL3 from *Bacillus stearothermophilus*, the analog of *E. coli* protein L2, participates in the formation of the triple complex [ribosome + GTP + EF-G] (*61*), and L11 is necessary for the factor-dependent hydrolysis of GTP (*49*, *62*). The similarity between proteins L7/L12 and L11 and L2 involved in the formation of the ribosomal peptidyltransferase center was supported by direct experiments in which proteins L7/L12 were covalently cross-linked with bifunctional agents to L2 (*63*) and L11 (*64*), showing that factor EF-Tu reacts with proteins L1, L5, L7/L12, L15, L20, L23, and L28, some of which also compose the peptidyltransferase center (*65*, *66*).

Certain proteins of the peptidyltransferase center participate as well in the formation of the binding site for the stringent factor (*52*).

Therefore, the active sites of the ribosome are interrelated in such a manner that binding of one of the components in the protein-synthesizing system regulates the interaction of other components.

Unfortunately, the above methods are not suitable for determining the fine structure of the active sites in the ribosome and their functional interrelationship. This problem requires different approaches; some of those that are now in use are described in the following sections.

II. Substrate Specificity of the Peptidyltransferase Center

Two main strategies are employed in studying the substrate specificity of the acceptor and donor sites of the ribosomal peptidyltransferase center. The first approach makes use of aminoacyl-tRNAs and peptidyl-tRNAs modified at their 3′ termini. However, in this case, it is difficult sometimes to assess the contribution of these substrate components, which interact directly with the peptidyltransferase center, to the overall process of binding. Indeed, the binding of aminoacyl-tRNA and peptidyl-tRNA to the ribosome involves anticodons (at least for aminoacyl-tRNA) and other regions of these molecules, e.g., the T-Y-C-G sequence (*67–72*). That is why studies on the substrate specificity of the acceptor and donor sites using modified aminoacyl-tRNA and peptidyl-tRNA molecules are of limited value, although they do provide some information about the peptidyltransferase center (*73*).

The second approach employing model substrates—namely, the amino acid or peptide (acylamino acid) derivatives of nucleosides, nucleotides, or dinucleoside phosphates—is more suitable for investigating the substrate specificity of the peptidyltransferase center. These derivatives are referred to as "minimal model" substrates. The definition implies that these molecules contain structural elements in quantities minimal but sufficient for manifesting the peptide acceptor or peptide donor properties.

The peptide acceptor activity of aminoacyl-tRNA fragments was noticed for the first time by Takanami (*74*) and that of adenosine amino acid esters by Waller (*75*). At the same time, starting from 1960, the mechanism of action of puromycin[2] became the subject of intensive studies (*76–78*). Puromycin acts as an aminoacyl-tRNA analog, participating in the one-step process of peptide bond formation. Once peptidylpuromycin is produced, it leaves the ribosome in the process of translocation, thus terminating the protein synthesis.

Formation of peptidylpuromycin on the ribosome requires neither GTP nor transfer factors nor a template. The reaction proceeds promptly on 70 S and 80 S ribosomes as well as on 50 S and 60 S subunits.

Numerous observations on the mechanism involved in the binding of puromycin and the acceptance of a peptide from peptidyl-tRNA suggest that the antibiotic is bound to the same site of the peptidyltransferase center that, normally, binds the aminoacyl-adenosine residue at the 3′ terminus of aminoacyl-tRNA.

In 1971 Raake (*79*) proposed a hypothesis whose basic principles were as follows. Puromycin is bound at special site in the peptidyltransferase center of ribosomes and differs from aminoacyl-tRNA in the mechanism of action; puromycin has a sandwich-like structure; the adenine base and the methoxybenzyl residue are in stacking interaction, and, during the reaction with peptidyl-tRNA, additional stacking occurs with the 3′-terminal adenosine and cytidine of peptidyl-tRNA. However, these principles are inconsistent with published data. Puromycin in crystals has an extended conformation (*80*), so that the two aromatic components are located at a distance that prevents any stacking interaction between them (the conformation of puromycin will be considered in detail below). The conformation of puromycin determined by X-ray analysis has been corroborated by quantum-chemical calculations (*81*) and NMR spectra of puromycin in solution (*82*). The hypothesis of Raake has been tested in experiments by Žemlička (*83–85*) and did not fit in with the results obtained (*85*).

The 3′-terminal fragments of peptidyl-tRNA, namely, C-A-A-C-C-A-(fMet), A-A-C-C-A-(fMet), A-C-C-A-(fMet), C-C-A-(fMet), and some other fragments can serve as peptide donors instead of peptidyl-tRNA (*86–88*). The reaction of these fragments with puromycin is catalyzed by *E. coli* 70 S ribosomes and their 50 S subunits (*86–88*), 80 S ribosomes and 60 S subunits from human glands (*89*), from rat and mouse liver (*90*), from *Saccharomyces cerevisiae* (*91*), from rabbit reticulocytes (*92*), and from wheat germ (*93*). It does not require a template, GTP, and transfer factors, and is stimulated by either ethanol or methanol. These results show definitely that the binding site for the 3′ terminus of peptidyl-tRNA is located on the large subunit of the ribosome.

Subsequently, Phe-tRNA or C-A-C-C-A-Phe was used as peptide acceptor instead of puromycin, and pA-fMet served as peptide donor (*94, 95*). This reaction could proceed in the absence of a template, transfer factors, and CTP; it was catalyzed by *E. coli* 70 S ribosomes and 50 S subunits (*96–100*) as well as by 80 S ribosomes from rat liver (*101*). The scheme of the reaction is as follows:

$$\text{pA-(fMet)} + \text{X-Phe} \xrightarrow[\text{ribosome}]{} \text{(fMet)-Phe-X}$$

where X = tRNA or C-A-C-C-A.

Puromycin could also play the role of a peptide acceptor; in that case, ribosomes catalyzed the reaction between pA-(fMet) and puromycin (*94, 102*). A-(fMet) also possessed donor properties (*102*).

Cytidine 5′-phosphate (pC) effectively stimulates the reaction of pA-(fMet) with Phe-tRNA or C-A-C-C-A-Phe (*103*). The mechanism of the stimulation consists, apparently, in the binding of pC to a region in the donor site of the peptidyltransferase center, which is occupied usually by the second nucleotide from the 3′ terminus of peptidyl-tRNA, namely, by the cytidylic acid residue (*98*, *104*). As soon as pC is bound by the appropriate region of the donor site, the adsorption of pA-fMet is stimulated, presumably by the allosteric mechanism. pC does not stimulate the activity of puromycin (*103*, *104*).

A. Studies of the Peptidyltransferase Center Using Modified Aminoacyl-tRNA and Peptidyl-tRNA

This approach is of limited application, since specific modification of the 3′ terminus of aminoacyl-tRNA is rather difficult. Moreover, interpretation of the results is complicated by the fact that the conformation of the 3′ terminus of aminoacyl-tRNA is not clarified in detail.

The conformation of the 3′ end of tRNA and aminoacyl-tRNA has been studied by various methods; the data thus obtained are considered in the recent review of Sprinzl and Cramer (*73*). These authors concluded that the C-C-A terminus in nonacylated tRNA is single-stranded and that any secondary structure of this part of the molecule is due only to stacking interactions between the bases. The closer to the 3′ end, the weaker is stacking interaction; the 3′-terminal adenosine can move rather easily in respect to the next (cytidylic) residue and, possibly, is not even involved in any stacking interaction.

The effect that introduction of an amino-acid residue produces on the conformation of tRNA has also been studied by various methods. Unfortunately, the results obtained are different and occasionally even contradictory. At any rate, it would be safe to state that introduction of an amino-acid residue does not cause major changes in the conformation of tRNA. However, this does not rule out the possibility of minor conformational changes at the 3′ terminus, as found by some authors (*73*). Apparently, thermodynamic barriers between different conformers are not high, and a dynamic equilibrium exists in solution that shifts easily in the process of tRNA functioning. It is likely that these conformational changes occur only at the 3′ end of tRNA and involve the geometry of the stacking interaction of the 3′-terminal N-C-C-A or N-C-C-A-aminoacyl.

Hardly anything is known about the effect of the peptide residue on the structure of the 3′ terminus in peptidyl-tRNA.

The 3′ end has a considerable influence on the conformational stability of tRNA. If several nucleotides are cleaved from the 3′ terminus of tRNA, its melting point drops (*105*), suggesting a decrease in the stability of the tRNA secondary structure.

There are few publications on the effect of the 3′-terminal structure in aminoacyl-tRNA and peptidyl-tRNA molecules on binding to the ribosome; the results obtained are concerned mainly with the position and the nature of amino-acid residues.

The location of the amino-acid residue at the 2′- or 3′-hydroxyl of the 3′-terminal adenosine has been the subject of several studies. We shall not discuss this aspect, as it is covered in reviews (*73, 106*). Yet, there are convincing lines of evidence to argue that the amino-acid residue in the complex [aminoacyl-tRNA + EF-Tu + GTP] is located at the 2′-hydroxyl of aminoacyl-tRNA and, as such, is found in the ribosome.

Aminoacyl-tRNAs modified at their 3′ termini are used to study the binding sites for amino-acid residues in the acceptor region of the peptidyltransferase center in the ribosome (*73*).

Several preparations of modified phenylalanyl-tRNA were obtained in which the amino-acid residue could not migrate between the 2′- and 3′-hydroxyl of the 3′-terminal adenosine (*107–111*). A high peptide-accepting activity close to that of Phe-tRNA was found in analogs in which phenylalanine occupied the 3′- position, i.e., tRNA-C-C-A(2′H)-Phe (*108, 109*) and tRNA-C-C-A-(3′NH-Phe) (*111*); the experiments were conducted in a cell-free system with *E. coli* ribosomes. The activity of 2′ isomers, i.e., tRNA-C-C-A(3′H)-Phe and tRNA-C-C-A_{oxred} (2′-Phe), was reduced to a half to a third (*73, 112–114*). In the latter case, the peptide bond was also formed at a lower rate.

Since compounds with amino acid either in 2′ or in 3′ positions possess peptide accepting activity, the NH_2 group of the amino acid residue from aminoacyl-tRNA in the acceptor site of the ribosome must be located near the symmetry axis normal to the C2′—C3′ bond of the terminal adenosine (Fig. 3). This conclusion has been corroborated by the data of X-ray diffraction analysis of puromycin.

Similar conclusions have been reached in work with ribosomes from rabbit reticulocytes in a cell-free system (*115*).

All of the above modified Phe-tRNAs were acetylated at the amino-acid residue. None of the preparations served as a peptide donor (*109, 115*); this can be attributed to a low activation energy of the residue AcPhe in tRNA-C-C-A(2′H)(AcPhe), tRNA-C-C-A(3′H)-(AcPhe), tRNA-C-C-A_{oxred}-(2′AcPhe) (*108, 116, 117*), and tRNA-C-C-A(3′NH-AcPhe) (*111*). Thus, there are no direct indications about the peptide residue position (3′ or 2′) in peptidyl-tRNA located in the donor site.

As was shown using DTyr-$tRNA^{Tyr}$, the residue of D-tyrosine is incorporated into the polypeptide chain in a cell-free system with *E. coli* ribosomes and poly(A-U) as a template (*118*). The rate of

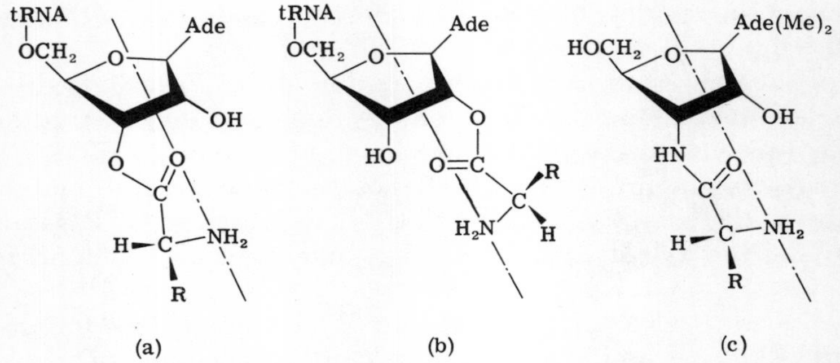

FIG. 3. Possible stereochemical conformation for the aminoacyl residue in 3′(a) or 2′(b) position of tRNA with identical position of the amino group. Conformation of puromycin in crystals (c).

D-tyrosine incorporation was 15–20% of that for L-tyrosine, but nevertheless D-tyrosine was discovered in the middle of the polypeptide chain as well as at its N- and C-termini.

Little progress has been made toward an understanding of how the nature of the amino-acid residue of aminoacyl-tRNA influences the peptide accepting activity. An explanation of this may be that other binding sites of aminoacyl-tRNA do not allow this contribution of the 3′ end of aminoacyl-tRNA to be apparent. However, the effect of the amino-acid residue of acetylaminoacyl-tRNA on the peptide donor activity in a system with *E. coli* ribosomes was studied in the absence of a template (*119*). AcLeu-tRNA served as a highly effective donor, and low peptide donor activity was displayed by AcGly-tRNA, AcPhe-tRNA, and AcPro-tRNA. Acetyldipeptidyl-tRNAs were slightly more active than acetylaminoacyl-tRNAs, and individual properties of their amino-acid residues were less pronounced. Such a great dependence of the peptide donor activity on the nature of the amino-acid residue of acetylaminoacyl-tRNA, in a system without a template, but in the presence of alcohol, shows that these substances apparently react with the ribosome only at the peptidyltransferase center.

An increase in the number of amino-acid residues in the peptidyl-tRNA enhances its affinity for the donor site of ribosomes (*120–122*). This cannot be attributed to a simple increase in the hydrophobicity of the peptide, since the capacity to be bound decreases as the hydrophobicity of the acetyl group rises in the series (*120*):

$$CH_3CO\text{-Phe-tRNA} > CH_3(CH_2)_6CO\text{-Phe-tRNA} > CH_3(CH_2)_{10}CO\text{-Phe-tRNA}$$

The peptide donor activity in a system without a template is manifested by dipeptidyl-tRNAs carrying, at the N-terminus of the peptide residue, bulky radicals of dyes (5-dimethylaminonaphthalenesulfo, anthracenyl, fluoresceinyl); this indicates that free spaces exist in the donor site (*123*). Apparently, the donor site of the peptidyltransferase center resides at the surface of the subunit, as peptidyl-tRNAs comprising up to 17 amino acid residues can be bound to the donor site of the ribosome from the cytoplasm (*18*). Moreover, interesting results have been obtained with a peptidyl-tRNA carrying a bis(chloroethyl)amino group and capable of covalent binding to ribosomes (*124*).

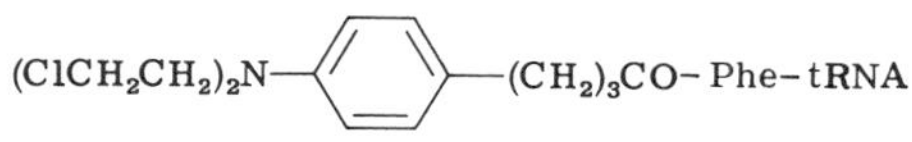

Once such a peptidyl-tRNA is bound covalently to the ribosome, mainly to 23 S RNA, it can react with Phe-tRNA, and the nascent peptide chain may contain 20–25 amino-acid residues.

Few researches have been published describing modification of the bases at the 3′ terminus of aminoacyl-tRNA. In two, the 3′-terminal adenosine was substituted by formycin (*125*) and tubercidin (*126*). Both modified tRNAs were capable of aminoacylating, binding to the acceptor site of *E. coli* ribosomes, accepting peptides, and participating in the overall process of polyphenylalanine synthesis, though at lower rates. In contrast, if one cytidine residue at the 3′ end of aminoacyl-tRNA was substituted by uridine, the tRNA-U-C-A-Phe entirely lost its ability to bind to the ribosome in the presence of [EF-Tu + GTP], and tRNA-U-C-A-(AcPhe) could not serve as a peptide donor (*127*).

B. Studies of the Peptidyltransferase Center Using Low Molecular Weight Substrates and Inhibitors

Minimal donors and acceptors and corresponding inhibitors of the peptidyltransferase center were much more informative for investigation of the role that individual components of these molecules play in the manifestation of biological activity.

First of all, it should be noted that minimal acceptors and donors are studied, as a rule, in model systems *in vitro*. Various systems with model acceptors (including puromycin) used at present are: systems with an endogenous template; polysomes on synthetic templates with oligopeptidyl-tRNA or acetylaminoacyl-tRNA; the reaction of acetylaminoacyl-tRNA or its 3′-terminal fragment in systems without a template in the presence of ribosomes. The activity of model donors is tested in systems without templates. Puromycin, aminoacyl-tRNA or its 3′-terminal pentanucleotide fragment serve as acceptors.

As has become evident from numerous publications, the activity of a model substrate depends on a testing system. The affinity of the substrate for the peptidyltransferase center changes, probably owing to complex mutual influences in the ribosome. For instance, the K_m for puromycin varies within the range of 10^{-4} to 10^{-6} M in passing from monosome to polysome (*128–130*). Therefore, caution should be exercised in interpreting the results of experiments in which the activity of a substrate has been determined in different systems.

During the last decade, minimal model acceptors have been tested in systems with ribosomes from *E. coli*, rat liver, pig uterus, rabbit reticulocytes, yeasts, etc. (*131–135*). Comparison of the results obtained with ribosomes from different sources shows that the behavior has the same general pattern irrespective of the type of ribosomes. Far less is known about minimal donors; they are active in systems with ribosomes from *E. coli* (*94–100*) and rat liver (*101*).

The general formula of compounds that serve as minimal substrates can be represented as follows:

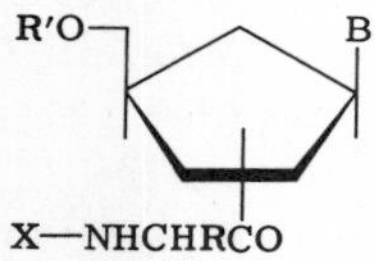

X = H for a minimal acceptor, and X = acyl or peptidyl for a minimal donor.

The base, the structure and configuration of the sugar residue, the substituent in the 5′ position and the amino-acid residue have been varied in the structures of minimal substrates. We follow this order in the presentation below.

1. The Base (B)

The effect of the base structure on the peptide acceptor activity of model substrates was studied with the phenylalanine esters of nucleosides (N-Phe) while its effect on the peptide donor activity was tested with the esters of *N*-formylmethionine and nucleoside 5′-phosphates [pN-(fMet)]. The substrate specificity of model substances depended to a significant extent on the nature of the base (Table II) (*95, 96, 136–140*).

As can be seen from Table II, the derivatives of adenine and hypoxanthine are most active among the substrates involving natural bases, which act either as peptide acceptors or as peptide donors. Apart from these features in common, some differences are note-

TABLE II
SUBSTRATE SPECIFICITY OF MODEL PEPTIDE ACCEPTORS (N-PHE) AND DONORS (pA-fMet) WITH DIFFERENT NUCLEIC BASES[a]

Base in substrate	Activity of	
	Acceptor	Donor
Ade	+++ (*136, 137*)	+++ (*95, 96*)
Hyp	++ (*136, 137*)	++ (*95, 96*)
Cyt	+ (*136, 137*)	− (*95, 96*)
Gua	− (*136, 137*)	+ (*95, 96*)
Ura	− (*136, 137*)	− (*95, 96*)
*br*Ade	+− (*138, 140*)	++ (*140*)
εAde	+++ (*139*)	− (*140*)
εCyt	++ (*139*)	
cS-Ade	++ (*139*)	− (*140*)
cO-Ade	++++ (*139*)	− (*140*)

[a] References are in parentheses.

worthy. εA-Phe and εC-Phe are active peptide acceptors; pεA-(fMet), however, lacks the donor activity and does not inhibit the reaction of pA-(fMet) with Phe-tRNA, which indicates that this compound does not interact at all with the donor site. Another difference between the substrates of the donor site and those of the acceptor site appears in nucleosides with restricted conformational flexibility. Compounds with an "anti-like" conformation (XX and XXI) are highly active peptide acceptors; XXI even exceeds A-Phe and puromycin in activity (S. G. Zavgorodny and B. P. Gottikh, personal communication). Yet, XX and XXI neither have the peptide donor activity nor react with the donor site [they do not compete with pA-(fMet)]. In contrast, p*br*A-(fMet) displays a high peptide donor activity (*140*). The bromine atom at C8 of the adenine does not interfere with the binding of p*br*A-to the acceptor site of the peptidyltransferase center, since p*br*A competes with the substrate of the acceptor site, C-A-C-C-A-Phe (*140*). It is known that *br*A and p*br*A have a "syn-like" conformation in crystals (*141*) and in aqueous solution (*142*). Incorporation of an acyl residue into the molecule of *br*A or p*br*A does not change the "syn-like" conformation, as was shown by NMR spectroscopy (*140*).

The amino group at C6 of the adenine in A-Phe can be substituted by H, $N(CH_3)_2$, CH_3O, and CH_3S groups (*138*) or by N_3 (*143*) without any noticeable loss of the activity. A high peptide acceptor activity is displayed by 3′(2′)-*O*-phenylalanyltubercidin (7-deaza-adenosine)[2] (*138*).

The following conclusions can be drawn from these results. The substrate specificity of both sites of the peptidyltransferase center toward the base is not absolute. Several purines and one pyrimidine, anomalous bi- and tricyclic structures are recognized. There is no specificity for any structural element of the base. However, the acceptor and donor sites differ in their specificity with respect to the conformation of model substrates.

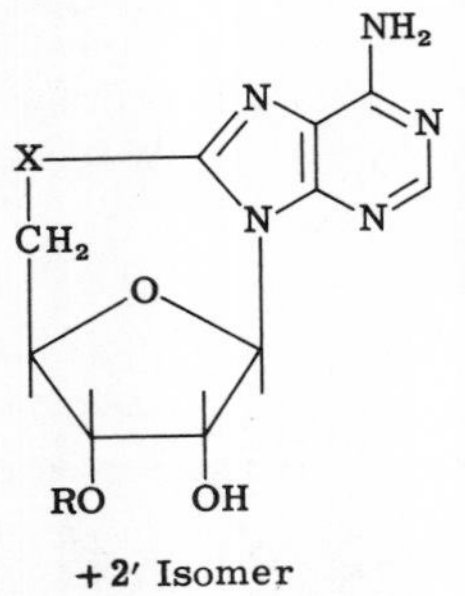

XX: X = S; a: R = Phe; b: R = fMet
XXI: X = O; a: R = Phe; b: R = fMet

2. The Sugar Residue

The role the structure of the sugar residue plays in manifesting the activity of peptide model acceptors has been reported in many papers; less evidence is available regarding substrates of the donor site. If the 2′-hydroxyl is methylated, as in the model acceptors A(2′OMe)-Phe and A(2′OMe)-Tyr (*144, 145*) or substituted by hydrogen, as in C-A(2′H)-Phe (*146–148*), the acceptor activity remains almost unchanged. However, the absence of the hydroxyl in A(2′H)-Phe (*137*), or its inversion as in araA-(3′Phe) (*138*), entirely deprives these compounds of the capacity to accept a peptide.

The type of bond (ester or amide) between the amino acid residue and ribose has only a slight effect on the peptide acceptor activity of model compounds (*131*).

Interesting results have been obtained with compounds in which an amino acid residue is located at the 2′-hydroxyl. C-A(3′H)-Phe, C-A(3′H)-Gly, C-A(3′OMe)-Phe, and C-A(3′OMe)-Gly can bind to the acceptor site of the peptidyltransferase center, whereas C-A(3′H), C-A(3′OMe), and C-A are not bound at the same concentrations (*117, 146*). These data may indicate that either a special region exists for binding an amino-acid residue in the 2′ position of the acceptor site, or a conformation of the nucleotide fragment more favorable for binding to the acceptor site is stabilized owing to the effect of the amino-acid residue. Apparently, C-A(3′H)-Phe and its analogs can bind to the ribosome because the NH_2 groups of their amino-acid residues react with the same locus of the acceptor site that fixes the NH_2 group of the amino-acid residue in the active acceptor, e.g., C-A(2′H)-Phe. However, only when the amino-acid residue occupies the 3′ position is the NH_2 group located in a way that favors the transfer of the peptide residue.

The following peptide model donors have been tested: pA(2′OMe)-(fMet), pA(3′OMe)-(fMet), C-A(2′OMe)-(fMet), and C-A-(3′OMe)-(fMet) (*95*, *149*). None of the compounds possesses the peptide donor activity or competes in the reaction with C-A-(fMet), i.e., they practically do not bind even to the donor site of the peptidyltransferase center.

A high peptide acceptor activity of A-(Phe)(Phe) and a noticeable activity of A-(Ala)(Ala) and A-(Val)(Val) are noteworthy (*150*). Thus, there must be a space (or binding loci?) in the acceptor site of ribosomes for amino-acid residues located at the 2′ position of the ligand.

The data on the effect of the 2′-substituent in the terminal adenosine are summarized in Table III.

The carbo analogs of puromycin[2] in which ribose is substituted by 2-hydroxy-3-aminocyclopentane (*151*) or 4-methyl-2-hydroxy-3-aminocyclopentane (*152*) display a high peptide acceptor activity in systems with *E. coli* ribosomes. Therefore, neither the endocyclic oxygen of ribose nor the 5′-hydroxymethyl group are of particular importance for binding a model acceptor to the ribosome.

A_{oxred}-(3′Phe) has a high peptide donor activity; its K_m is only one order of magnitude higher than the K_m of puromycin in systems

TABLE III

SUBSTRATE ACTIVITY OF MODEL ACCEPTORS AND DONORS MODIFIED AT SUGAR RESIDUES[a]

Acceptor site		Donor Site	
Compound	Activity	Compound	Activity
A-Phe	++	pA-fMet	++
A(2′OMe)-Phe, A(2′OMe)-Tyr	+	pA(2′OMe)(fMet)	−
A-(Phe)(Phe)	++	pA(3′OMe)(fMet)	−
A(2′H)-Phe	+−		
araA-(3′Phe)	−		
C-A-Phe	+++	C-A-fMet	+++
C-A(2′OMe)-Phe	++	C-A(2′OMe)-(fMet)	−
C-A(2′H)-Phe	+++	C-A(3′OMe)-(fMet)	−

[a] Abbreviations are defined in footnote 1 (p. 2).

with *E. coli* monosomes and polysomes (*153, 154*). However, A_{oxred}-(2′Phe) does not accept peptides (*146, 151*), while pA_{oxred}-(3′fMet) and pA_{oxred}-(2′fMet) do not display peptide donor activity (*95*).

3. Substituents in the 5′-Position

As noted in 1967, only the last compound in the series A-Gly, pA-Gly, and C-A-Gly has substrate activity, being slightly inferior to puromycin (*155*). U-A-Gly, U-U-Gly, and C(2′H)-A-Gly do not accept peptides (*155*). It follows, therefore, that the nature of the 5′-substituent is essential for the manifestation of the substrate activity.

The role of this substituent has been the object of further studies. pA-Phe is more active than A-Phe (*156*); Cm-A-Phe[3] in a system with *E. coli* ribosomes (*157*) and C-A(3′NH-Phe) in systems with ribosomes from *E. coli* and rat liver (*131, 133*) are rather more active than A(3′NH-Phe) or puromycin. Yet, C2′ → 5′A(3′NH-Phe), an analog with the abnormal bond 2′ → 5′, accepts a peptide almost two orders of magnitude less than does C-A(3′NH-Phe) in a system with ribosomes from rabbit reticulocytes (*158*). Um-A-Phe[3] and Am-A-Phe[3] do not accept peptides (*156*). U-A(3′NH-Gly), C-A(3′NH-Gly), and A(3′NH-Gly) are also not active (*131, 133, 159*).

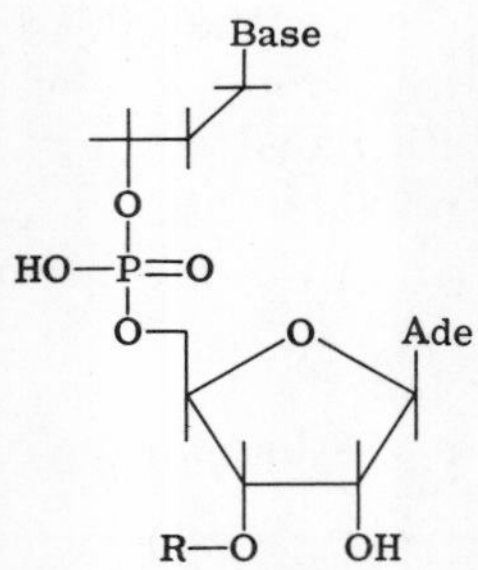

+2′ Isomer

R = Phe
XXIIa: Base = Cyt Cm–A–Phe
XXIIb: Base = Ura Um–A–Phe
XXIIc: Base = Ade Am–A–Phe

R = fMet
XXIId: Base = Cyt Cm–A–(fMet)
XXIIe: Base = Ura Um–A–(fMet)

The substituent in the 5′ position of a minimal donor also has a considerable effect on the manifestation of peptide donor activity (see Table IV) (*97, 102*). The 5′-phosphate group in pA-(fMet) increases the activity two- to threefold, although this effect can be caused by a considerable rise in the solubility of pA-(fMet), as compared to A-(fMet), in the incubation medium. The residue of cytidylic acid in the 5′

[3] As shown by CD spectroscopy, Cm-A-Phe, Um-A-Phe, and Am-A-Phe do not differ significantly in conformation in solution from C-A-Phe, U-A-Phe, and A-A-Phe, respectively.

TABLE IV
DEPENDENCE OF THE PEPTIDE DONOR ACTIVITY AND THE PEPTIDE ACCEPTOR ACTIVITY OF MODEL SUBSTRATES ON THE NATURE OF THE 5′-SUBSTITUENT[a]

Acceptor site		Donor site	
Compound	Activity	Compound	Activity
A-Phe, pA-Phe	++	A-(fMet)	+
C-A-Phe	+++	pA-(fMet)	++
Cm-A-Phe	+++	C-A-(fMet)	+++
		Cm-A-(fMet)	+
C-A-Gly, C-A(3′NH-Gly)	+		
pA(3′NH-Gly)	−		
A-A(3′NH-Gly), G-A(3′NH-Gly)	−		
U-A(3′NH-Gly), C(2′H)-A-Gly	−	C(2′H)-A-(fMet)	++
Um-A-Phe, Am-A-Phe	−	Cm-A-(fMet)	−
C(2′-5′)A-Phe	+−		

[a] Abbreviations are defined in footnote 1 (p. 2).

position raises the activity by nearly two orders of magnitude as compared to that of pA-(fMet). C(2′H)-A-(fMet) is also very effective.

Important information about the effect of the 5′-substituent on the peptide donor activity of pA-(fMet) has been obtained from studies of the stimulation of its reaction. As has been mentioned above, the most probable mechanism of stimulation consists of binding the stimulant to a region of the donor site normally occupied by a nucleotide second from the 3′-terminus of peptidyl-tRNA. The effect the nature of a stimulant exerts on its function is presented in Table V (*97, 104, 160*).

TABLE V
EFFECT OF THE STRUCTURE OF COMPOUNDS ON THEIR STIMULATING ACTIVITY[a,b]

Compounds with high stimulating activity	Compounds with low stimulating activity	Compounds without stimulating activity
pC, p_2C, p_3C	C, Cyt	pA, pU
cm^5C	pC	
p-s^4U		
C-C		A-C, U-C
pC(2′H)	C(2′H)	
pC(3′NH_2), pC(2′NH_2)	C(3′H)	pC_{oxred}, (C2′p + C3′p)
p-m^4C	p-m_2^4C	araC, pεC

[a] From Kotusov *et al.* (*97, 104*) and Cerna *et al.* (*160*).
[b] Abbreviations are defined in footnote 1 (p. 2).

As can be seen from Table V, 5′-phosphate increases the activity of a stimulant, and the activity decreases slightly in the series $p_2C > pC > p_3C$. The activity of cytosine is roughly half that of cytidine and one-fifth that of pC. Besides pC, pG is also active but not in all variants of the system; pA and pU do not stimulate the reaction, nor do C3′p, C2′p, pC_{oxred} or araC. The stimulating activity of $pC(3'NH_2)$ and $pC(2'NH_2)$ is even higher than that of pC. The NH_2 group of cytosine is essential for the activity: its methylation to a monomethylamino group halves the stimulating activity; dimethylation decreases the activity to an eighth; introduction of an ethene group results in a complete loss of the pC activity.

4. The Amino-Acid Residue

The dependence of the peptide acceptor activity on the nature and position of an amino-acid residue in minimal peptide acceptors has been studied in detail in various cell-free systems containing ribosomes from *E. coli*, yeasts, rat liver, and rabbit reticulocytes (*131, 133, 135–137, 159, 161*). These studies show that the amino-acid residue must be located at the 3′ hydroxyl rather than at the 2′ or 5′ hydroxyl. It must be an α-amino acid since compounds with a more remote NH_2 group do not possess acceptor activity. The amino-acid residue must have the L- rather than the D-configuration. Moreover, the sidechain of the amino-acid residue is also essential for the peptide acceptor activity of adenosine or adenylic esters. A-Phe and A-Tyr are very effective acceptors of peptides in all of the studied systems. The esters of A-Ser, A-Leu, and A-Met are comparatively good peptide acceptors. The activity of A-Glu, A-Pro, and A-Val is low. A-Gly and A-Trp practically do not accept peptides. The activity of the A-Lys and A-Ala changes to a great extent depending on the system being used (Table VI).

The contribution of the amino-acid residue in C-A-aminoacyl activity is manifested also, though to a lesser degree. C-A-Phe is a good peptide acceptor whereas C-A-Pro and C-A-Asp are less effective in a system with ribosomes from pig uterus (*162*). Binding to the accepter site decreases in the series C-C-A-Phe > C-C-A-Leu > C-C-A-Lys > C-C-A-Ser > C-C-A-Glu > C-C-A-Ala (*163*).

Many authors have speculated on what causes the binding and the peptide acceptor activity of model substrates to depend on the amino-acid residue. Rychlik was the first to suggest that the acceptor site has a hydrophobic pocket for fixing the aromatic radicals of phenylalanine and tyrosine as well as an anionic center for binding the side radical of lysine (*137, 164*). This suggestion is inconsistent with

TABLE VI
DEPENDENCE OF SUBSTRATE ACTIVITY ON THE NATURE AND CONFIGURATION OF AMINO-ACID RESIDUE

Compounds with typical substrate activity	Acceptor site[a]	Donor site[a]
High	C-A-Phe, A-Phe, A-Tyr A(3′NH-Phe), A(3′NH-Tyr)	pA-(fMet), pA-(fLeu) pA-(fLLeu-LLeu) pA-(fDLeu-LLeu)
Moderate	C-A-Pro, C-A-Asp, A-Lys A-Met, A-Ala, A-Leu, A-Val, A-Ser, A-Pro, and respective 3′-amides	pA-(fPhe) pA-(fLLeu-DLeu) pA-(fDLeu-DLeu)
Low or absent	A-Gly, A-Trp, A-(D-Phe), and respective 3′ amides	pA-(fGly)

[a] The abbreviations in column 3 are explained in footnote 1 (p. 2).

the fact that A(3′NH-Trp) (also having an aromatic system) is completely, or almost completely, inactive as a peptide acceptor (*131–133, 159*). Moreover, it does not fit with comparatively high activities of puromycin analogs carrying the hydrophilic amino acid residue *S*-(ω-hydroxyhexyl)cysteine (XXIII) (*165*), a ureido analog of A-Lys (XXIV) in which the cationic amino group has been converted into a neutral ureido group (*166*), and a hexahydro analog of A-Phe (XXV) (*167*).

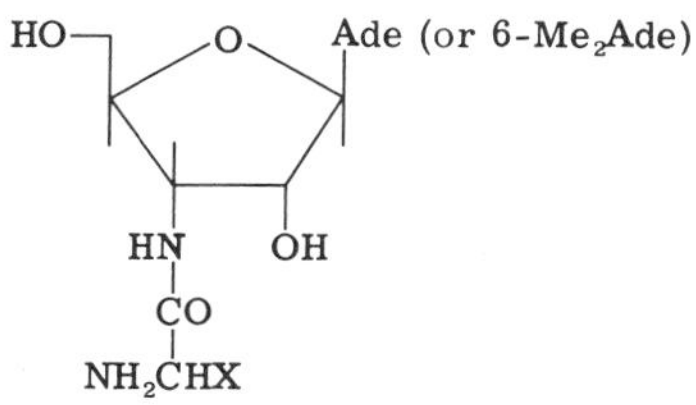

XXIII: X = $CH_2S(CH_2)_6CH_2OH$
XXIV: X = $(CH_2)_4NHCONH_2$
XXV: X = $CH_2C_6H_{11}$

According to Raake, there is a stacking interaction between the 3′-terminal C-A in peptidyl-tRNA and the aromatic ring of the amino-acid residue in puromycin (*79*). At present, however, this hypothesis seems to be highly unlikely.

The nature of the amino-acid residue influences the thermodynamically favored conformation of minimal acceptors and donors (*95*). This effect of the amino-acid residue on the activity is most pronounced for mononucleotide acceptors and weakens as the nucleotide chain lengthens in C-A and C-C-A. It is the polynucleotide chain rather than the amino-acid residue that determines the conformation of the 3′ end in aminoacyl-tRNA.

The effect of the C-terminal amino acid on the peptide donor properties of the amino acid and adenosine 5′-phosphoric esters has been tested for several amino acids (*168;* see *95, 102*). The donor activity decreases in the series pA-(fMet) > pA-(fLeu) > pA-(fPhe). Hardly any activity has been found for pA-(fGly) (see Table VI). These results agree with the data obtained for oligonucleotide fragments (*87*).

The activity of minimal donors with L-amino-acid residues is several times higher than these with D residues, as was found for the esters of acylleucine and adenosine 5′-phosphate (*169*).

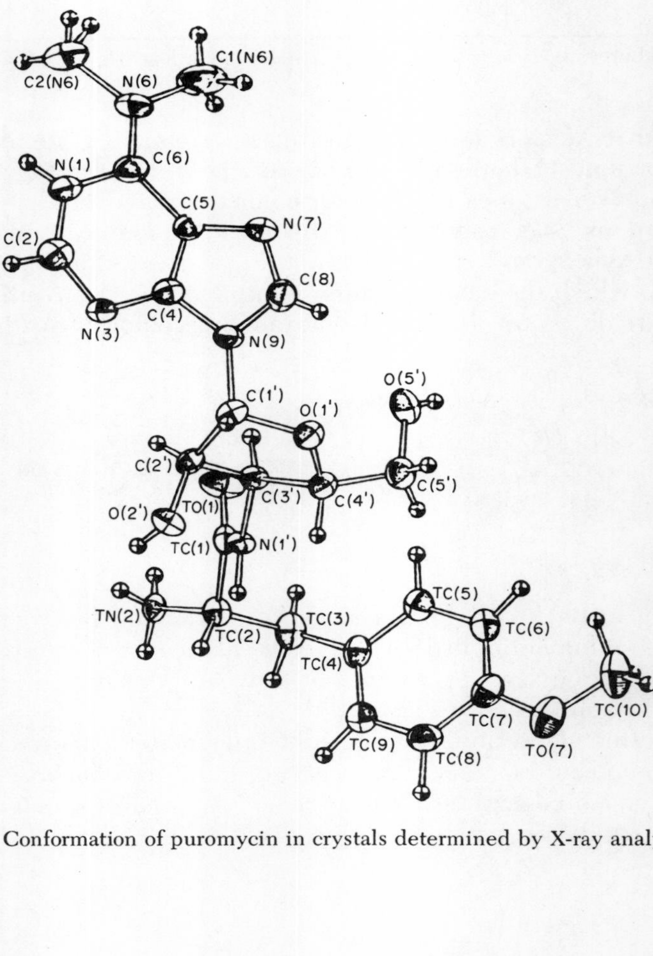

FIG. 4. Conformation of puromycin in crystals determined by X-ray analysis (*80*).

5. The Acyl Group

The activity of minimal donors depends to a great extent on the nature of the acyl group. It was established for the 2′(3′)-*O*-acylmethionyladenosine 5′-phosphates that their peptide donor activity decreased in the series CH_3SO_2— > HCO— > CH_3CO— ≫ C_6H_5CO— (*95*, *102*). The esters of adenosine 5′-phosphate and *N*-formyldipeptides were more active than those of *N*-formylamino acids (*169*).

The data presented in this section have led to a better understanding of the binding areas in the acceptor and donor sites of ribosomes. Their analysis shows that both the chemical nature and conformation of model substrates are essential for the activity. It is not yet possible to interpret the structure of the peptidyltransferase center in terms of conformational analysis, although this strategy seems to be most suitable for the next developments to be made. The first advances are associated with the determination of puromycin conformation (Fig. 4). What is known already about the conformation of model acceptors in the ribosome (the position of the NH_2 group in the amino acid residue and the "anti-like" position of the base) fits well with the puromycin conformation. The conformation of the donor site has been hardly analyzed at all so far. We think that the conformation of pA-(fMet) and C-A-(fMet) in crystals should be determined first of all; the conformation of pC in crystals has been established (*151*).

III. Mechanism of Peptide Bond Formation

Each peptide bond in the ribosome is formed as a result of electrophilic attack at the NH_2 group of the amino-acid residue in aminoacyl-tRNA by the carbonyl group of the peptide residue in peptidyl-tRNA. The energy required for this act is accumulated in aminoacyl-tRNA synthesis coupled with ATP hydrolysis. Hydrolysis of 1 mol of ATP to AMP and pyrophosphate releases about 8 kcal. All, or nearly all, of this energy is transferred after the enzymic act to the aminoacyl-tRNA being formed. However, only about 0.5 kcal/mol is required for amide bond formation. Therefore, the overall balance of energy can be expressed by the following reaction (additional energy expended for the transport of aminoacyl-tRNA to the ribosome and for the functioning of the ribosome is not considered).

$$\text{ATP} + \text{amino acid} \rightarrow \text{peptide bond} + \text{AMP} + \text{PP}_i + 7.5\ \text{kcal}$$

The exothermic nature of the reaction, which commences with

aminoacyl-tRNA ligases and terminates on ribosomes, makes the overall process irreversible.

A. Catalytic Functions of the Peptidyltransferase Center

1. Effect of Ions

Several conditions must be met for the ribosome to carry out the peptidyltransferase reaction. The first is the presence of monovalent metal or ammonium ions. NH_4^+ and K^+ are active cations for *E. coli* polysomes whereas Na^+, Li^+, and Cs^+ inhibit the biosynthesis (*170, 171*).

The peptidyltransferase center of *E. coli* ribosomes can exist in two states, active and inactive (*172–175*). The removal of NH_4^+ ions from *E. coli* ribosomes completely inactivates them. The ribosomes can be activated by heating at 40°C in the presence of NH_4^+ or K^+; 50 S, but not 30 S, subunits are activated, as was found in experiments with *E. coli* mutant ribosomes capable of catalyzing the reaction of AcPhe-tRNA with puromycin only at a high concentrations of K^+ or NH_4^+ (*173*). One K^+ or NH_4^+ ion per ribosome must be bound to peptidyltransferase to become capable of catalyzing the reaction. Polysomes from higher organisms, e.g., from rat liver or brain, require K^+ or Na^+ rather than of NH_4^+ or Li^+ (*176*). Likewise, one ion of K^+ or Na^+ is sufficient for activation of the ribosome. Conversion of the inactive form of *E. coli* ribosomes into the active one, and vice versa, is inhibited by the antibiotics streptomycin, neomycin, erythromycin, vernamycin, lincomycin, and chloramphenicol. Kanamycin inhibits only the activation, and kasugamycin prevents the inactivation (*174*).

The presence of bivalent metal ions, particularly Mg^{2+} (which occasionally can be substituted by Mn^{2+}), is also necessary (*170, 171, 176*). Ca^{2+} is not effective whereas Be^{2+}, Zn^{2+}, and Hg^{2+} inhibit the activity of peptidyltransferase. Apparently, Mg^{2+} ions are required for substrate binding rather than for catalysis. For example, NH_4^+ and Mg^{2+} ions are required for binding aminoacyl-tRNA or C-A-C-C-A-Phe to the acceptor site (*122, 172, 175, 177*) and NH_4^+ seems to be necessary for the catalysis (*122, 172, 177*). However, only Mg^{2+} is needed for binding AcPhe-tRNA or C-A-C-C-A-(AcLeu) to the donor site (*122, 175*).

In systems without templates, alcohols (at a concentration of 30 to 50%) are very effective, particularly in stimulating the fragment reaction and the reaction with a minimal donor (*178*). Alcohols favor substrate binding to the donor site and, to a lesser extent, to the acceptor site (*177, 179–181*). The mechanism of their action is not clear; presumably, alcohols modify the spatial organization of the peptidyltransferase center. The activity of alcohols decreases in the series methanol > ethanol > isopropanol. The reaction is also stimulated by acetone and 3.2 M NH_4^+ (*182*).

2. pH-Dependence

The reaction of peptidyltransferase depends on the pH of the medium. The optimum pH of the reaction lies within the range of 8.5 to 9.0 when the complex [fMet-tRNA + ribosome + A-U-G] reacts with puromycin, methanol, and water (in the presence of RF factors or acetone) in a system with *E. coli* monosomes (*183*), when peptidylpuromycin is synthesized by rat-liver polysomes (*176*) and *E. coli* polysomes (*184*), and when the ester bond is formed with the hydroxy analog of puromycin (*185*). This optimal pH value of the transferase reaction is the closest to the pK of the imidazole group or the α-amino group of the protein.

Transition of the ribosome from the form that exists at neutral pH values to the catalytically active state is accompanied by a loss of a proton (*176*). When the pH increases, a second inflection appears in the curve for the activity vs. pH corresponding to a group with the pK = 9.2 (*175*), which is similar in ionization constant to the hydroxyl of tyrosine or the ϵ-amino group of lysine. Therefore, two functions of peptidyltransferase with the pK of ca. 7.5 and 9.5 are involved in manifestation of the catalytic activity of the ribosome; the first one must be deprotonated and the second functional group has to be protonated.

3. Types of Bond Whose Formation Is Catalyzed by Ribosomes

The natural peptide acceptor in the ribosome is either the NH_2 group of an amino acid residue in aminoacyl-tRNA or water (in termination of peptide bond formation). However, *E. coli* ribosomes can catalyze the transfer of a peptide to other nucleophilic groups as well.

As has been shown for the hydroxy analog of puromycin, ribosomes can catalyze ester bond formation (185). Hydroxypuromycin reacts with the nascent polyphenylalanine chain in a system that contains ribosomes carrying poly(U) as a template; as the result, polyphenylalanyl-oxypuromycin is formed.

Likewise, ribosomes can catalyze polycondensation of phenyllactic acid if phenyllactyl-tRNA is used instead of phenylalanyl-tRNA, or form ester bonds in the presence of RNA from a mutant strain of bacteriophage A_MB_2 as a template (*186*).

If a thio analog of puromycin containing an SH group that substitutes for the NH_2 group of puromycin is used as a peptide acceptor, ribosomes catalyze synthesis of the thioester bond (*187*).

Escherichia coli ribosomes also catalyze thioamide bond formation when 2′(3′)-*O*-(*N*-acetylthioleucyl)adenosine 5′-phosphate is used as a peptide donor while Phe-tRNA or C-A-C-C-A-Phe serves as a peptide acceptor (*188*).

Thus, *E. coli* ribosomes are capable of catalyzing synthesis of several types of bond: amide (—CO—NH—), ester (—CO—O—), thioester (—CO—S—), thioamide (—CS—NH—). One may anticipate that other types of bond can also be formed in ribosomes.

Peptidyl-tRNA in ribosomes can react with alcohols such as methanol or ethanol (*189*). If the complex [fMet-tRNA + 70 S ribosome + A-U-G] is prepared, the methyl ester of formylmethionine, $fMetOCH_3$, is formed in 20% methanol, and $fMetOC_2H_5$ in 20% ethanol, in the presence of free tRNA or its 3′-terminal fragment C-C-A, which serves as a sort of cofactor. Such a reaction has been described for ribosomes from rabbit reticulocytes or rat liver (*190*).

This capacity of peptidyltransferase to selectively channel the reaction into hydrolysis of alcoholysis is found while studying the effect of certain organic solvents on artificial termination. As has been already mentioned, fMet-tRNA in the complex [fMet-tRNA + 70 S ribosome + A-U-G] reacts in alcohol-aqueous solutions, in the presence of free tRNA or C-C-A, with alcohols yielding fMetOR (where R = CH_3, C_2H_5, or CH_3CHCH_3). This complex is hydrolyzed in solutions containing 30% acetone to produce formylmethionine (*183*).

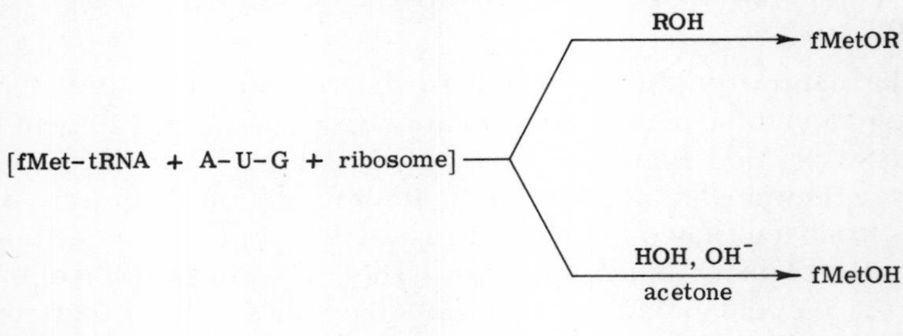

B. Attempts to Inactivate Selectively Different Functions of the Peptidyltransferase Center

Several attempts have been made to inactivate selectively one of the functions of the peptidyltransferase center by the action of various chemical agents.

If *E. coli* ribosomes or 50 S subunits are treated with T1-ribonuclease under strictly specified conditions (the ratio between ribosomes and ribonuclease, temperature, time, etc.), the donor site is almost entirely inactivated while the acceptor site remains active (*191*). A method for selective inactivation of the acceptor site in the peptidyltransferase center has not been elaborated so far.

Attempts have been made to modify chemically one of the protein functions responsible for catalyzing peptide bond formation. However, treatment of ribosomes with reagents for the OH group of serine (diisopropyl fluorophosphate, etc.), the SH group of cysteine (*p*-chloromercuribenzoate, *N*-ethylmaleimide, etc.), or the imidazole group of histidine (chloromethyl ketones of N-protected amino acids) does not inactivate them (see *102*). The binding of C-A-C-C-A-Leu and C-A-C-C-A-(Ac-Leu), respectively, to the acceptor and donor sites of the peptidyltransferase center in ribosomes of higher organisms, as well as the transferase activity, is even stimulated by reagents for SH-groups (*92, 192*).

The transferase activity of *E. coli* 50 S subunits is entirely inhibited by photochemical oxidation; only one of the protein functional groups is modified here (*193*). This group seems to be the imidazole residue of histidine as follows from the pH-dependence (the optimal pH for modification is 7.5 to 7.7) and the absence of protection with mercaptoethanol. It has been stated that the treatment does not interfere with the binding of Phe-tRNA and AcPhe-tRNA to the modified 50 S subunits. However, experiments conducted with C-A-C-C-A-Phe and C-A-C-C-A-(AcPhe) show that the binding of C-A-C-C-A-Phe to ribosomes is upset upon photooxidation of 50 S subunits, and therefore the reaction is not selective (J. Černa, private communication). The erroneous statement of the above authors (*193*) stems, apparently, from wrong testing of Phe-tRNA binding.

C. Mechanism of Catalysis

Many publications are concerned with the mechanism of catalysis in the peptidyltransferase center of ribosomes (see *194*). Two hypotheses have been proposed. One postulates nucleophilic or acid–base catalysis by the functional groups of the peptidyltransferase center. Several attempts have been made to isolate an acyl-ribosomal derivative (*84, 85, 195*) according to the equation

$$\text{C-A-C-C-A-Ac}[^3\text{H}]\text{Leu} + \text{ribosome} \rightleftharpoons \text{C-A-C-C-A} + \text{Ac-}[^3\text{H}]\text{Leu-ribosome}$$

However, no one has succeeded in isolating the complex Ac[^{3}H]Leu-ribosome. We too have failed to do this.

According to the other hypothesis, peptidyltransferase catalyzes peptide bond formation merely by sterical orientation of the reaction components, and the peptide bond is synthesized spontaneously. This hypothesis has been proposed after unsuccessful attempts to inhibit the action of peptidyltransferase using inhibitors for different functions of proteins.

A number of compounds in which a model donor and a model acceptor are covalently bound have been synthesized and tested (*84, 85, 195*). If the 3′-terminal fragments of peptidyl-tRNA and aminoacyl-tRNA are located in close vicinity in the peptidyltransferase center, one may expect that condensed model substrates will display a high affinity for the peptidyltransferase center due to additive binding of the two components of the molecule to the acceptor and donor sites simultaneously. Compounds XXV and XXVI were tested; neither had an elevated affinity for the peptidyltransferase center of ribosomes. Compound XXVI possessed a moderate peptide acceptor activity close to that of puromycin (*85*). However, these results do not rule out the hypothesis of sterical orientation since the conformation of XXVI and XXVII may not meet the requirements imposed by the peptidyltransferase center. It has been found, in particular, that the bases of the adenosine residues in XXVII are in the intramolecular stacking interaction (*196*).

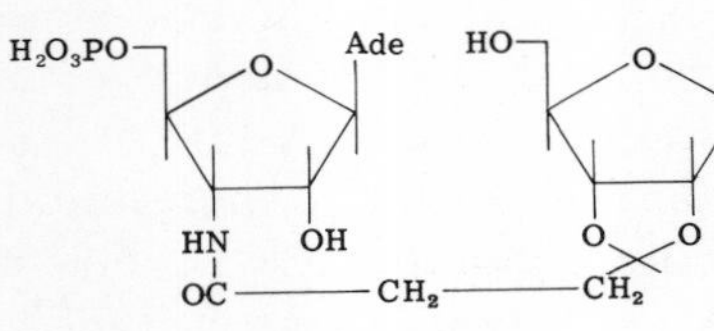

XXVI

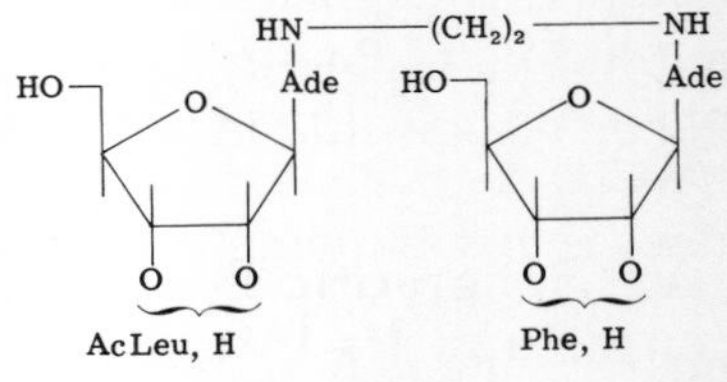

XXVII

The hypothesis of spatial orientation cannot account for the catalytic role of ribosomes in alcoholysis and hydrolysis of peptidyl-tRNA. Actually, only negligible, if any, alcoholysis of aminoacyl-tRNA occurs in 20 to 40% alcohols in the absence of ribosomes although one would expect this process to take place. Ribosomes catalyze this reaction under identical conditions.

Therefore, the catalytic mechanism of peptide bond formation in ribosomes remains obscure.

D. Are Cytoplasmic Proteins Necessary for the Functioning of the Peptidyltransferase Center?

Apparently, no external protein factors from the cytoplasm are required for peptide bond formation in the ribosome. This conclusion has been corroborated by studying the reaction of a peptide donor with puromycin, the reaction of the fragments of aminoacyl-tRNA and peptidyl-tRNA, the reaction with a minimal donor, and the fact that the process does not need extra energy from the cytoplasm.

However, a protein isolated from the cytoplasm of *E. coli*, referred to as EFP (M_r ca. 50,000), accelerates 10- to 20-fold the reaction of puromycin with the complexes [ribosome + A-U-G + fMet-tRNA] and [50 S subunit + A-U-G + fMet-tRNA]; likewise, it stimulates several fold the translation of the templates poly(U) and poly(A) by 70 S ribosomes (*197–199*). The protein is most effective at low concentrations of puromycin. The affinity of puromycin for ribosomes increases in the presence of EFP by an order of magnitude (without EFP, $K_m = 2.4 \times 10^{-5}$ M; in the presence of EFP, $K_m = 2.0 \times 10^{-6}$ M). EFP has no effect on fMet-tRNA binding; it neither binds puromycin nor causes GTP hydrolysis, and does not require the presence of proteins L7/L12. Its function remains obscure. Presumably, it is involved in the formation of the complex [ribosome + template + fMet-tRNA] by creating a conformation of the ribosome suitable for the subsequent translation.

It follows from this that the catalytic mechanism of peptide bond formation in ribosomes needs further study. It remains to be found out whether there are any groups of protein or rRNA involved in the catalysis and to clarify whether an individual component of the ribosome is responsible for this function.

IV. Schematic Presentation of the Peptidyltransferase Center and Its Interrelation with Other Functional Sites of Ribosomes

One of the goals in studies on the substrate specificity of the peptidyltransferase center is composing its three-dimensional structure. At present, this problem cannot be entirely solved. Yet, the data available have inspired some authors to publish two schemes for the peptidyltransferase center.

A. Schemes of the Peptidyltransferase Center

The first scheme for the peptidyltransferase center was proposed by Harris and Symons in 1973 (*200*). The main theoretical premise underlying this scheme is as follows. Since the C-C-A fragment is identical in all the tRNAs, there must be binding sites for each of its components, i.e., three bases, ribose residues, and phosphate group(s). The scheme is also based on different peptide acceptor activities of model acceptors; a high, or sufficiently high, acceptor activity is displayed by A-Phe, A-Tyr, A-Leu, A-Met, A-Ser, A-Lys, and A-Ala. Since the amino-acid residues in these compounds contain hydrophobic or cation-bearing radicals, the acceptor site has been presumed to have a hydrophobic pocket and a cation-binding center (for lysine). The binding of the remaining amino-acid derivatives has not been discussed by the authors.

As can be seen in Fig. 5, the peptidyltransferase center comprises in the acceptor site, according to Harris and Symmons, areas for binding the 3′-terminal adenosine (III) and the two cytidine residues (IV and V), the phosphate residue(s) between these nucleosides (IVa), as well as a hydrophobic pocket for aromatic amino acids (I) and an anionic center for the ϵ-amino group of lysine (II). The donor site contains an area for binding the C-C-A fragment (IX), a hydrophobic pocket for the side radicals of the C-terminal amino acids from the

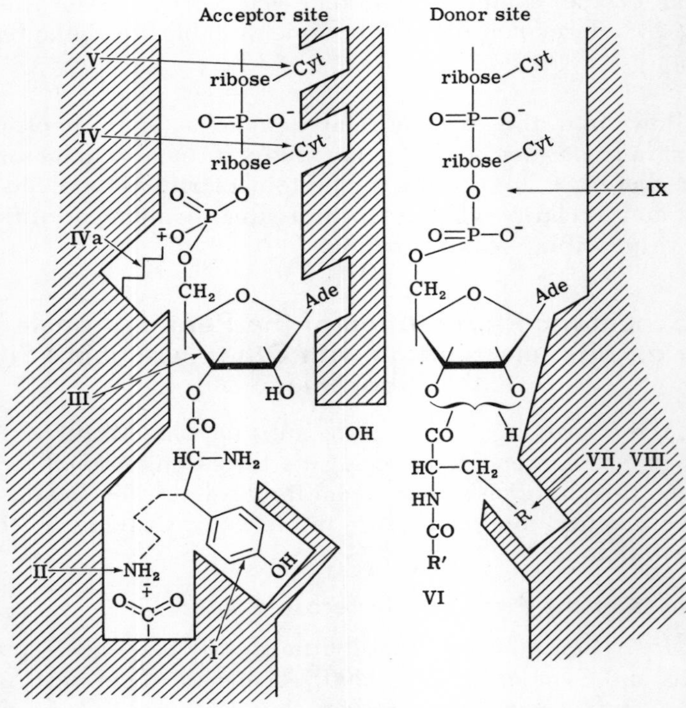

FIG. 5. Schematic model of the peptidyltransferase center of *Escherichia coli* ribosomes according to Harris and Symons (*200*). The binding areas are discussed in the text.

nascent peptide (VII and VIII) if they are hydrophobic, or a hydrophilic pocket for the side radicals if the amino acid is hydrophilic. An area for binding the remote peptide fragment (VI) is also proposed without any details.

The second detailed scheme for the peptidyltransferase center was put forward by the authors of this review in collaboration with Gottikh in 1975 (*95*). The overall sequence of events occurring in the peptidyltransferase center of ribosomes is supposed to be the following. As has been shown earlier, substrates of this center have different structures in their particular parts (amino acid or peptide) at each moment during the formation of a peptide chain. Thus, one may anticipate that these two sites of the peptidyltransferase center contain regions for recognizing those components of substrates whose structure does not change. However, not all of the constant components in the structure of substrates must necessarily be bound to the regions of the peptidyltransferase center.

In accordance with this hypothesis, the acceptor site of the peptidyltransferase center possesses areas for binding the adenine of the 3′-terminal adenosine and the amino group of the amino-acid residue. For a model substrate to display the acceptor activity, its molecules in solution must have a conformation that corresponds to the structure of the acceptor site. The role of the next two nucleotides (pC) is to fix a conformation of adenosine favorable for the acceptor site of the peptidyltransferase center.

The situation is somewhat different for the donor site. There is direct evidence, such as stimulation of the activity of minimal substrates by pC, that an additional region exists for binding a substrate besides the terminal peptidyladenosine. Therefore, the donor site must comprise two areas binding nucleotide components; these are referred to as D_A and D_C. Apparently, there is also a third area, D_P, as peptidyl-tRNA containing several amino-acid residues has a higher affinity for the donor site than acylaminoacyl-tRNA (see Section I).

The scheme of the peptidyltransferase center implied that interactions of this center with the 3′ termini of aminoacyl-tRNA and peptidyl-tRNA must be of the protein–nucleic acid type. This statement soon received support, for the acceptor site, from experiments in which εA-Phe and εC-Phe (whose complementary pairing with rRNA according to the Watson–Crick principle is ruled out) showed a high peptide acceptor activity (*139*).

The schematic presentation of the peptidyltransferase center from bacterial ribosomes, in conformity with the above hypothesis, is given in Fig. 6.

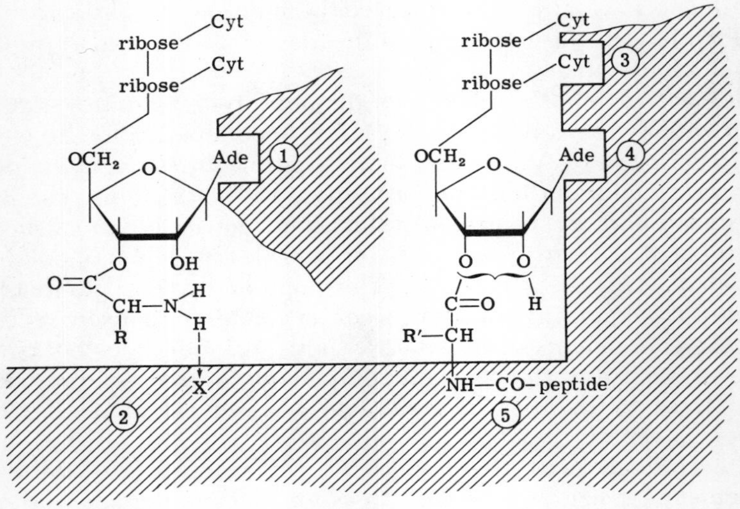

FIG. 6. Schematic model of the peptidyltransferase center of *Escherichia coli* ribosomes (95). Binding areas of the acceptor terminal adenine (1), NH_2 group of amino-acid residue (2), the donor cytosine (3), the donor adenine (4), and peptide residue (5).

The presence of the NH_2-, OH-, or SH-group in the acyl residue of a minimal acceptor is necessary not only for accepting a peptide but also for binding to the acceptor site of the peptidyltransferase center. For example, compounds XXVIII and XXIX do not react with the acceptor site at all.

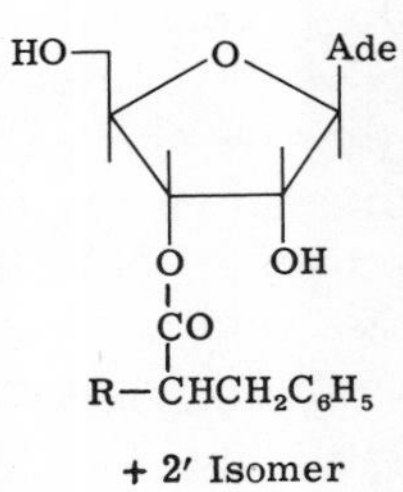

XXVIII: R = H (*138*)
XXIX: R = OCH_3 (*201*)

The mechanism of peptide bond formation specifies that the NH_2-group is the nucleophilic component of the reaction. That is why the functionality of the protein in the acceptor site of the peptidyltransferase center would increase the nucleophilic nature of the NH_2 group being bound or, at least, would not decrease it. This necessitates the presence of a hydrogen bond in which the group of the ribosomal protein or nucleic acid serves as a proton acceptor while the NH_2 group of the amino acid residue in aminoacyl-tRNA acts as a proton donor.

B. Studies of Binding Sites in the Peptidyltransferase Center by Means of Binding Constants

The most reliable approach for determining the number of binding sites in the peptidyltransferase center is to compare the binding constants for puromycin, aminoacyl esters of adenosine, and oligonucleotides. However, this strategy is complicated by the fact that these constants depend, to a great extent, on a system in which they are assayed. It has not been found out why the values of equilibrium and Michaelis constants differ among various systems. Yet, it is obvious that these values depend on the state of ribosomes (polysomes or monosomes), on the presence in the medium of methanol or ethanol which considerably increase binding of substrates to the peptidyltransferase center, etc.

Table VII (*40, 94, 128, 129, 131, 153, 165, 170, 176, 185, 202–205*) presents the values of K_s and K_m reported for the acceptor site as well as the values of K_s and K_m determined in this laboratory for the donor site. It should be noted, first of all, that the K_s and K_m for various model acceptors have not been studied systematically. Second, interrelationships exist in the peptidyltransferase center. For instance, the K_m of puromycin for monosomes differs from that for polysomes by two orders of magnitude. Moreover, the difference in the K_s values for C-A-C-C-A-Phe and puromycin (or A-Phe) measured in the absence of a peptide donor is about 10^4 whereas the K_m for A-C-C-A-Tyr differs from that for A-Tyr only by about 30-fold. One may expect that the substrate binding areas in the acceptor site of the ribosome are organized differently when the donor site is vacant and when it is occupied. The situation has been studied to a lesser extent for polysomes.

However, the K_s of minimal substrates in free ribosomes differs from the K_s of oligonucleotide fragments by about 10^4 (Table VII). Apparently, additional binding areas exist in both sites apart from the binding areas for the terminal adenosine and an amino acid or peptide residue. There is a qualitative difference between the properties of these binding areas in the acceptor and donor sites. As has been noted above, cytidylic acid stimulates the activity of a minimal donor but not that of a minimal acceptor.

The data of Table VII suggest also that studies of the peptidyltransferase center must take into account the interrelationship of the sites, by comparing systems with monosomes and polysomes.

TABLE VII

EQUILIBRIUM CONSTANTS AND MICHAELIS CONSTANTS OF MODEL SUBSTRATES

	Acceptor site				Donor site		
Substrate	K_{diss} (M)	References	K_m (M)	References	Substrate (50% ethanol)[e]	$K_{diss}(204)$ (M)	$K_m(205)$
					Monosomes		
C-A-C-C-A-Phe	3.6×10^{-8}[a]	*202*			C-A-C-C-A-(AcLeu)	5.8×10^{-7}	6.8×10^{-8}
C-A-C-C-A-Phe	1.0×10^{-8}[a]	*129*			C-C-A(3′NH-(MsLeu)	3.8×10^{-6}	
C-A-C-C-A-Phe	1.1×10^{-6}[b]	*129*			C-A(3′NH)-(fMet)	1.8×10^{-4}	1.1×10^{-4}
Z[d]-Tyr			2.5×10^{-5}[a]	*128*	C-A-(fMet)	5.0×10^{-4}	1.3×10^{-4}
A-C-C-A-Tyr			8.0×10^{-6}[a]	*131*	pA(3′NH)-(MsMet)	2.8×10^{-3}	0.5×10^{-3}
C-A(3′NHPhe)			1.7×10^{-4}[b]	*131*	pA-(fMet)	2.3×10^{-3}	3.6×10^{-3}
Puromycin	1.4×10^{-4}[a]	*185, 203*	$(1.7\text{–}3.1) \times 10^{-4}$[a,b]	*131, 94,*	pA-(fPhe)		1.9×10^{-3}
				185, 165			
				128, 153			
Puromycin	7.1×10^{-4}[b]	*40*					
A-Phe			2.0×10^{-4}[a]	*153*			
A-Tyr			3.3×10^{-4}[a]	*131*			
A(3′NH-Phe)			2.8×10^{-4}[a]	*131*			
					Polysomes		
C-A-C-C-A-Phe			2.5×10^{-4}[b]	*129*			
A-C-C-A-Tyr			3.0×10^{-6}[b]	*128*			
Z[d]-Tyr			4.0×10^{-6}[b]	*128*			
Puromycin			$(2.4\text{–}7.5) \times 10^{-6}$[a,b]	*131, 170*			
				153, 128			
			8.0×10^{-6}[c]	*176*			
A-Phe			1.9×10^{-6}	*153*			
A-Tyr			4.5×10^{-6}	*128*			

[a] In 20 to 30% alcohols. [b] Without alcohol. [c] Ribosomes from rat liver and brain; otherwise, *E. coli* ribosomes.
[d] Z = A-A-U-C-C-U-U-C-C-C-C-C-A-C-C-A-C-C-A-. [e] The representation is explained in footnote 1, p. 2 (last paragraph). Ms = $CH_3SO_2^-$.

Once a new peptide bond is formed, the situation in the peptidyltransferase center changes. Apparently, the binding area for the amino group in the acceptor site disappears, as this group is converted into an amide group. Furthermore, it remains unclear whether peptidyl-tRNA in the acceptor site and free tRNA in the donor site are still in contact with the respective binding areas for nucleotide components.

C. Substrates and Products of the Peptidyltransferase Center

The essence of changes occurring in the peptidyltransferase center after transpeptidation can be interpreted in the following terms (*206*). In the posttranslocational state of the ribosome, aminoacyl-tRNA and peptidyl-tRNA are the substrates of the acceptor and donor sites, respectively. However, as soon as the reaction is over, the newly formed peptidyl-tRNA in the acceptor site and the free tRNA in the donor site, both being the reaction products, may have a lower affinity for their respective sites. Then the free tRNA would readily leave the donor site of the peptidyltransferase center, and the peptidyl-tRNA the acceptor site. As the result, the donor site becomes accessible to the peptidyl-tRNA, which forms new contacts translocating to this site. Therefore, the translocation is thermodynamically favored. If this is true, the peptidyltransferase center in the ribosome plays a more significant role than has been believed heretofore.

The data on the competition of C-A-C-C-A-(AcLeu) with C-A-C-C-A-Phe for binding to the acceptor site and that of tRNA with C-A-C-C-A-(AcLeu) for binding to the donor site are in line with this idea (*207*, *208*). The incubation mixtures contained ribosomes and 30 to 50% alcohol, but no templates. C-A-C-C-A-(AcLeu) was bound to the acceptor site less than 1% as strongly as C-A-C-C-A-Phe (*207*); the binding of tRNA was weaker by over 10^{-3} than that of C-A-C-C-A-(AcLeu), a substrate of the donor site (*208*). Likewise, tRNA weakly competed for the acceptor site, and C-A-C-C-A-Phe for the donor site, when their concentrations were raised 100-fold.

The above data indicate that, in the absence of a template, there is a distinct specificity in the binding of fragments to the acceptor or donor site, depending on whether the amino acid residue is acetylated or not.

D. Sequence of Events in the Peptidyltransferase Center during Translocation

The following sequence of events seems to occur in the peptidyltransferase center of ribosomes. As soon as transpeptidation is over, the binding of tRNA to the donor site and that of peptidyl-tRNA to the acceptor site are weakened. That is why peptidyl-tRNA is transferred from the acceptor site to the donor site, gaining by one to two orders of magnitude in the binding constant. Therefore, translocation of peptidyl-tRNA from the acceptor site to the donor site can be attributed to diffusion and does not require additional energy from the outside. The latter statement fits in well with the data of Pestka (*209*) as well as with the findings of Gavrilova and Spirin (*210*) showing that nonenzymic translation is possible in the absence of GTP and transfer factors.

Synthesis of oligopeptides in a system with *E. coli* ribosomes in the absence of a template and transfer factors has been found in this laboratory using pA-(fMet) as a model peptide donor and Phe-tRNA or C-A-C-C-A-Phe as a model peptide acceptor. Apart from formyldipeptidyl-tRNA or a formyldipeptidyl fragment, tripeptide and tetrapeptide derivatives have been isolated from the incubation mixture (*100*). The mechanism for the formation of these peptide derivatives is not yet known. It may be translocation of formyldipeptidyl-tRNA produced in the first transpeptidation reaction and the subsequent transpeptidation to a new molecule of the acceptor bound to the acceptor site, etc. Or, it may be transfer of formyldipeptidyl-tRNA or formyldipeptidyl oligonucleotide through a solution by a mechanism of dissociation–association. Anyway, transfer of peptidyl-tRNA or peptidyl oligonucleotides from the acceptor site to the donor site does take place, and therefore this transfer is thermodynamically favored.

Polypeptide synthesis is also catalyzed by 50 S subunits with the participation of aminoacyl-tRNAs (*88*, *100*) or aminoacyl oligonucleotides (*100*).

Apparently, the sources of energy for translocation are contained in the substrates, and are somehow realized in the process of translocation.

E. Conformation of Model Substrates

As has been noted in Section II, model substrates with an "anti-like" conformation in the ribosome are highly active as peptide acceptors whereas model substrates with a "syn-like" conformation in solution are effective peptide donors. These data suggest a probable conformation of the 3′ termini of aminoacyl-tRNAs and peptidyl-tRNAs.

Another piece of evidence indicating a possible change in the conformation of the 3′ end of aminoacyl-tRNA as compared to tRNA, and that of tRNA as compared to peptidyl-tRNA, is provided by PMR spectroscopy studies of model substrates for the acceptor and donor sites, and adenosine 5′-phosphate, in aqueous medium (*211*). The ribose residue in adenosine 5′-phosphate is found predominantly in the

N-conformation ($C_{3'}$-endo) whereas this residue in 3′-aminoacyl- or 3′-peptidyladenosine 5′-phosphate is mainly in the S-conformation ($C_{2'}$-endo) (*211*). This difference in conformation indicates that, presumably, the conformation of the acceptor end of aminoacyl-tRNA changes compared to that of tRNA. It follows therefore that the binding of a substrate to the acceptor site is determined by the presence of an amino-acid residue, which interacts directly with the acceptor site of the peptidyltransferase center, as well as by a favorable conformation of the 3′-terminal adenosine.

The reverse sequence of conformational changes in the ribose residue may occur when the peptide residue leaves peptidyl-tRNA upon transpeptidation. Thereupon tRNA loses its contacts with the donor site and leaves it.

F. Interdependence in the Peptidyltransferase Center and Its Functional Interrelation with Other Sites

The presence of peptidyl-tRNA in the donor site stimulates binding of the complex [aminoacyl-tRNA + EF-Tu + GTP] to *E. coli* ribosomes (*212*), and that of the complex [aminoacyl-tRNA + EF-I + GTP] to ribosomes from rabbit reticulocytes (*115, 213, 214*). The stimulating action is not displayed if the donor site is vacant or occupied with nonacylated tRNA or tRNA carrying a peptide at the 2′ hydroxyl of the terminal adenosine, tRNA-C-C-A(3′H)-(Gly_3Phe) (*115*). Therefore, the peptidyltransferase center is directly involved in the stimulation of complex binding; the peptide residue in peptidyl-tRNA at the donor site must be located at the 3′ hydroxyl. Thus, there is a direct functional interrelation between the peptidyltransferase center and the binding sites for the complex [EF-Tu + GTP + aminoacyl-tRNA] or [EF-I + GTP + aminoacyl-tRNA].

The nature of peptides is also essential for the stimulation of substrate binding to the acceptor site and its reaction. Endogenous peptidyl-tRNA reacts with puromycin in a system with rat-liver ribosomes (and 60 S subunits of ribosomes) and 60 mM K^+, whereas AcPhe-tRNA requires 300 mM K^+ (*177*). As has been found in studies with ribosomes from pig uterus, the reaction with puromycin decreases in the series of peptidyl-tRNAs coded by the templates: endogenous template > poly(C-U) > poly(G-U) > poly(U) (*162*).

A number of antibiotics[2] (erythromycin, carbamycin, tylosin, niddamycin, spiramycin III, lincomycin, streptogrammin A) effectively inhibit the reaction of synthetic peptidyl-tRNAs with puromycin; however, they are poor inhibitors of a similar reaction with endogenous peptidyl-tRNAs (*215*).

The effect of pC binding on the reactivity of a minimal donor, such as pA-(fMet), is discussed in Section II.

The direct functional relationship of the peptidyltransferase center with the binding site for the stringent factor is also obvious. The binding of the stringent factor to the ribosome and synthesis of ppG(3′)pp or pppG(3′)pp commence in a protein-synthesizing system only when the acceptor site of the ribosome is occupied with a codon-specific deacylated tRNA. Therefore, the absence of an amino-acid residue from the acceptor site of the peptidyltransferase center activates the ribosome. Moreover, tRNAs modified at their 3′ termini, e.g., tRNA-C-C-C, tRNA-C-C-A_{ox}, tRNA-C-C-A_{oxred} (*216, 217*), tRNA-C-C-A(3′H), and tRNA-C-C-A(3′NH_2) (*218*), do not promote synthesis of ppG(3′)pp and pppG(3′)pp catalyzed by ribosomes. The 2′ hydroxyl of adenosine is not essential (*218*). Apparently, the 3′ hydroxyl of aminoacyl-tRNA has an allosteric effect on the stringent factor involved in catalysis (*218*). This effect is believed to be realized through protein L11, as ribosomes of *E. coli* mutant in L11 (rel C) bind the stringent factor but do not synthesize the guanosine polyphosphates (*52*).

The effect of antibiotics[2] on the binding of substrate to the peptidyltransferase center has been shown by several workers. We shall not consider here the inhibition, since this effect can be realized by the mechanism of direct competition as well as by the allosteric mechanism (see reviews *219, 220*). It would be relevant to mention, as an example, the stimulating effect of sparsomycin, chloramphenicol, erythromycin, PA-114A, etc., on the binding of C-A-C-C-A-(AcLeu), C-A-C-C-A-(AcPhe) and C-A-A-C-C-A-(fMet) to the donor site of *E. coli* ribosomes and 50 S subunits (*207, 221, 222*).

The stimulating action of peptidyl-tRNA (at the donor site) on the binding of antibiotics has been demonstrated for chloramphenicol and lincomycin (*223*). If the donor site is occupied by peptidyl-tRNA, the affinity of the ribosome for chloramphenicol is 4 times higher, and that for lincomycin 1.5 to 2 times greater, than in the case of free ribosomes.

Finally, it must be mentioned that the complex [EF-G + GTP] in a system with *E. coli* ribosomes, or the complex [EF-II + GTP] in a system with eukaryotic ribosomes, is attached with a high capacity to ribosomes that carry peptidyl-tRNA at the acceptor site and free tRNA at the donor site. However, the complex tends to dissociate as soon as tRNA leaves the ribosome, and peptidyl-tRNA is translocated from the acceptor site to the donor site. It follows therefore that the presence of a peptide residue in peptidyl-tRNA at the donor site decreases the affinity of the ribosome for the translocation factor, but increases it for the factor that catalyzes the binding of aminoacyl-tRNA to the ribosome.

It is noted in Section III that the binding of tRNA or C-C-A to the complex [ribosome + A-U-G + fMet-tRNA] in the presence of alcohols causes transfer of fMet to alcohols catalyzed by ribosomes (*189, 190*). These facts indicate that there is a direct correlation between the reaction of transpeptidation and the functional state of the acceptor site.

The experimental results presented in this section and their theoretical interpretation suggest that the peptidyltransferase center plays a more general role in the ribosome than only catalysis of peptide bond formation. However, they do not yet clear up the role of GTP and elongation factors, though they demonstrate the correlation between the functional state of the peptidyltransferase center and the capacity of ribosomes to interact with elongation factors.

Conclusions

The evidence presented in this survey gives sufficient grounds to believe that the role of the peptidyltransferase center cannot be regarded solely as catalysis of peptide bond formation. The following functions of this center should be mentioned: (*a*) dependence of the functional state of the peptidyltransferase center on the binding of transfer factors; (*b*) synthesis of guanosine polyphosphate dependent on the stringent factor; (*c*) apparently, the direct participation of the peptidyltransferase center in the process of translocation.

The peptidyltransferase center can be regarded as a dynamic system that undergoes reversible changes in the processes of transpeptidation and translocation.

$$\text{posttranslocational state} \underset{\text{translocation}}{\overset{\text{transpeptidation}}{\rightleftarrows}} \text{pretranslocational state}$$

The energy required for these changes is supplied by the substrate and realized in the course of peptide bond formation. Conformational transitions of the peptidyltransferase center seem to be related to various subsequent events occurring in the ribosome and catalyzed by protein factors.

Such formulation of the problem necessitates a complex approach in studying the peptidyltransferase center as a polyfunctional and dynamic system.

Acknowledgment

The authors would like to express their gratitude to Marina Verkhovtseva for preparation of the english version of the manuscript.

References

1. B. S. Cooperman, *Bioorg. Chem.* **4**, 81–115 (1978).
2. R. Brimacombe, G. Stöffler, and H. G. Wittmann, *ARB* **47**, 217–249 (1978).
3. F. Hernandez, D. Vazquez, and J. P. C. Ballesta, *EJB* **78**, 267 (1977).
4. E. S. Bochkareva, V. G. Budker, A. S. Girshovich, D. G. Knorre, and N. M. Teplova, *FEBS Lett.* **19**, 121 (1971).
5. C. R. Cantor, M. Pellegrini, and H. Oen, *in* "Ribosomes" (N. Nomura, A. Tissières and P. Lengyel, eds.), p. 573. Cold Spring Harbor Lab., Cold Spring Harbor, New York, 1974.
6. H.-G. Wittmann, *EJB* **61**, 1(1976).
7. G. Stöffler and H. G. Wittmann, *in* "Molecular Mechanisms of Protein Biosynthesis" (H. Weissbach and S. Pestka, eds.), p. 117. Academic Press, New York, 1977.
8. M. Pellegrini, H. Oen, D. Eilat, and C. R. Cantor, *JMB* **88**, 809 (1974).
9. H. Oen, M. Pellegrini, and C. R. Cantor, *FEBS Lett.* **45**, 218 (1974).
10. J. B. Breitmeyer and H. F. Noller, *JMB* **101**, 297 (1976).
11. D. Eilat, M. Pellegrini, H. Oen, N. de Groot, Y. Lapidot, and C. R. Cantor, *Nature* **250**, 514 (1974).
12. M. Sopori, M. Pellegrini, P. Lengyel, and C. R. Cantor, *Bchem.* **13**, 5432 (1974).
13. E. Collatz, E. Küechler, G. Stöffler, and A. P. Czernilofsky, *FEBS Lett.* **63**, 283 (1976).
14. M. Yukioka, T. Hatayama, and S. Morisava, *BBA* **390**, 192 (1975).
15. A. P. Czernilofsky, E. E. Collatz, G. Stöffler, and E. Küechler, *PNAS* **71**, 230 (1974).
16. K. Bauer, A. P. Czernilofsky, and K. Küechler, *BBA* **395**, 146 (1975).
17. A. P. Czernilofsky, G. Stöffler, and E. Küechler, *ZpChem* **355**, 89 (1974).
18. D. Eilat, M. Pellegrini, H. Oen, Y. Lapidot, and C. R. Cantor, *JMB* **88**, 831 (1974).
19. N. Hsiung, S. A. Reines, and C. R. Cantor, *JMB* **88**, 841 (1974).
20. N. Hsiung and C. R. Cantor, *NARes* **1**, 1753 (1974).
21. A. S. Girshovich, E. S. Bochkareva, V. A. Kramarov, and Yu. A. Ovchinnikov, *FEBS Lett.* **42**, 213 (1974).
22. N. Sonenberg, M. Wilchek, and A. Zamir, *PNAS* **72**, 4332 (1975).
23. N. Sonenberg, M. Wilchek, and A. Zamir, *EJB* **77**, 217 (1977).
24. L. Bispink and H. Matthaei, *FEBS Lett.* **37**, 291 (1973).
25. A. Barta, E. Küechler, C. Branlant, J. Sriwidada, A. Krol, and J. P. Ebel, *FEBS Lett.* **56**, 170 (1975).
26. C.-J. Lai, J. E. Dahlberg, and B. Weisblum, *Bchem.* **12**, 457 (1973).
27. C.-J. Lai, B. Weisblum, S. R. Fahnestock, and M. Nomura, *JMB* **74**, 67 (1973).
28. G. W. Tischendorf, H. Zeichhardt, and G. Stöffler, *PNAS* **72**, 4820 (1975).
29. H. Hatayama, M. Yukioka, and S. Morisawa, *Mol. Biol. Rep.* **2**, 181 (1975).
30. N. Sonnenberg, M. Wilchek, and A. Zamir, *PNAS* **70**, 1423 (1973).
31. O. Pongs and W. Messer, *JMB* **101**, 171 (1976).
32. O. Pongs, R. Bald, and V. Erdmann, *PNAS* **70**, 2229 (1973).
33. O. Pongs, R. Bald, T. Wagner, and V. Erdmann, *FEBS Lett.* **35**, 137 (1973).
34. P. Greenwell, R. J. Harris, and R. H. Symons, *EJB* **49**, 539 (1974).
35. R. Harris, P. Greenwell, and R. Symons, *BBRC* **55**, 117 (1973).
36. A. Nicholson and B. Cooperman, *FEBS Lett.* **90**, 203 (1978).

37. B. Cooperman, E. Jaynes, D. Brunswick, and M. Luddy, *PNAS* **72**, 2974 (1975).
38. S. Dietrich, I. Schrandt, and K. Nierhaus, *FEBS Lett.* **47**, 136 (1974).
39. G. Stöffler, *Abstr. Int. Congr. Biochem., 10th, 1976* p. 87 (1976).
40. E. Jaynes, P. Grant, R. Wieder, G. Giangrande, and B. Cooperman, *Bchem.* **17**, 561 (1978).
41. J. Stahl, K. Dressler, and H. Bielka, *FEBS Lett.* **47**, 167 (1974).
42. M. Minks, M. Ariatti, and A. O. Howtrey, *ZpChem* **356**, 109 (1975).
43. K. Nierhaus and V. Montejo, *PNAS* **70**, 1931 (1973).
44. K. Nierhaus, V. Montejo, D. Nierhaus, S. Dietzich, and I. Schrandt, *Acta Biol. Med. Ger.* **33**, 613 (1974).
45. G. A. Howard and J. Gordon, *FEBS Lett.* **48**, 271 (1974).
46. J. P. G. Ballesta and D. Vazquez, *FEBS Lett.* **48**, 266 (1974).
47. C. Bernabeu, D. Vazquez, and J. Ballesta, *EJB* **79**, 469 (1977).
48. I. L. Armstrong and W. P. Tate, *JMB* **120**, 155 (1978).
49. B. I. Schrier and W. Möller, *FEBS Lett.* **54**, 130 (1975).
50. J. H. Highland, G. A. Howard, E. Ochsner, G. Stöffler, R. Hasenbank, and J. Gordon, *JBC* **250**, 1141 (1977).
51. M. Kazemie, *EJB* **67**, 373 (1976).
52. L. Christiansen and K. H. Nierhaus, *PNAS* **73**, 1839 (1976).
53. C. Bernabeu, D. Vazquez, and J. P. G. Ballesta, *EJB* **69**, 233 (1976).
54. F. Dohme and K. H. Nierhaus, *PNAS* **73**, 2221 (1976).
55. H. Teraoka and K. Nierhaus, *FEBS Lett.* **88**, 223 (1978).
56. R. Reyes, D. Vazquez, and J. P. G. Ballesta, *BBA* **435**, 317 (1976).
57. R. Reyes, D. Vazquez, and J. P. G. Ballesta, *EJB* **73**, 25 (1977).
58. P. Westermann, B. Gross, and W. Haumann, *Acta Biol. Med. Ger.* **33**, 699 (1974).
59. G. Stöffler, *in* "Ribosomes" (N. Nomura, A. Tissières and P. Lengyel, eds.), p. 615. Cold Spring Harbor Lab., Cold Spring Harbor, New York, 1974.
60. W. P. Tate, C. T. Caskey, and G. Stöffler, *JMB* **93**, 375 (1975).
61. S. Fahnestock, *Bchem.* **14**, 5321 (1975).
62. J. A. Maassen and W. Möller, *JBC* **253**, 2777 (1978).
63. R. R. Traut, R. L. Heimark, T. T. Sun, A. Bollen, and J. W. B. Hershey, *in* "Ribosomes" (M. Nomura, A. Tissières and P. Lengyel, eds.), p. 271. Cold Spring Harbor Lab., Cold Spring Harbor, New York, 1974.
64. A. Expert-Bezançon, D. Barritault, M. Milet, and D. H. Hayes, *JMB* **108**, 781 (1976).
65. C. S. José, C. G. Kurland, and G. Stöffler, *FEBS Lett.* **71**, 133 (1976).
66. U. Fabian, *FEBS Lett.* **71**, 256 (1976).
67. J. Ofengand and K. Hanes, *JBC* **244**, 6241 (1969).
68. D. Richter, V. A. Erdmann, and M. Sprinzl, *Nature NB* **246**, 132 (1973).
69. J. Ofengand, *in* "Molecular Mechanisms of Protein Biosynthesis" (H. Weissbach and S. Pestka, eds.), p. 8. Academic Press, New York, 1977.
70. M. Glukhova, N. Belitsina, and A. Spirin, *EJB* **52**, 197 (1975).
71. M. Sprinzl, T. Wagner, S. Lorenz, and V. A. Erdmann, *Bchem.* **15**, 3031 (1976).
72. S. V. Kirillov, V. I. Makhno, V. B. Odinzov, and Y. P. Semenkov, *EJB* **89**, 304 (1978).
73. M. Sprinzl and F. Cramer, This series **22**, 1 (1978).
74. M. Takanami, *PNAS* **52**, 1271 (1964).
75. J. P. Waller, T. Erdös, F. Lemoine, S. Guttemann, and E. Sandrin, *BBA* **119**, 566 (1966).
76. D. Nathans, *in* "Antibiotics" (D. Gottlieb and P. D. Shaw, eds.), Vol. 1, p. 251. Springer-Verlag, Berlin, and New York, 1967.

77. S. Pestka, *ARB* **40**, 697 (1971).
78. S. Pestka, *Annu. Rev. Microbiol.* **25**, 487 (1971).
79. I. D. Raake, *BBRC* **43**, 168 (1971).
80. M. Sundaralingam and S. K. Arora, *JMB* **71**, 49 (1972).
81. N. Yathindra and M. Sundaralingam, *BBA* **308**, 17 (1973).
82. H. P. M. de Leeuw, J. R. de Jager, H. J. Koeners, J. H. van Boom, and C. Altona, *EJB* **76**, 209 (1977).
83. C. Li and J. Žemliča, *J. Org. Chem.* **42**, 706 (1977).
84. P. Bhutta, C. Li, and J. Žemlička, *BBRC* **77**, 1237 (1977).
85. C. Li, P. Bhutta, and J. Žemlička, *Bchem.* **17**, 2537 (1978).
86. R. E. Monro and K. A. Marcker, *JMB* **25**, 347 (1967).
87. R. E. Monro, J. Černa, and K. A. Marcker, *PNAS* **61**, 1042 (1968).
88. R. E. Monro, *Nature* **223**, 903 (1969).
89. R. Neth, R. E. Monro, C. Heller, E. Battaner, and D. Vazquez, *FEBS Lett.* **6**, 198 (1970).
90. A. K. Falvey and T. Staehelin, *JMB* **53**, 1 (1970).
91. E. Battanez and D. Vazquez, *BBA* **254**, 316 (1971).
92. L. Carrasco and D. Vazquez, *EJB* **50**, 317 (1975).
93. M. M. Sikorski, J. Černa, I. Rychlik, and A. B. Legocki, *BBA* **475**, 123 (1977).
94. J. Černa, I. Rychlik, A. A. Krayevsky, and B. P. Gottikh, *FEBS Lett.* **37**, 188 (1973).
95. A. A. Krayevsky, M. K. Kukhanova, and B. P. Gottikh, *NARes* **2**, 2223 (1975).
96. J. Černa, I. Rychlik, A. A. Krayevsky and B. P. Gottikh, *Acta Biol. Med. Ger.* **33**, 877 (1974).
97. V. V. Kotusov, M. K. Kukhanova, V. S. Victorova, A. A. Krayevsky, A. D. Treboganov, N. V. Gnutchev, V. L. Florentiev, and B. P. Gottikh, *Mol. Biol.* (*Moscow*) **10**, 1394 (1976).
98. V. V. Kotusov, M. K. Kukhanova, A. A. Krayevsky, and B. P. Gottikh, *Mol. Biol. Rep.* **3**, 151 (1976).
99. V. V. Kotusov, M. K. Kukhanova, N. E. Sal'nikova, L. V. Nikolayeva, A. A. Krayevsky, and B. P. Gottikh, *Mol. Biol.* (*Moscow*) **11**, 671 (1977).
100. V. V. Kotusov, L. V. Nikolayeva, M. K. Kukhanova, A. A. Krayevsky, and B. P. Gottikh, *Mol. Biol.* (*Moscow*) **12**, 628 (1978).
101. M. K. Kukhanova, A. A. Krayevsky, B. P. Gottikh, and I. Stahl, *FEBS Lett.* **101**, 225 (1979).
102. M. K. Kukhanova, A. A. Krayevsky, and B. P. Gottikh, *Mol. Biol.* (*Moscow*) **11**, 1357 (1977).
103. J. Černa, *FEBS Lett.* **58**, 94 (1975).
104. A. A. Krayevsky, L. S. Victorova, V. V. Kotusov, M. K. Kukhanova, A. D. Treboganov, N. B. Tarussova, and B. P. Gottikh, *FEBS Lett.* **62**, 101 (1976).
105. B. Beltchev, M. Yaneva, and D. Staynov, *EJB* **64**, 507 (1976).
106. S. Hecht, *Tetrahedron* **33**, 1671 (1977).
107. M. Sprinzl, K. H. Scheit, H. Sternbach, F. von der Haar, and F. Cramer, *BBRC* **51**, 881 (1973).
108. M. Sprinzl and F. Cramer, *Nature NB* **245**, 3 (1973).
109. S. M. Hecht, J. W. Kozarich, and F. J. Schmidt, *PNAS* **71**, 4317 (1974).
110. F. Cramer, F. von der Haar, and E. Schlimme, *FEBS Lett.* **2**, 136 and 354 (1968).
111. T. H. Fraser and A. Rich, *PNAS* **70**, 2671 (1973).
112. G. Chinali, M. Sprinzl, A. Parmeggiani, and F. Cramer, *Bchem.* **13**, 3001 (1974).
113. S. M. Hecht and A. C. Chinault, *PNAS* **73**, 405 (1976).
114. S. M. Hecht, *Acc. Chem. Res.* **10**, 239–245 (1977).
115. E. Baksht, N. de Groot, M. Sprinzl, and F. Cramer, *Bchem.* **15**, 3639 (1976).

116. H. G. Zachau and W. Karau, *Chem. Ber.* **93**, 1830 (1960).
117. J. Ofengand and C.-M. Chen, *JBC* **247**, 2049 (1972).
118. R. Calender and P. Berg, *JMB* **26**, 39 (1967).
119. J. C.-H. Mao, *BBRC* **52**, 595 (1973).
120. A. Panet, N. de Groot, and Y. Lapidot, *EJB* **15**, 222 (1970).
121. N. de Groot, A. Panet, and Y. Lapidot, *EJB* **15**, 215 (1970).
122. S. Watanabe, *JMB* **67**, 443 (1972).
123. M. K. Kukhanova, A. A. Krayevsky, A. A. Alexandrova, S. A. Streltsov, and B. P. Gottikh, *Mol. Biol. Rep.* **1**, 397 (1974).
124. E. S. Bochkareva, V. G. Budker, A. S. Girshovich, D. G. Knorre, and N. M. Teplova, *Dokl. Akad. Nauk SSSR* **201**, 353 (1971).
125. E. Baksht, N. de Groot, M. Sprinzl, and F. Cramer, *FEBS Lett.* **55**, 105 (1975).
126. S. C. Uretsky, G. Acs, E. Reich, M. Mori, and L. Altwerger, *JBC* **243**, 306 (1968).
127. R. M. Sundari, H. Pelka, and L. H. Schulman, *JBC* **252**, 3941 (1977).
128. Z. Hussain and J. Ofengand, *BBRC* **49**, 1588 (1972).
129. S. Pestka, R. Vince, S. Daluge, and R. Harris, *Antimicrob. Agents & Chemother.* **4**, 37 (1973).
130. S. Pestka, *in* "Methods in Enzymology" (L. Grossman and K. Moldave, eds.), vol. 30, Part F, pp. 470 and 479. Academic Press, New York, 1974.
131. E. F. Vanin, P. Greenwell, and R. H. Symons, *FEBS Lett.* **40**, 124 (1974).
132. R. J. Harris, J. E. Hanlon, and R. H. Symons, *BBA* **240**, 244 (1971).
133. D. J. Eckermann, P. Greenwell, and R. H. Symons, *EJB* **41**, 547 (1974).
134. S. Harbon and F. Chapeville, *EJB* **13**, 375 (1970).
135. A. A. Krayevsky and B. P. Gottikh, *in* "Molecular Bases of Protein Synthesis" (V. A. Engelhardt, ed.), p. 241. "Nauka," USSR, 1971.
136. I. Rychlik, J. Černa, S. Chladek, J. Žemlička, and Z. Haladova, *JMB* **43**, 13 (1969).
137. J. Černa, I. Rychlik, J. Žemlička, and S. Chladek, *BBA* **204**, 203 (1970).
138. J. Žemlička, S. Chladek, D. Ringer, and K. Quiggle, *Bchem.* **14**, 5239 (1975).
139. S. Chladek, D. Ringer, and E. M. Abraham, *NARes* **3**, 1215 (1976).
140. S. V. Popovkina, S. G. Zavgorodny, A. V. Azhayev, V. V. Kotusov, L. S. Victorova, M. K. Kukhanova, A. A. Krayevsky, and B. P. Gottikh, *Mol. Biol. (Moscow)* **12**, 397 (1978).
141. S. S. Tavale and H. M. Sobell, *JMB* **48**, 109 (1970).
142. M. Ikehara, S. Uesugi, and K. Yoshida, *Bchem.* **11**, 830 (1972).
143. K. Quiggle, M. L. Wejrowski, and S. Chladek, *Bchem.* **17**, 94 (1978).
144. V. A. Pozdnyakov, Yu. V. Mitin, M. K. Kukhanova, L. V. Nikolayeva, A. A. Krayevsky, and B. P. Gottikh, *FEBS Lett.* **24**, 117 (1972).
145. M. K. Kukhanova, V. A. Pozdnyakov, A. A. Krayevsky, Yu. V. Mitin, and B. P. Gottikh, *Mol. Biol. (Moscow)* **8**, 389 (1974).
146. D. Ringer, K. Quiggle, and S. Chladek, *Bchem.* **14**, 514 (1975).
147. D. Ringer and S. Chladek, *FEBS Lett.* **39**, 75 (1974).
148. D. Ringer and S. Chladek, *BBRC* **56**, 760 (1974).
149. L. A. Alexandrova and J. Smrt, *Collect. Czech. Chem. Commun.* **42**, 1694 (1977).
150. J. Černa, S. Chladek, I. Rychlik, and J. Žemlička, *BBA* **199**, 291 (1970).
151. R. Vince and S. Daluge, *J. Med. Chem.* **17**, 578 (1973).
152. R. G. Almquist and R. Vince, *J. Med. Chem.* **16**, 1396 (1973).
153. H. Hussain and J. Ofengand, *BBRC* **50**, 1143 (1973).
154. S. Chladek, D. Ringer, and J. Žemlicka, *Bchem.* **12**, 5135 (1973).
155. I. Rychlik, S. Chladek, and J. Žemlička, *BBA* **138**, 640 (1967).
156. B. P. Gottikh, L. V. Nikolayeva, A. A. Krayevsky, and L. L. Kisselev, *FEBS Lett.* **7**, 112 (1970).

157. N. B. Tarussova, L. V. Victorova, T. L. Tsilevich, M. K. Kukhanova, A. A. Krayevsky, and B. P. Gottikh, *Bioorg. Khim.* **2**, 69 (1976).
158. E. J. Hengesh and A. J. Morris, *BBA* **299**, 654 (1973).
159. R. H. Symons, R. J. Harris, L. P. Clarke, J. F. Wheldrake, and W. H. Elliott, *BBA* **179**, 248 (1969).
160. J. Černa, A. Holy, and I. Rychlik, *Collect. Czech. Chem. Commun.* **43**, 3279 (1978).
161. D. Nathans and A. Neidle, *Nature* **197**, 1076 (1963).
162. W. B. Butler and N. R. Maledon, *BBA* **454**, 329 (1976).
163. J. L. Lessard and S. Pestka, *JBC* **247**, 6901 (1972).
164. I. Rychlik, J. Černa, S. Chladek, P. Pulkrabek, and J. Žemlička, *EJB* **16**, 136 (1970).
165. R. Vince and K.-L. Fong, *BBRC* **81**, 559 (1978).
166. F. W. Lichtenthaller, T. Morino, and I. Rychlik, unpublished results.
167. M. Ariatti and A. O. Hawtrey, *BJ* **145**, 169 (1975).
168. M. K. Kukhanova, A. A. Krayevsky, R. Ja. Vigestane, and B. P. Gottikh, *Biofizika* **22**, 719 (1977).
169. M. A. Kharshan, R. Ja. Vigestane, S. V. Popovkina, M. K. Kukhanova, A. A. Krayevsky, and B. P. Gottikh, *Bioorg. Khim.* **3**, 494 (1977).
170. S. Pestka, *PNAS* **69**, 624 (1972).
171. R. Miskin, A. Zamir, and D. Elson, *JMB* **54**, 355 (1970).
172. R. Miskin, A. Zamir, and D. Elson, *BBRC* **33**, 551 (1968).
173. H. Teraoka and K. Tanaka, *BBA* **247**, 309 (1976).
174. R. Miskin and A. Zamir, *JMB* **87**, 121 (1974).
175. J. Černa, *FEBS Lett.* **15**, 101 (1971).
176. S. Pestka, R. Goorha, H. Rosenfeld, C. Neurath, and H. Hintikka, *JBC* **247**, 4258 (1972).
177. B. Edens, H. A. Thompson, and K. Moldave, *Bchem.* **14**, 54 (1975).
178. H. A. Thompson and K. Moldave, *Bchem.* **13**, 1348 (1974).
179. M. L. Celma, R. E. Monro, and D. Vazquez, *FEBS Lett.* **6**, 273 (1970).
180. R. E. Monro, T. Staehelin, M. L. Celma, and D. Vazquez, *CSHSQB* **34**, 357 (1969).
181. D. Vazquez, E. Battaner, R. Neth, G. Heller, and R. E. Monro, *CSHSQB* **34**, 369 (1969).
182. E. Silverstain, *BBA* **186**, 402 (1969).
183. C. T. Caskey, A. L. Beaudet, E. Scolnick, and M. Rosman, *PNAS* **68**, 3136 (1971).
184. E. Cundliffe and K. McQuillen, *JMB* **30**, 137 (1967).
185. S. Fahnestock, H. Neumann, V. Shashoua, and A. Rich, *Bchem.* **9**, 2477 (1970).
186. S. Fahnestock and A. Rich, *Nature* **229**, 8, 10 (1971); *Nature NB* **229**, 8 (1971).
187. J. Gooch and A. O. Hawtrey, *BJ* **149**, 209 (1975).
188. L. S. Victorova, L. S. Kotusov, A. V. Azhayev, A. A. Krayevsky, M. K. Kukhanova, and B. P. Gottikh, *FEBS Lett.* **68**, 215 (1976).
189. E. Scolnick, G. Milman, M. Rosman, and T. Caskey, *Nature* **225**, 152 (1970).
190. V. T. Innanen and D. M. Nicholls, *BBA* **361**, 221 (1974).
191. J. Černa, I. Rychlik, and J. Jonak, *EJB* **34**, 551 (1973).
192. V. G. Moore, R. E. Atchison, G. Thomas, M. Moran, and H. F. Noller, *PNAS* **72**, 844 (1975).
193. K. K. Wan, N. D. Zahid, and R. M. Baxter, *EJB* **58** 397 (1975).
194. R. J. Harris and S. Pestka, *in* "Molecular Mechanism of Protein Biosynthesis" (H. Weisbach and S. Pestka, eds.), p. 413. Academic Press, New York, 1977.
195. F. Seela and V. A. Erdmann, *BBA* **435**, 105 (1976).
196. J. Žemlicka and J. Owens, *J. Org. Chem.* **42**, 517 (1977).
197. M. C. Ganoza and J. L. Fox, *JBC* **249**, 1037 (1974).

198. B. R. Glick and M. C. Ganoza, *PNAS* **72**, 4257 (1975).
199. B. R. Glick and M. C. Ganoza, *EJB* **71**, 483 (1976).
200. R. J. Harris and R. H. Symons, *Bioorg. Chem.* **2**, 286 (1973).
201. J. Žemlicka and J. Owens, *BBA* **442**, 71 (1976).
202. R. Harris and S. Pestka, *JBC* **248**, 1168 (1973).
203. S. Pestka, *ABB* **136**, 80 (1970).
204. M. K. Kukhanova, S. A. Streltsov, L. S. Victorova, A. V. Azhayev, B. P. Gottikh, and A. A. Krayevsky, *FEBS Lett.* **102**, 198 (1979).
205. S. V. Streltsov, A. V. Kosenuk, M. K. Kukhanova, A. A. Azhayev, A. A. Krayevsky, and B. P. Gottikh, *FEBS Lett.* **104**, 279 (1979).
206. P. Leder, *Adv. Protein Chem.* **27**, 213 (1973).
207. M. Yukioka and S. Morisawa, *BBA* **254**, 304 (1971).
208. A. A. Krayevsky, M. K. Kukhanova, A. V. Azhayev, and B. P. Gottikh, *in* "tRNA, Chemistry and Synthesis" (M. Wiěworowsky, ed.), p. 249. Kiekrz, Poland, 1976.
209. S. Pestka, *JBC* **244**, 1533 (1969).
210. L. P. Gavrilova and A. S. Spirin, *in* "Methods in Enzymology" (L. Grossman and K. Moldave, eds.), Part F., Vol. 30, p. 452. Academic Press, New York, 1974.
211. A. V. Azhayev, S. V. Popovkina, N. B. Tarussova, A. A. Krayevsky, M. K. Kukhanova, and B. P. Gottikh, *NARes* **4**, 2223 (1977).
212. M. Springer and M. Grunberg-Manago, *BBRC* **47**, 477 (1972).
213. D. P. Leader, J. G. Wool, and J. J. Castles, *BJ* **124**, 537 (1971).
214. E. Baksht and N. de Groot, *Mol. Biol. Rep.* **1**, 493 (1974).
215. S. Pestka, H. Rosenfeld, R. Harris, and H. Hintikka, *JBC* **247**, 6895 (1972).
216. G. Chinali, R. Liou, and J. Ofengand, *Bchem.* **17**, 2761 (1978).
217. D. Richter, *BBRC* **81**, 359 (1978).
218. M. Sprinzl and D. R. Richter, *EJB* **71**, 171 (1976).
219. S. Pestka, This Series **17**, 217 (1976).
220. S. Pestka, *in* "Molecular Mechanisms of Protein Biosynthesis" (H. Weissbach and S. Pestka, eds.), p. 467. Academic Press, New York, 1977.
221. T. Hishizawa, J. L. Lessard, and S. Pestka, *PNAS* **66**, 523 (1970).
222. B. Ulbrich, G. Mertens, and K. H. Nierhaus, *ABB* **190**, 149 (1978).
223. A. Contreras and D. Vazquez, *EJB* **74**, 539 (1977).

Patterns of Nucleic Acid Synthesis in *Physarum polycephalum*

GEOFFREY TURNOCK

Department of Biochemistry
University of Leicester
Leicester, England

I. Introduction

A. Life Cycle

The slime mold *Physarum polycephalum* is classified as a Myxomycete, a type of organism that has affinities with both protozoa and fungi (*1*). It has a life cycle (Fig. 1) that includes two vegetative growth phases, one of which, the plasmodial stage, constitutes the most notable feature of the Myxomycetes. The transition from the microscopic, uninucleate amebae to the macroscopic, multinucleate

Progress in Nucleic Acid Research
and Molecular Biology, Vol. 23

ISBN 0-12-540023-3

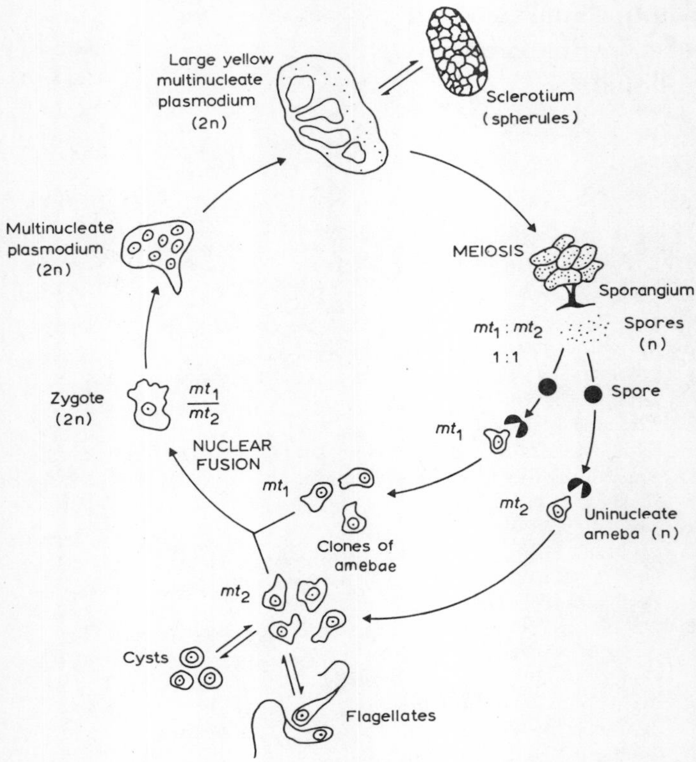

FIG. 1. The life cycle of *Physarum polycephalum*, illustrating the formation of a heterothallic plasmodium by the mating of amebae of different mating types (mt_1, mt_2) (n = haploid, 2n = diploid). From Ashworth and Dee (*1*).

plasmodium is achieved by sexual fusion of amebae carrying different alleles at a mating type locus (*2*), although there is one allele of the same locus that does permit plasmodium formation to occur within clones (*3*). Maintaining a plasmodium under conditions of starvation and exposure to light leads to the production of spores, which hatch in a favorable environment to yield amebae. In contrast, starvation of a plasmodium in the dark results in the production of a dry, resistant structure called a sclerotium. This, when placed in contact with growth medium, hatches to regenerate a plasmodium.

This life cycle has several aspects that might be of interest to a cell biologist; for example, sporulation, sclerotia formation, and the amebal–plasmodial transition. Of these, the first two have analogies in other lower eukaryotes; the third constitutes an interesting example of the differentiation of one type of cell to another, but study of the biochemical details of the transition has been limited until recently by the inability of amebae to grow in the absence of a bacterial food source. However, this problem has now been solved (*4*), and a biochemical and genetical analysis of the amebal–plasmodial transition should soon be possible.

As stated in the first paragraph, it is the plasmodial form that is the unique feature of the Myxomycetes and that has largely stimulated interest in the study of *P. polycephalum* and related organisms. This is so because a plasmodium, which lacks a cell wall and is indefinite in form, grows without cell division, and the nuclei in the common cytoplasm divide with a high degree of synchrony. That this synchrony is natural in origin and does not require the chemical or physical perturbation of the organism makes the plasmodial form of *Physarum* an ideal system in which to analyze biochemical processes in relation to the nuclear division cycle. The purpose of this article is to review the work that has been done on the synthesis of DNA and RNA during the cycle, correlating wherever possible both *in vivo* and *in vitro* experiments.

B. Growth of Plasmodia

The plasmodial form of *Physarum*, for which an axenic medium was developed some years ago (*5*), can be grown in a number of different ways: (*a*) on medium solidified with agar (Fig. 2); (*b*) on filter paper supported on a stainless steel grid in a vessel containing sufficient growth medium to maintain contact with the underside of the paper; and (*c*) in suspension culture obtained by transferring a piece of plasmodium to liquid medium in a conical flask. Agitation on a reciprocating shaker or vigorous stirring produces spontaneous fragmentation of the so-called microplasmodia as growth proceeds, and cultivation in this form is analogous to that of many other microorganisms.

The ability to grow plasmodia in a variety of different modes provides flexibility in the design of experiments for different purposes. For example, suspension cultures of microplasmodia provide the most convenient way of maintaining stocks; in our laboratory, a new line of microplasmodia is initiated from a stock of sclerotia at intervals of four to five months in order to minimize any selective changes that may be produced by continuous growth in liquid culture. Although plasmodia are easily maintained on agar plates, the time for which a particular line may be subcultured by serial transfer at regular intervals is limited by a process of aging (*6*). The length of time for which a plas-

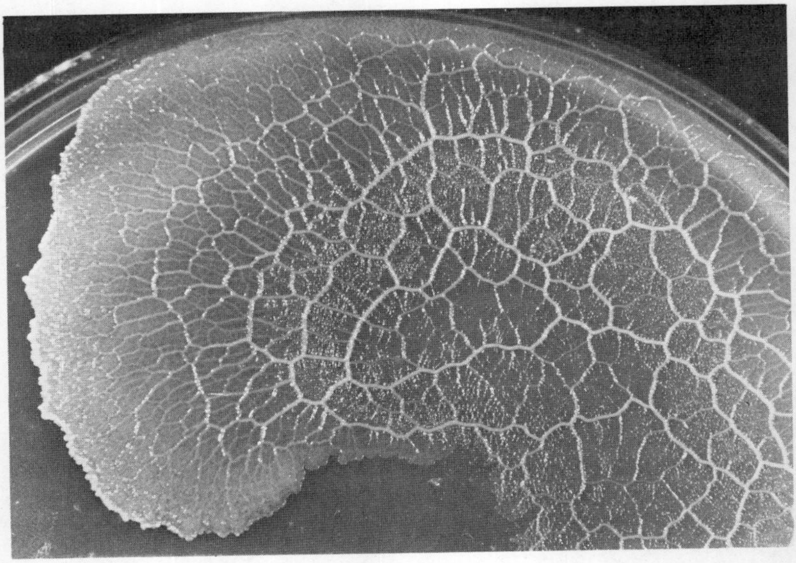

FIG. 2. Photograph (magnification ×2) of a plasmodium of *Physarum* growing on an agar plate. The network of veins that extends from the growing edge of the plasmodium is a characteristic morphological feature; vigorous cytoplasmic streaming occurs within the veins.

modium may be maintained varies from strain to strain, but is usually several months; the biochemical basis of the ultimate death of a plasmodium has not been established, although the progressive accumulation of abnormal nuclei has been observed in one instance (7).

New plasmodia are easily initiated on agar plates, but the close juxtaposition of the plasmodium to the surface makes it difficult to take samples for biochemical analysis without contamination of the material with solidified medium. Accordingly, most experiments designed to exploit the synchronous nature of the nuclear division in plasmodia utilize cultures growing on a filter paper or membrane filter support. To obtain such cultures, the procedure used is briefly as follows. A culture of microplasmodia in the early exponential phase of growth is harvested by brief centrifugation and resuspended in a small volume of sterile distilled water, and portions of the concentrated suspension of microplasmodia are pipetted onto the dry paper (5). After 30 to 45 minutes to allow fusion of the microplasmodia, the growth medium is gently layered under the paper until contact between paper and liquid is attained. The cultures are then incubated in the dark at 26°C, the optimum temperature for growth.

C. Nuclear Division in *Physarum*

Details of the nuclear division cycle in growing plasmodia have been provided by both light (*8–10*) and electron microscopy (*9–11*). The salient features of mitosis are that it is synchronous within one plasmodium and that each nucleus divides within a closed nuclear envelope. The approximate times occupied by the successive stages of the nuclear-division cycle are as follows: early prophase (15 to 20 minutes), prophase (5 minutes), metaphase (7 minutes), anaphase (3 minutes), telophase (5 minutes), reconstruction (75 minutes), and interphase (480 minutes). A single, large nucleolus is the dominant, cytological feature of the interphase nucleus, and it is the characteristic changes undergone by this organelle that identify in particular the early stages of mitosis.

The observation that the division of nuclei in the common cytoplasm of a plasmodium is synchronous immediately suggests the involvement of cytoplasmic factors in the regulation of nuclear division, and a number of studies have been designed to investigate the nature of the nuclear–cytoplasmic interactions (*12, 13*). Because many experiments have used plasmodia that have only undergone three or four nuclear divisions subsequent to the fusion of microplasmodia, it is important to appreciate that the synchronous nuclear divisions observed under these conditions are in no sense artificial. The synchrony may be observed to continue for many generations of growth if a plasmodium is subdivided at intervals, thus keeping the size of the "cell" within reasonable limits. However, in a plasmodium established in a large tray apparatus and allowed to grow to the third postfusion mitosis, nuclei from opposite sides may be out of phase by 20 to 40 minutes (*8*). Within an intermitotic time of about nine hours, this is an acceptable degree of uncertainty and does not outweigh the advantages, for many biochemical experiments, of being able to sample from a single culture.

D. Isolation of Nuclei

The absence of a cell wall makes the isolation of nuclei by gentle procedures a relatively easy matter (*14*). Also, the restriction of mitosis as an intranuclear event means that it is possible to isolate nuclei at any stage of the division cycle, although the nuclear membrane is more fragile during the period from prophase to telophase. Nuclei are separated from other cell material by differential centrifugation, and are thus readily prepared for *in vitro* biochemical studies.

E. Preparation of Nucleic Acids

DNA and all the major classes of RNA may be isolated from plasmodia and subcellular fractions provided rigorous conditions to minimize damage by nucleases are maintained (*15–17*). Polysaccharides are a major contaminant of preparations of nucleic acids. However, DNA may be selectively precipitated with streptomycin (*18*) and RNA may be purified by patient reprecipitation with alcohol from dilute solution.

F. Radioactive Labeling of Nucleic Acids

Radioactive nucleic acid precursors, principally nucleosides, have been used to label nucleic acids in *Physarum* since studies of the organism began in the early 1960s. However, only recently has a systematic investigation of the metabolism of nucleosides and of its significance for labeling experiments been carried out.

The uptake and metabolism of uridine has been analyzed using both microplasmodia and surface cultures (*19*). Uptake by microplasmodia occurs via a saturable transport system with a K_m of approximately 30 μM. Other ribo- and deoxyribonucleosides are competitive inhibitors of the uptake of uridine and vice versa, and it is surmised that there is a common uptake system for nucleosides. Uridine can be detected in the intracellular pool at a concentration higher than that in the medium, suggesting that it enters microplasmodia by a process of active transport.

Of the radioactivity (derived from [5,6-^{3}H]- or [2-^{14}C]uridine) that is incorporated into macromolecules, 90 to 95% is found in nucleic acids, the ratio between the radioactivity in RNA and that in DNA being about 30 : 1 (*20*). However, uridine, as well as being incorporated into the nucleotide pool, is also catabolized via uracil to yield $^{14}CO_2$ from [2-^{14}C]uridine. Although the enzyme that can convert uracil to UMP has been detected in cell-free extracts of *Physarum* (*20a*), this reaction does not appear to constitute a major pathway, because uracil rapidly becomes detectable in the acid-soluble pool and progressively accumulates in the medium (*19*). A practical consequence of the operation of the catabolic pathway for uridine is that continuous radioactive labeling of nucleic acids cannot be achieved without resupplying the radioactive precursor at appropriate intervals.

The rate of uptake of uridine by surface plasmodia is a linear function of its concentration in the medium, providing an interesting contrast with the saturable transport system detected in microplasmodia (*19*). Entry of uridine into a plasmodium must occur either by passive diffusion or pinocytosis or a combination of both processes. It makes the discovery of active uptake in microplasmodia rather more surprising in that this mode of growth does not represent a "natural" form of the organism. However, it should be recalled that, subsequent to the fusion of two amebae, the size of the plasmodium will for some time approximate that of the microplasmodia generated in shake culture. It is feasible that the requirement to take advantage of any soluble nutrients changes as a function of plasmodial growth.

The utilization of thymidine by microplasmodia is interesting because it is subject to extensive catabolism even when supplied at low concentrations. Table I summarizes the distribution of radioactivity derived from [2-^{14}C]-, [6-^{3}H]-, and [Me-^{3}H]thymidine in microplasmodia growing in mycological peptone, a medium that fortuitously contains only small amounts of nucleosides (*19*). The extent to which RNA and protein can be labeled relative to DNA is striking. Thymidine is rapidly taken up and the radioactivity that remains in the medium after 24 hours of growth represents products of catabolism that cannot readily be utilized. The data presented for [2-^{14}C]-thymidine show that the pattern of metabolism can change with the concentration of nucleoside in the medium. This emphasizes the caution that must be exercised in attempting to use radioactive

TABLE I

UTILIZATION OF THYMIDINE BY MICROPLASMODIA OF *Physarum*[a]

	Concentration of added thymidine (μM)				
	[2-^{14}C]-			[6-^{3}H]-	[Me-^{3}H]-
	0.3	10	50	10	10
	Distribution (%) of radioactivity after 24 hours of growth in mycological peptone (*19*)				
Cells	38	8	6	15	15
Medium	9	18	33	85	85
Lost (as CO_2)	53	74	61	n.a.	n.a.
	Distribution (%) of radioactivity inside the cells				
Acid-soluble	5	8	5	69	43
Protein	24	30	28	7	27
RNA	2	18	16	5	13
DNA	69	44	50	19	17

[a] Unpublished data of L. Hall and G. Turnock.

thymidine to label DNA, especially in other peptone- and tryptone-based media that can contain substantial quantities of nucleosides. Fink and Nygaard (*20a*, and personal communication) have studied the catabolism of both thymidine and other deoxy- and ribonucleosides by *Physarum* and have established the existence of a number of the relevant enzymes by *in vitro* experiments.

G. Pulse-Labeling and Isotope-Dilution

Since the introduction of isotopically labeled precursors, it has been common practice to assess the rate of synthesis of macromolecules by measuring the incorporation of a pulse of radioactivity from an appropriate labeled compound. In most instances, however, no correction is made for dilution of the specific activity of the precursor by endogenous synthesis, so that comparison of the radioactivity incorporated by different samples can only be made on a qualitative basis. This follows because fluctuations in the pool size of the precursor will directly affect the specific activity it attains during the fixed time of the pulse, and this will be reflected in the amount of radioactivity incorporated into the macromolecule. This can of course be taken into account by measuring the specific activity of the precursor (*21*), although there is still the uncertainty in the case of eukaryotes as to whether the specific activity of, say, the total UTP of a cell will necessarily be equal to that of the triphosphate in the nucleus (*22*).

Pulse-labeling experiments with *Physarum* have not so far involved any attempt to relate the specific activity attained by an immediate precursor to the radioactivity incorporated into a macromolecule. For nucleic acids the complexity of the metabolism of nucleosides seems to make the assessment of such a relationship a necessary basis for any quantitative pulse-labeling studies. In a general context, Fink (*23*) has studied the pool sizes of both ribo- and deoxyribonucleoside triphosphates during the mitotic cycle in *Physarum* (Fig. 3). The amounts of all eight compounds showed variation during the cycle, the pools of the deoxyribonucleoside triphosphates expanding just before mitosis, whereas the ribonucleoside triphosphates increased during mitosis and also one hour and five hours after mitosis. The changes are large enough to make clear the importance of taking into account the specific activity of, say, UTP in measurements of RNA synthesis during the mitotic cycle by pulse-labeling with radioactive uridine.

As an alternative to the uncertainties of pulse-labeling, an isotope-dilution method for studying the synthesis of nucleic acids during the mitotic cycle has been developed (*20*). A culture of microplasmodia is given a small amount of radioactive nucleoside, which is

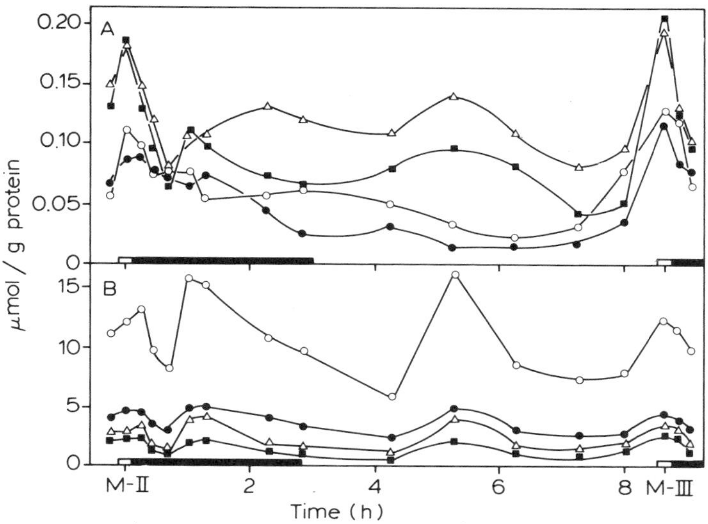

FIG. 3. Changes in the deoxyribo- and ribonucleoside triphosphate pools during the mitotic cycle. The figures show data from a single experiment in which a plasmodium was labeled with $^{32}P_i$. The ^{32}P activity of the triphosphates was determined after two-dimensional chromatography in the presence of standards. For each extract two or three thin-layer chromatography plates were developed. The open and black bars along the abscissas are a schematic representation of the duration of mitosis and S phase, respectively. (A) Deoxyribonucleoside triphosphates: ○, dATP; ●, dGTP; △, dTTP; ■, dCTP. (B) Ribonucleoside triphosphates: ○, ATP; ●, GTP; △, UTP; ■, CTP. From Fink (23).

rapidly utilized by incorporation into nucleic acids and also by catabolism, as described in Section I,F. After a period of growth sufficient to ensure that any radioactivity in metabolically unstable nucleic acids is recycled into stable species, a surface plasmodium is formed from the labeled microplasmodia and the specific activity of the nucleic acid concerned is determined from the second postfusion mitosis (M-II) onward. As the specific activity is inversely proportional to the amount of the nucleic acid in the culture, the relative accumulation curve for the nucleic acid is easily calculated.

The isotope-dilution method, in its simple form, is applicable only to metabolically stable nucleic acids, namely DNA, tRNA, and rRNA. However, elucidation of the patterns of synthesis of these macromolecules is of basic importance in the biochemical characterization of the synchronous nuclear division cycle. Also, once the accumulation curve for a particular molecule has been defined it is possible to make use of it to determine the characteristics of synthesis of other components that may not be metabolically stable. The synthesis of protein during the mitotic cycle has been measured in this way (*24*).

A further advantage of the isotope-dilution method for analyzing the synthesis of macromolecules in *Physarum* stems from the form of growth of a plasmodium as a single cell. For example, by comparison with the ease with which samples of defined size may be taken from suspension cultures of microorganisms or animal or plant cells, accurate "normalization" of samples from a plasmodium is difficult. The isotope-dilution method obviates this problem, because it requires only the isolation of a sample (suitably purified) of the macromolecule concerned, not its quantitative recovery from a defined amount of plasmodium.

II. Synthesis of DNA

A. The Genome of *Physarum*

1. Organization of the DNA

In the G2 phase of the mitotic cycle, a nucleus of a heterothallic strain of the Wisconsin isolate of *Physarum* has a DNA content (4C; C = DNA content of haploid set of chromosomes) of 1.0 to 1.2 pg (*25*), while germinating spores and amebae possess an amount of DNA close to the 2C value. This appreciable quantity of genetic material is distributed among a large number of small chromosomes, about 70 for the diploid plasmoidal nucleus (*25*), and evaluation of the karyotype has therefore been technically difficult. Chromosome spreads from amebae of heterothallic strains show approximately 40 chromosomes.

Measurements of the nuclear DNA content and chromosome number (*25*) of the Colonia strain, which is unusual in that plasmodia form within single clones of amebae (*3*), are consistent with the view that it is haploid throughout its life cycle. However, a plasmodium of a Colonia strain always contains a few diploid nuclei, and it is possible that these are involved in a conventional alternation of ploidy during sporulation (*26*).

The nuclear DNA of *Physarum*, like that of many eukaryotes, includes a significant proportion of repetitive sequences interspersed with nonrepetitive sequences. By measuring the kinetics of reassociation of denatured DNA, the proportion of repeated sequences has been calculated to be 45% with a $C_0t_{\frac{1}{2}}$ of 0.07 to 0.7 mol $L^{-1}s$. The nonrepeated base sequences reassociate with a $C_0t_{\frac{1}{2}}$ of 500 to 1100 mol $L^{-1}s$ and represent a complexity that is 130 times greater than that of the genome of *Escherichia coli* (*27*). A detailed study has been made of the class of repeated sequences organized in inverted register in the genome (*28*). Under appropriate conditions, such inverted repeat sequences can form so-called foldback or hairpin loops, and in *Physarum* DNA, the sequences that generate foldback DNA are 150 to 3000 nucleotides in length with a number average of 340. About half of the sequences involved form looped structures, rather than simple hairpins, with loop sizes averaging 1200 nucleotides in length. Foldback sequences appear to be widely distributed throughout the genome in that 60 to 80% of large DNA molecules (2.2×10^4 nucleotides in length; selected from alkaline sucrose gradients) contain such sequences (*29*). Moreover, the size of individual foldback duplexes and the length of sequences that separate them are nonrandom. The functional significance of inverted repeat sequences in the DNA of eukaryotes is as yet unknown. Some at least are represented in the hnRNA of mammalian cells (*30*), and it is consequently suggested that they may be involved in the regulation of gene expression. This idea is reinforced by the demonstration that mRNAs are probably generated by cleavage and ligation of different regions of individual molecules of hnRNA [reviewed by Williamson (*31*)].

The DNA that contains the genes for rRNA has been the subject of intensive study, and each unit of this DNA has been shown to possess a rotational axis of symmetry near its center, constituting an example of an inverted repeat structure (*32, 33*). Initially, it was demonstrated that the genes for rRNA, as in other eukaryotes, are located in the nucleolus (*34*). The nucleolar DNA has a greater density (1.714 g cm^{-3}) than the bulk of the nuclear DNA (1.702 g cm^{-3}) and forms a separate satellite band, containing about 1 to 2% of the DNA, when centrifuged to equilibrium in CsCl (*34*). Characterization of the nucleolar DNA by electron microscopy and restriction endonuclease mapping shows that it consists of linear molecules of a discrete size, $M_r = 37 \times 10^6$ (*32, 33*). The sequences that code for the 18 S, 26 S, and 5.8 S ribosomal RNAs are represented twice per intact molecule, once at each end. In addition, blocks of short, inverted repetitious sequences are located around the rotational axis of symmetry of the molecule. Although most of the investigations of rDNA in *Physarum* have employed the plasmodial form of the organism, the nucleolar DNA molecules extracted from amebae are known to be of a similar size (*16*). The data for amebae were obtained using cells growing axenically with a doubling time of 72 to 96 hours as compared to 8 to 10 hours for plasmodia. In both amebae and plasmodia, hybridization measurements indicate 275 genes for each of the rRNAs per haploid genome, and it thus seems clear that the amebal–plasmodial transition is not associated with any selective change in the amount of rDNA. Equally, rDNA as a percentage (0.16 to 0.18%) of total nuclear DNA does not change during spherulation (*35*) or in strains of different ploidy (*18*), and a similar value has recently been obtained with DNA from spores (*36*). These observations point to the existence of a control mechanism that can titrate the number of rDNA molecules in relation to the whole genome.

The observed number of extrachromosomal, nucleolar DNA molecules might represent a balance between synthesis and breakdown. However, isotope-dilution experiments show that both the rDNA and the rest of the nuclear DNA are metabolically stable during active growth of plasmodia (*37*). The question then arises as to how the partition of the molecules between the daughter nuclei during mitosis is organized. This has been studied by electron microscopy and high-resolution autoradiography (*11, 38*). The interphase nucleolus has a structure similar to that in other eukaryotes, consisting of large particulate zones surrounding fibrillar, denser regions. After fixing with osmium tetroxide, the fibrillar regions exhibit opaque granules approximately 25 nm in diameter. The opaque granules serve as a diagnostic feature within the globular bodies into which the nucleolus fragments during late prophase. The fragments persist during mitosis and are observed to contribute to the nucleoli that form within the daughter nuclei. Autoradiography has demonstrated the presence of DNA in the nucleolar remnants (*11*) so that, linking together the biochemical and cytological observations, it is reasonable to suppose that they represent the means by which the rDNA is distributed during nuclear division. Whether or not there is a mechanism that ensures that the distribution is equal cannot be determined from the electron micrographs. In a plasmodium, of course, the nuclei remain within a common cytoplasm, and so some variation in the number of rRNA genes per nucleus may not be important.

Whether any copies of the genes for rRNA are integrated into the chromosomal DNA of *Physarum* has not yet been ascertained. That such integration may occur is suggested by studies of the rDNA of *Tetrahymena pyriformis*. In the macronucleus of this protozoan, which divides amitotically, the rDNA also exists as a component of multiple copies of linear, independent molecules ($M_r = 13 \times 10^6$) (*39*). However, the formation of a macronucleus after conjugation is accompanied by an increase in the fraction of rDNA from 0.03% in the micronucleus to 0.3% in the macronucleus, and a single copy of rDNA is detectable in the chromosomal DNA of the micronucleus (*40*).

Our considerable knowledge of the genes that code for rRNA in

Physarum is not matched by equivalent information concerning other specific genes. Hall and Braun (*16*) have shown that, as in other eukaryotes, the genes for the 5 S rRNA are not linked to the genes for the other RNAs of the ribosome. When tested against DNA fractionated by centrifuging to equilibrium in CsCl, 5 S RNA hybridized to a component of the main band of the nuclear DNA, not to the nucleolar satellite. Similarly, unfractionated tRNA hybridized only with main-band DNA. The extent of the hybridization in the two cases showed that there are 685 genes for 5 S RNA, as compared with 275 genes for the other rRNAs, and 1050 genes for tRNAs.

2. CHROMATIN STRUCTURE

A ubiquitous feature of chromatin in eukaryotes is the arrangement of the DNA in repeating nucleoprotein subunits or nucleosomes, in which about 200 base-pairs of DNA are associated in a regular manner with histones. An assay frequently used for demonstrating this type of structure in chromatin involves partial digestion with micrococcal nuclease, which produces a characteristic repeat pattern among the DNA fragments; this technique shows that chromatin from *Physarum* is organized into nucleosomes (*41*, *42*).

The ease with which nucleoli may be isolated and the extra-chromosomal nature of the rDNA have provided the stimulus for investigating the structure of this specific fraction of the genome. Grainger and Ogle (*42a*) treated purified nucleoli with micrococcal nuclease and obtained a pattern of DNA fragments identical with that of the bulk of the chromatin from the nucleus, indicating that the rDNA is also organized in nucleosomes. A different approach exploited the fact that nucleolar DNA may be differentially labeled in G2 phase (Section II,B) (*42b*), and showed that nucleolar and main-band DNA generate identical repeat lengths (165 ± 5 base-pairs) when nuclei are digested with micrococcal nuclease. These results (*42b*) are important because they demonstrate that the structure of the nucleolar DNA is organized in a manner similar to that of the chromosomes, despite the fact that it consists of several hundred molecules that are independent of the remainder of the genome. The nucleoli, and chromatin derived therefrom, thus constitute a good system in which to study the relationship between structure, particularly the role of individual proteins, and function.

B. S Phase

The period occupied by the synthesis of the nuclear DNA is a major landmark in the progression of the cell cycle. In many types of cell, there is a gap (G1) between the completion of mitosis and the onset of S phase, and a second gap (G2) between the end of DNA synthesis and the next nuclear division (*43*). However, in the case of plasmodia of *Physarum*, pulse-labeling experiments (Fig. 4) with radioactive nucleic acid precursors indicate that the synthesis of DNA begins immediately after nuclear division (*44, 45*), that is to say, without the imposition of a detectable G1 phase. S phase itself appears to last two to three hours out of the eight to nine-hour intermitotic period.

Although the technique of pulse-labeling has the attraction of being very sensitive, it can give only a qualitative picture of the synthesis of the macromolecule that is being examined, unless the radioactivity incorporated can be related to the specific activity of the immediate precursor (Section I,G). Depending upon the rate of entry of radioactivity into the precursor pool, the specific activity of the precursor may increase continuously throughout the period of the pulse; see, for example, the kinetics of labeling of dTTP by radioactive thymidine supplied to plasmodia of *Physarum* (*46*). Rigorous numerical analysis of labeling experiments of this type is, therefore, difficult and has rarely been attempted; two examples of what may be achieved may be seen in studies of nucleic acid biosynthesis in *E. coli* (*21, 47*).

The uncertainties involved in the evaluation of pulse-labeling experiments have led us to apply the technique of isotope-dilution (Sec-

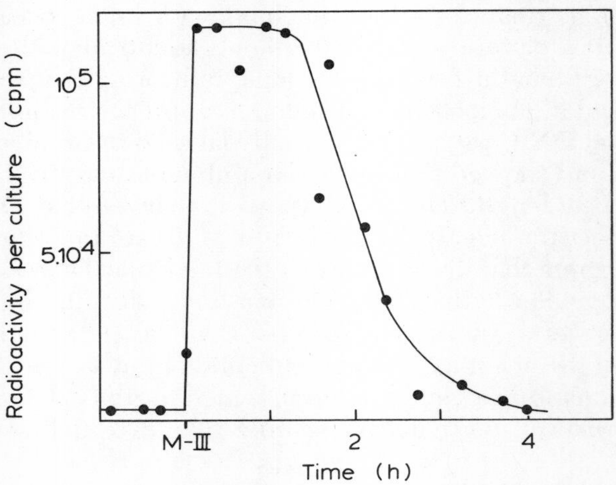

FIG. 4. Incorporation of [^{3}H]thymidine into DNA. Plasmodia were labeled for 10 minutes at different times during the mitotic cycle. After extraction with cold acid and treatment with RNase, the radioactivity of the insoluble residue was determined. From Braun *et al.* (*45*).

tion I,G) to the analysis of S phase in plasmodia of *Physarum*. Initial experiments (*20*) with strain M_3C-VIII showed rapid DNA synthesis during the first two hours of the mitotic cycle, so that by the end of this period approximately 80% of the DNA had replicated. Subsequently, there was a large decrease in the rate of DNA synthesis, which continued to the end of the cycle. An increase of 90 to 100% in nuclear DNA between postfusion mitoses II and III was routinely observed, the uncertainty in individual measurements being ±3 to 4%.

Figure 5 shows in detail the synthesis of DNA in early S phase in relation to the identifiable stages of nuclear division. It is clear that a G1 phase, if present, is too short to be detected within the limits of the method, so that the early conclusion, based on pulse-labeling experiments, that a G1 phase is absent from the nuclear division cycle of *Physarum* is substantiated. It would be interesting to see if a G1 phase

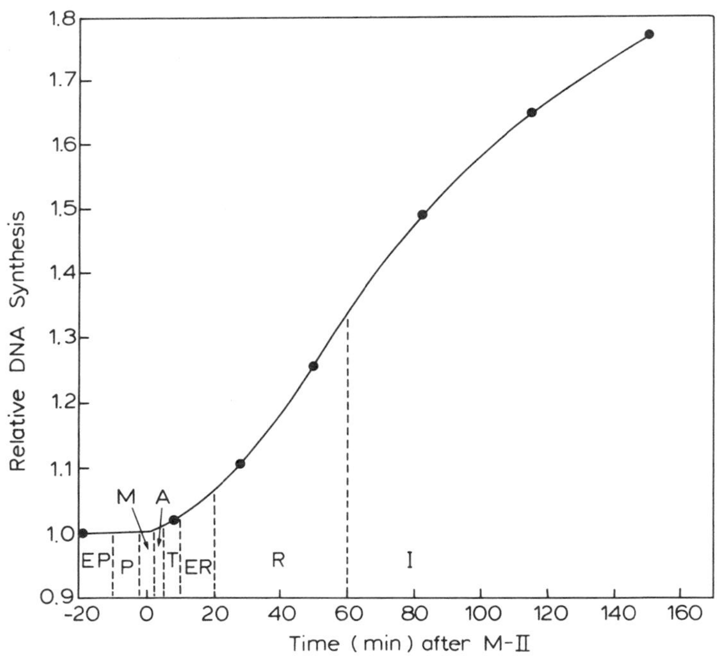

FIG. 5. S phase, determined by the isotope-dilution method (Section I,G), in relation to the stages of nuclear division. Nuclei were isolated from samples of a large plasmodium labeled with [^{3}H]thymidine, and the specific activity of the DNA (*20*) was determined. EP, early prophase; P, prophase; M, metaphase (M-II); A, anaphase; T, telophase; ER, early reconstruction; R, reconstruction; I, interphase. M. Cunningham and G. Turnock, unpublished results.

is introduced into the cycle when the growth rate is reduced. Unfortunately, although microplasmodia can be grown satisfactorily in defined media with growth rates three to four times less than those characteristic of peptone-based media (*15*), coherent plasmodia with good synchrony are not formed in surface culture, so experiments at slow rates of growth have not yet been possible.

Although 70 to 80% of the nuclear DNA replicates during the first two hours of the cycle, the remainder appears to be synthesized at a much lower rate during G2 phase (*20, 47a*). Figure 6 presents the results of two experiments in which the isotope-dilution method was used to assay DNA synthesis subsequent to the first 100 minutes of the mitotic cycle. In the first (Fig. 6A) a single plasmodium in a large tray apparatus was used, and in the second (Fig. 6B) a replicate series of plasmodia in Petri dishes was employed; in both cases DNA synthesis

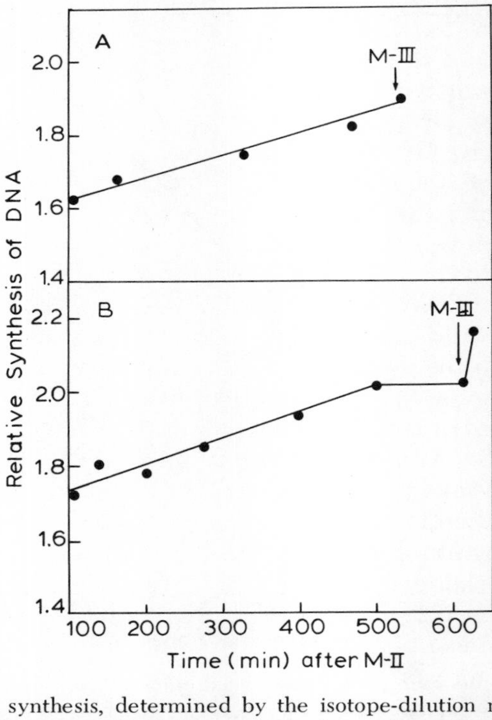

FIG. 6. DNA synthesis, determined by the isotope-dilution method, during G2 phase. In (A) nuclei were isolated from a single, large plasmodium and in (B) from a series of replicate plasmodia in Petri dishes. DNA was extracted from each sample, and its specific activity was determined (*20*). M. Cunningham and G. Turnock, unpublished results.

continued until virtually the next nuclear division. There is no obvious reason why the isotope-dilution method should yield misleading data with respect to the synthesis of DNA, and it has in fact been known for some time that replication of the nucleolar DNA, which includes the genes for rRNA, takes place during both S phase and G2 phase (*34*, *48*, *49*). Replication of the nucleolar DNA must therefore contribute to the slow, continual synthesis of DNA subsequent to S phase itself, although this DNA comprises only 1 to 2% of the nuclear DNA (*34*).

Does some of the synthesis of DNA in G2 phase result from asynchrony within the population of nuclei? There is little support for this possibility; autoradiography after pulse-labeling with [^{3}H]thymidine shows that at least 99% of the nuclei incorporate radioactivity at the beginning of the S phase (*45*), and even in large plasmodia at M-III, nuclei from different areas divide within 15 to 20 minutes of each other (*8*, and our own observations). However, other autoradiographic experiments (*50*) suggest that some of the DNA synthesis in G2 phase might be due to "abnormal" nuclei, as opposed to simple asynchrony. The small amount of incorporation of [^{3}H]thymidine detected in G2 phase was of two types: (*a*) for the majority of nuclei, with normal morphology, the silver grains were located predominantly over the nucleoli; and (*b*) for a small proportion (0.5%) of nuclei that were abnormally large, the distribution of silver grains suggested extensive incorporation into extranucleolar chromatin. The large, presumably polyploid nuclei were heavily labeled, even in late G2 phase. It is not immediately obvious why a polyploid nucleus should not follow the same kinetics of DNA synthesis as a normal nucleus. Nevertheless, the phenomenon could contribute to the synthesis of DNA during G2 phase, although the amount of DNA involved is difficult to quantify. The results obtained by a cytofluorometric method (*47a*) for assaying the DNA content of individual nuclei suggest that the increase in DNA during G2 phase (10%) cannot in fact be due to synthesis by a small, special class of nuclei.

It appears, therefore, that the simple definitions of S and G2 phases given at the beginning of Section II,B are blurred to a small extent in *Physarum*. Further evidence that some nucleoplasmic DNA synthesis extends into the G2 phase comes from work on the details of DNA chain growth, described in the next section. DNA chains initiated late in S phase do not achieve their final size until well on into G2 phase (*50a*). Although ligation of smaller chains is involved in this increase in size, it is possible that some net synthesis of DNA is also required. At a time in the mitotic cycle when the demand for substrates for DNA synthesis must be low, by comparison with early S phase, it may not be possible to assay this synthesis reliably by exogenous labeling with radioactive thymidine. On the other hand, nucleolar DNA can apparently be successfully labeled during G2 phase (*42b*); this contradiction could be explained if nucleoplasmic and nucleolar DNA synthesis draw on physically distinct deoxynucleoside triphosphate pools, with different kinetics of labeling from radioactive thymidine supplied in the growth medium.

C. Ordered Replication of Nuclear DNA

The high degree of natural synchrony characteristic of nuclear division in *Physarum* makes it an ideal organism in which to analyze the organization of the replication of the nuclear DNA during S phase. A fundamental contribution to this area of research was made by Braun and Wili in 1969 (*51*), when they showed, using a combination of radioactive and density labeling of DNA, that DNA molecules that replicate in a small fraction (about 20%) of the S period of one interphase are replicated at the same relative time in the following interphase. This suggests that there is a fixed, temporal order for the replication of the genome.

Since this early work, and against the background of the demonstration that initiation of bidirectional synthesis begins at many points along a molecule of DNA in a mammalian nucleus (*52*), research with *Physarum* has been directed toward defining the average size and properties of replicons in the nuclear DNA. Experiments with isolated nuclei have proceeded in parallel with *in vivo* studies, and the knowledge gained by both approaches is discussed.

Elucidation of the details of the replication of nuclear DNA in *Physarum* has inevitably involved experiments in which plasmodia have been labeled with radioactive thymidine. The radioactivity incorporated into DNA is then analyzed by zone sedimentation at neutral or alkaline pH. In the former case, it is hoped that the molecules of DNA will be equivalent to those in the chromosome, and in the latter case, single-stranded units of DNA are examined. The apparent simplicity of the overall procedure is deceptive, and there are technical problems that can hinder the interpretation of the results, especially when comparing data obtained by slightly different methods in independent laboratories. For instance, as indicated in Section I,F, thymidine is subject to extensive catabolism with concomitant randomization of radioactivity into other macromolecules. This has not usually been taken into account in the design of labeling procedures, so that it is often not clear, for example, whether "continuous labeling" really was continuous.

The extraction and analysis of nuclear DNA has as its primary problem the difficulty of designing a procedure least likely to yield DNA damaged either by the action of enzymes or by shear during the isolation. In parallel with this, there is considerable uncertainty in defining the sizes of very large molecules. Empirical relationships between sedimentation coefficient and molecular weight (*53*) are used, but it is a common practice to use only one type of molecule as a reference standard, and errors in the sedimentation coefficients and molecular weights, calculated by reference to the standard, may be substantial, although this is rarely acknowledged.

From what has just been said, the need for reasonable caution in interpreting the results both of labeling and of the physical analysis of DNA should be apparent. Nevertheless, this should not lessen the interest in the results so far obtained; rather it should sharpen the desire to refine the available techniques and improve the theoretical basis for the analysis of the data, particularly in the case of kinetic experiments.

Brewer (*54*) found that uniformly labeled DNA from nuclei isolated in G2 phase and lysed directly on top of a sucrose-density gradient had an average, apparent molecular weight of 2.3×10^8 at neutral pH and 4×10^7 at alkaline pH. In contrast, the radioactivity in DNA labeled for only 10 minutes sedimented in positions equivalent to average molecular weights of 3.6×10^8, with a broad size distribution, and 1.5×10^7 in the two types of gradient, respectively. The pulse-labeled molecules observed at neutral pH were considered to be replicating (forked) structures: treatment with Pronase and prior labeling of plasmodia with [^{14}C]lysine showed that such molecules were not held together by proteins. Labeling for 10 minutes with radioactive thymidine, followed by growth of the plasmodium in unlabeled medium, demonstrated that the very large, replicating molecules started to give rise to molecules of parental size within 30 minutes. Separation in density gradients at alkaline pH indicated that the molecules with a molecular weight of 1.5×10^7 took up to 90 minutes to achieve the value of 4×10^7 characteristic of nonreplicating DNA isolated in G2 phase.

Subsequent work (*55*) showed that, although most double-stranded DNA molecules initiate synthesis early in S phase, synthesis of the subunits within them examined under alkaline conditions, is initiated throughout at least the first two hours of this period. These observations are compatible with work on the synthesis of DNA in mammalian cells in which synthesis is initiated at multiple sites and at different times within a given DNA molecule (*52*).

Funderud and Haugli (*56*) also analyzed DNA labeled 30 minutes after M-II by zone sedimentation in alkaline sucrose-density gradients, using periods of labeling as short as 15 seconds. Their results suggest that the major initial product is a single-stranded DNA of 5 to 7 S (molecular weight 1.4×10^5), with the corollary that DNA synthesis is discontinuous on both strands (*56*). By 10 minutes, and with the probable involvement of several discrete intermediates, the primary products of replication are incorporated into DNA molecules of molecular weight of approximately 10^7. Molecules of this size probably correspond to the pulse-labeled DNA, with a molecular weight estimated to be 1.5×10^7, detected in Brewer's experiments (*54*). Thereafter, in prolonged chase experiments, these molecules were observed to accumulate in single-stranded molecules at least 20 to 40 times larger, a size that begins to approach that anticipated (a maximum M_r of 2.5×10^9) for a single-stranded DNA from a whole chromosome. It is not clear why Brewer (*54*) did not detect any single-stranded molecules larger than 4×10^7; possibly the conditions of lysis inadvertently facilitated degradation of the chromosomal DNA to pieces of that size.

More recent work in Haugli's laboratory has involved a further extensive series of pulse-chase experiments during S phase and G2 phase and analysis of the DNA by sedimentation in alkaline sucrose gradients (*50a*). Synthesis of 5 to 7 S primary replication units ("Okazaki fragments"; molecular weight revised to 6×10^4) occurs throughout the S period. Concomitantly, during the first hour of S phase [corresponding to the maximum net rate of DNA synthesis (Fig. 5)], the 5 to 7 S molecules are joined together at a rate of 6 to 8 units per minute. It is proposed that the 35 to 45 S molecules ($M_r = 1$ to 2×10^7) constitute the fundamental units (replicons) of replication of DNA in *Physarum*. Further increase in size appears to occur in a stepwise manner, possibly reflecting clustering of isochronous replicons along a chromatid.

Haugli *et al.* have also been able to demonstrate that replication of DNA in *Physarum* is bidirectional (*57*) by exploiting their findings that it occurs by discontinuous synthesis of "Okazaki fragments" on both strands and that little or no ligation of adjacent replicons takes place during the first hour after initiation (*56*). A plasmodium was pulse-labeled with bromodeoxyuridine at the beginning of S phase to provide sites in the DNA that could be broken by exposure to UV light (*58*). The plasmodium was then pulse-labeled with [^{3}H]deoxyadenosine and then allowed to grow in unlabeled medium. Nuclei were isolated at different times, some portions exposed to UV-light to permit photolysis of the DNA, and the size of irradiated, radioactive DNA compared to that of nonirradiated DNA by zone sedimentation in alkaline sucrose gradients. The reduction in size of the [^{3}H]DNA produced by photolysis suggested a mechanism for the replication of DNA in which initiation of synthesis begins in the center of a replicon and proceeds bidirectionally at equal rates for both forks (*57*).

A major goal of research into DNA synthesis in *Physarum* is the development of *in vitro* conditions capable of sustaining extensive, net replication of nuclear DNA. Significant progress has been made using both simple homogenates (*59*) and isolated nuclei. Following the discovery that addition of dextran to homogenates enhances the ability of the nuclei to synthesize DNA (*60*), homogenates prepared 30 minutes after M-III were shown to sustain for about an hour a rate of DNA synthesis calculated to be 25% of the rate *in vivo*. Overall, about 15% of the genome is replicated in this system (*60*). Although carried out prior to the introduction of the use of dextran to stabilize homogenates (*59*), the experiment illustrated in Fig. 7, which compares the activity of homogenates prepared at different times during G2 and S phases, clearly relates the amount of DNA synthesized *in vitro* to the phase of maximum synthesis in the plasmodium (compare with Figs. 4 and 5). Extensive synthesis of DNA occurs only in homogenates pre-

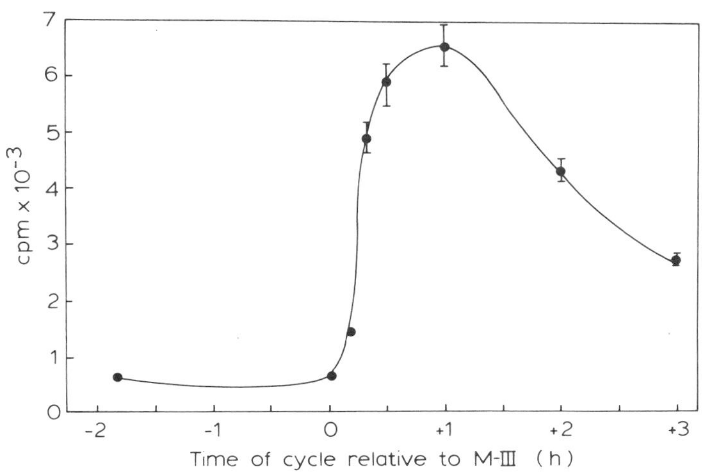

FIG. 7. DNA synthesis in homogenates obtained from plasmodia at various times in the mitotic cycle. Sectors of individual plasmodia were homogenized, debris was removed, and the incorporation of [^{3}H]dATP was determined as an index of DNA synthesis. Bars indicate the average deviation of duplicate assays. From Brewer and Ting (*59*).

pared during S phase, suggesting that it must relate closely to the replication of DNA as it occurs *in vivo*.

The synthesis of nuclear DNA in homogenates (*60*) requires all four deoxynucleoside triphosphates, Mg^{2+}, ATP, and the chelating agent EGTA [ethylene glycol bis(β-aminoethyl ether)-*N*,*N*,*N*′,*N*′-tetraacetic acid]. It is inhibited by Triton X-100, presumably because the nonionic detergent interferes with the integrity of the nuclear membrane. Addition of dextran to homogenates prepared in its absence, or to recombined nuclear and postnuclear supernatant fractions, gives a significant increase in the amount of DNA synthesized, and it is suggested (*60*) that the polysaccharide may function by providing an environment that stabilizes replication complexes. Another, possibly related, observation is the partial purification (E. N. Brewer, personal communication) of a substance, tentatively identified as a glycoprotein with a molecular weight of 30,000, from plasmodia that increases the rate and extent of DNA synthesis in nuclei isolated in S phase, although it does not initiate DNA synthesis in nuclei isolated from plasmodia in G2 phase.

The size of nuclear DNA synthesized in homogenates has been compared with that of nonreplicating (G2) DNA by zone sedimentation in alkaline, sucrose-density gradients (*60*). About half of the radioactivity incorporated into DNA *in vitro* was associated with molecules that increased in size with time, and the remainder had a sedimentation coefficient of about 10 S.

Funderud and Haugli (*61*) also studied the synthesis of nuclear DNA *in vitro*, using a modification (*62*) of an established (*14*) method for isolating nuclei. Combining density labeling and radioactive labeling, they showed that DNA made in isolated nuclei is a continuation of synthesis begun prior to isolation. Pulse-chase experiments demonstrate some incorporation of single-stranded molecules of molecular weight of about 50,000 into molecules approaching the size of an average replicon ($M_r = 10^7$; *50a*).

In summary, exploitation of the natural synchrony of nuclear DNA synthesis in *Physarum* has already provided valuable results concerning the organization of DNA replication during S phase. For the future, one might hope to see the development of conditions that will permit nuclei, isolated at a defined time in S phase, to complete replication of their DNA *in vitro*. An interesting question here is whether it will be necessary to incubate nuclei in a system supplemented with all the requirements for protein synthesis; the interrelationship between DNA replication and protein synthesis is discussed in Section II,E. The ultimate objective is the development of an *in vitro* system that will allow not only the completion of DNA replication but also the subsequent division of the nuclei. In addition to the natural synchrony of this process, *Physarum* possesses two other advantages for the biochemical analysis of nuclear division, namely, extracts can be prepared by gentle procedures, and cytokinesis does not occur, thereby reducing the complexity of the system.

D. Replication of rDNA

Unlike the bulk of the nuclear DNA, which replicates during the first two to three hours of the mitotic cycle, the nucleolar DNA has been shown to replicate during both S phase and G2 phase (*34, 48*), with the possible exception of the first hour after mitosis (*34*). Subsequent to the characterization of the structural organization of the genes that code for rRNA (Section II,A), the replication of rDNA has been examined in more detail (*63*) in order to assess the process in relation to individual molecules. To recall the salient features of nucleolar DNA, each molecule ($M_r = 37 \times 10^6$), of which there are approximately 150 per haploid genome, has a rotational axis of symmetry and carries two sets of genes for rRNA; nucleolar DNA is metabolically stable.

Vogt and Braun (*63*) used alternate periods of density labeling with bromodeoxyuridine and radioactive labeling with [^{3}H]thymidine to investigate the pattern of replication of nucleolar DNA, which was prepared from nuclei by a selective extraction procedure and fractionated by centrifugation to equilibrium in CsCl. They found that probably all the molecules can participate in replication, which is an important result in relation to the control of the process because it clearly eliminates reliance on a small number of "master" copies. The replication of rDNA was found to be unscheduled in that a molecule that replicates at a particular time in one cycle has an equal probability of replicating again in any time interval in the subsequent cycle. A consequence of this random selection of molecules for replication is that a molecule can replicate more than once per cycle or not at all. The probability that a given molecule will not replicate during a particular cycle is nearly 0.5 (*63*); in experiments in which bromodeoxyuridine was present from M-III to M-IV, 40 to 60% of radioactive DNA (labeled in the previous G2 phase) had a hybrid density in good agreement with the value expected on the basis of random selection.

The mechanism of replication of nucleolar DNA was investigated using electron microscopy to examine isolated molecules for putative replication forks (*63*). Such structures were found in molecules from the "heavy" side of the nucleolar peak in zone sedimentation analysis in a sucrose-density gradient. Most of the molecules with forks possessed two such structures connected by an "eye" of variable size. On the basis of the examination of a number of molecules, it is tentatively proposed that replication can initiate at 0.33, 0.45, 0.55, or 0.67 unit from one end. The points at 0.33 and 0.67, and at 0.45 and 0.55, are identical because the DNA is symmetrical around its center (*32*). Replication then proceeds bidirectionally at approximately equal rates from the origin. For a rate of movement of a replication fork of 1 μm min^{-1}, a single nucleolar DNA molecule should replicate in 10 minutes.

E. Dependence of DNA Synthesis on Protein Synthesis

A perennial question of interest in trying to dissect the relationship between the synthesis of different classes of macromolecules is that of identifying rate-limiting, and therefore potentially regulatory, interactions. For example, in a gross sense, it is obvious that the ability of a cell to replicate its DNA must depend on its capacity to synthesize proteins, so that if the latter process is inhibited, the former must ultimately be affected. However, the kinetics with which the synthesis of DNA responds to inhibition of protein synthesis may provide information about the requirement for proteins for particular components of the reaction, for example, the initiation of DNA synthesis or the rate of chain extension.

Cycloheximide is one of a number of inhibitors of protein synthesis in eukaryotes [reviewed in Stimac *et al.* (*64*)] that have been employed to investigate the dependence of DNA replication on the continued ability of a cell to synthesize proteins. Stimac *et al.* (*64*) used incorporation of [^{3}H]thymidine to measure DNA synthesis in animal cells in culture, testing the validity of this procedure by comparing the pool size and specific activity attained by dTTP in control and inhibited cultures. They found that eight different treatments, including addition of cycloheximide, all produced a similar reduction in the rate of DNA synthesis to a minimum value in about the same time (one to two hours). The importance of this result is that, because of the diverse nature of the treatments used, it suggests a fundamental dependence of DNA synthesis on protein synthesis. When only one drug is used, even though a major site of action may have been identified, there is always the uncertainty that a secondary response may be quite unrelated to the primary inhibition.

DNA fiber autoradiography showed that the rate of movement of replication forks is reduced within 15 minutes of the onset of inhibition of protein synthesis to an extent sufficient to account for the reduction in the incorporation of thymidine (*64*). For times greater than one hour, there appears to be an additional effect due to inhibition of initiation of new replicons.

Cycloheximide inhibits the incorporation of thymidine into the DNA of *Physarum* (*65*). When the inhibitor is added just before prophase, neither mitosis nor DNA synthesis occurs; addition during mitosis or at different times during S phase allows only a small amount of DNA synthesis. A more detailed examination of the effect of adding cycloheximide at different times during S phase led to the claim (*66*) that the genome of *Physarum* consists of at least ten replicative units controlled by proteins synthesized at defined times during the S period. However, the nature of the replicative units and their associated specific proteins do not appear to have been characterized any further.

Other investigations have tested the effect of cycloheximide in terms of the mechanism of DNA synthesis. Figure 8 shows that elongation of DNA strands stops within 15 minutes of addition of the drug (*46*). In contrast, alternate periods of labeling with [^{14}C]thymidine and [^{3}H]deoxyadenosine showed that the initiation of new replicons is not adversely affected by cycloheximide. The DNA synthesized in the presence of the antibiotic and labeled with [^{3}H]deoxyadenosine has been characterized (*67*). Very short periods of labeling [as in Funderud and Haugli (*56*); see also Section II,C] led to the claim that cycloheximide inhibits the formation of 5 to 7 S "Okazaki fragments," slows down the joining of these to 30 to 35 S molecules, and prevents the slow maturation of the latter to higher molecular weight DNA.

A notable feature of these experiments (*67*) is the use of times of treatment with cycloheximide as short as two minutes, which still allowed observation of differences in the sedimentation profiles of pulse-labeled DNA. This raises the question of whether the effects observed were actually due to inhibition of protein synthesis. To counter criticism of this nature, a mutant with ribosomes resistant to inhibition by cycloheximide (*68*) was also tested (*67*); it gave patterns of labeling of DNA that were very similar in the presence and in the absence of the drug.

The kinetics of labeling of dTTP by exogenous [^{3}H]thymidine are altered in the presence of cycloheximide, making it difficult to quantify the amount of DNA synthesized after addition of the drug (*46*). This is another situation in which the isotope-dilution method is of value; Fig. 9 illustrates the use of this procedure to determine the kinetics of inhibition of DNA synthesis by cycloheximide added 30 minutes after mitosis. It is apparent that inhibition of DNA replication

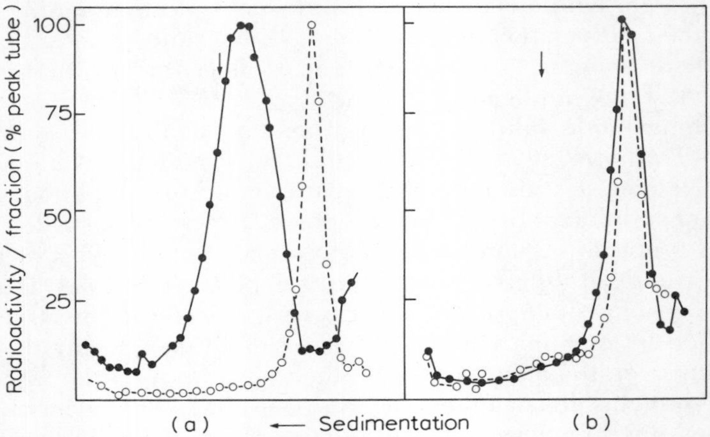

FIG. 8. Effect of cycloheximide on the elongation of progeny DNA strands. Five minutes after metaphase, plasmodial halves were transferred to medium containing [^{3}H]thymidine ± cycloheximide (10 μg/ml). After 15 minutes, half of each section was harvested and the other half was rinsed and placed in medium containing nonradioactive thymidine (10^{-4} M) ± cycloheximide and harvested 90 minutes later. Nuclei were isolated and lysed and the DNA analyzed by zone sedimentation in alkaline, sucrose-density gradients (*54*). Fractions from the gradients were assayed for ^{3}H and ^{14}C. Samples were centrifuged in separate tubes, but pairs have been plotted on the same graph; ^{14}C-labeled, nonreplicating DNA (60 S; $M_r = 4 \times 10^7$) from *Physarum* was included in each tube as a reference, and its position is indicated by an arrow. ○---○, Nuclei isolated from plasmodia incubated for 15 minutes in the presence of [^{3}H] thymidine; ●——●, nuclei isolated from plasmodia incubated for 15 minutes in the presence of [^{3}H]thymidine and for 90 minutes in nonradioactive medium. (a) Control. (b) Cycloheximide present during both pulse and chase. From Evans *et al.* (*46*).

becomes effective within about 10 minutes and rapidly increases in severity.

The mechanism by which DNA synthesis is reduced by cycloheximide is still unknown, although the comparative study of inhibitors (*64*), referred to earlier, substantiates the view that it acts by inhibiting of protein synthesis. Accepting this hypothesis, there are two principal ways in which inhibition of protein synthesis could produce an effect on DNA replication. First, net DNA synthesis might require a continual supply of either structural proteins or regulatory proteins or both (*64, 66*). Second, the response may reflect a control system that titrates the general capacity of the cell to synthesize proteins. Unless the amounts of free, "essential" proteins are extremely small, the rapid response of DNA synthesis to cycloheximide (Fig. 9; *46, 67*) would tend to favor the second possibility. A precedent for this

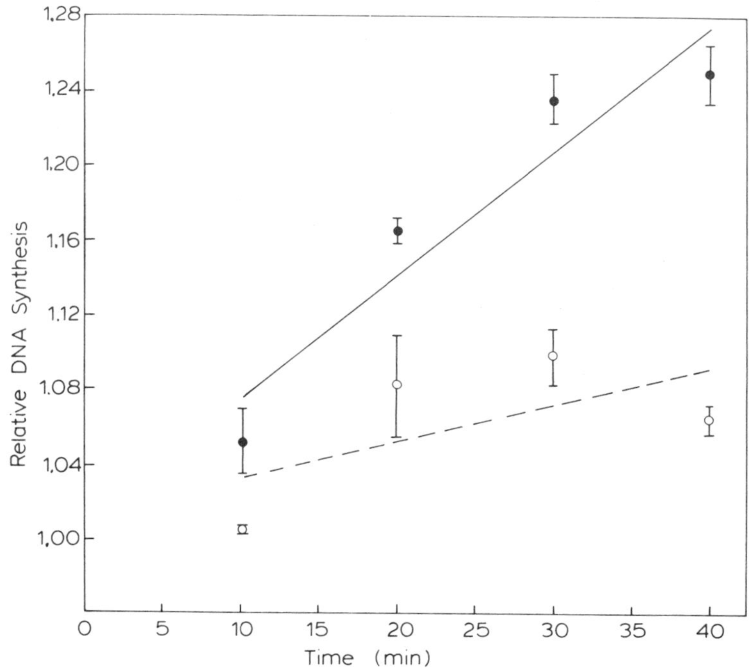

FIG. 9. Inhibition of DNA synthesis by cycloheximide. A replicate series of plasmodia, prelabeled with [^{3}H]thymidine so that DNA synthesis could be followed by the isotope-dilution method, was used for this experiment. Thirty minutes after M-III 3 samples were taken from each plasmodium to provide zero-time DNA specific activities, while three sectors were transferred to plates containing 20 μg of cycloheximide per milliliter and three others were transferred to normal medium. At intervals of 10 minutes, sets of three treated and three untreated sectors were harvested, total DNA was extracted (insufficient material was available to permit isolation of nuclei), and the specific activity of each sample was determined. The bars indicate the standard errors of the means for each set of three samples, normalized to a value of 1.0 for the zero-time samples from the initial cultures. ●, Control; ○, treated with cycloheximide. M. Cunningham and G. Turnock, unpublished results.

type of control is the link, mediated by a guanosine tetraphosphate, between protein synthesis and transcription in bacteria (*69*).

III. Synthesis of RNA

A. General Analysis of Transcription

The design of experiments for the investigation of the synthesis of RNA in any organism is as complex as that required for the study of

DNA synthesis. Unambiguous analysis of data obtained by labeling with radioactive precursors is a problem common to both and has already been discussed (Sections I, F and I, G). Measurements of the synthesis of RNA also have to take into account its varied composition and differentiation of function and, because of the ubiquity of post-transcriptional processing, possible differences between the primary rate of synthesis on the genome and the rate of synthesis of the mature, functional molecule in the cytosol. Any statement about "the rate of RNA synthesis . . ." needs to be carefully qualified if it is to have a clear meaning, especially for purposes of comparison with experiments involving a different technical base.

The earlier investigations of the synthesis of RNA during the mitotic cycle in *Physarum* were carried out before many of the problems alluded to in the previous paragraph were fully appreciated, but they are still of interest. Pulse-labeling with [^{3}H]uridine as an assay for total RNA synthesis (*70*) yielded the visually dramatic results reproduced in Fig. 10, which shows a biphasic pattern of incorporation of the precursor into RNA. The minima in the pattern of incorporation occur at mitosis and in mid-interphase, one maximum coinciding with the period of nucleolar reconstruction and another in G2 phase. Pulse-labeling of plasmodia was carried out for periods of 10 minutes, and throughout interphase approximately 70% of the radioactivity incorporated was found in the nuclear fraction (*71*).

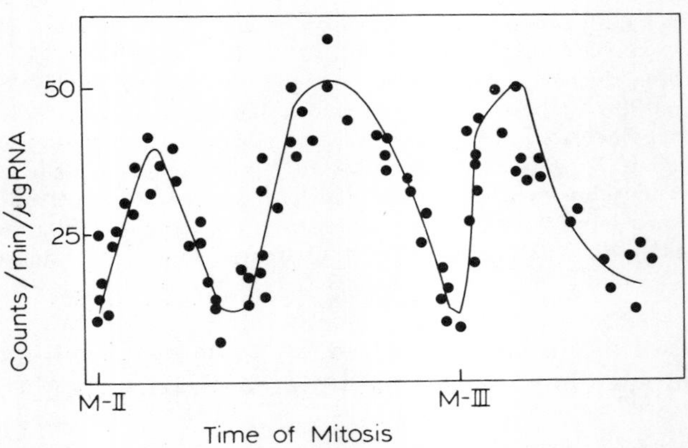

FIG. 10. Uptake of 10-minute pulses of [^{3}H]uridine into RNA at different times of the mitotic cycle. The time between M-II and M-III was 8 to 10 hours. Each point represents the specific activity of RNA in a single plasmodium. From Mittermayer *et al.* (*70*).

Although variations in the size of the nucleoside triphosphate pools during the mitotic cycle can be measured (Fig. 3), no attempt has yet been made to correlate changes in the rate of uridine incorporation into RNA (Fig. 10) with possible variation in the specific activity achieved by UTP in short periods of labeling. To do so would perhaps not be very rewarding, given that it is changes in the rates of synthesis of individual RNAs that are of much greater interest, and that this information can be obtained using other techniques. It should also be noted that the specific activity of UTP extracted from whole cells may not be equal to that of UTP in the nucleus, especially after short periods of labeling (*22*). In one study of mRNA and rRNA synthesis in animal cells, it has even been found that there is a permanent, systematic difference in the specific activity of UTP used for the synthesis of the two types of RNA when cells are labeled continuously with [^{3}H]uridine (*72*).

Despite the uncertain relationship between the incorporation of [^{3}H]uridine and any variation in the actual rate of RNA synthesis during the mitotic cycle, it is interesting to note that nuclei isolated from plasmodia at different times during the cycle can also exhibit a biphasic pattern in their ability to incorporate radioactive ribonucleoside triphosphates into RNA (*73, 74*). Although it may be argued that any systematic changes in the intranuclear concentrations of the triphosphates during the cycle should be reflected in the *in vitro* system, the endogenous concentrations appear to be low because the incorporation of [^{3}H]UTP by isolated nuclei is dependent upon the presence of the other three substrates (*17, 73*). However, the ability of isolated nuclei to synthesize RNA is affected by the ionic conditions used for the assay (cf. *74, 75*), so that the similarity between the behavior of nuclei (*73, 74*) and intact plasmodia (*70*) does not rigorously prove that the rate of total RNA synthesis *in vivo* follows the biphasic pattern of Fig. 10.

B. RNA Polymerases in *Physarum*

Multiple forms of RNA polymerase have been detected in the nuclei of both lower and higher eukaryotes, and there is good evidence that the different enzymes exhibit differentiation of function with respect to the genes that they transcribe (*76*).[1] The three main classes of enzyme characterized can be separated by ion-exchange chromatography, and they differ from each other also in their response to the

[1] See the article by Biswas *et al.* in Vol. 15 of this series.

polypeptide antibiotic α-amanitin. This fungal toxin has been particularly useful in helping to establish different functional roles for the three classes of RNA polymerase.[1] Class I (A) enzymes are unaffected by quite high concentrations of α-amanitin, are found within the nucleolus, and are responsible for the synthesis of rRNA, whereas class II (B) activity is very sensitive to inhibition by the drug, is located in the nucleoplasm and catalyzes the synthesis of hnRNA. Class III (C) enzymes, depending on the organism, have an intermediate or quite high resistance to inhibition by α-amanitin, and there is evidence that they transcribe the genes for 5 S RNA and tRNAs.

In work with *Physarum* the first evidence for the existence of more than one type of RNA polymerase came from the study of RNA synthesis in isolated nuclei and isolated nucleoli (*74*). Nuclei isolated at different times during the mitotic cycle show a differential response to addition of Mn^{2+} and $(NH_4)_2SO_4$ and to α-amanitin (Table II). The results demonstrate that over half the RNA polymerase activity in S-phase nuclei is inhibited by α-amanitin, with little response to change in ionic conditions, and that the reverse is true for nuclei isolated in G2 phase. α-Amanitin had almost no effect on the synthesis of RNA by nucleoli, supporting the conclusion that the data obtained with nuclei (Table II) reflect a change in the proportion of nucleolar (rRNA) compared to nucleoplasmic (α-amanitin-sensitive) RNA synthesis during the mitotic cycle.

Subsequently, the existence of class I and class II RNA polymerases in *Physarum* has been established in several laboratories, using a variety of methods to fractionate crude extracts (*77–79*), and

TABLE II
EFFECT OF α-AMANITIN ON RNA SYNTHESIS IN ISOLATED NUCLEI[a]

Concentration of α-amanitin (μg/ml)	Radioactivity incorporated from [^{3}H]UTP (cpm/μg protein)			
	S nuclei		G2 nuclei	
	Mg^{2+}	Mn^{2+} + $(NH_4)_2SO_4$	Mg^{2+}	Mn^{2+} + $(NH_4)_2SO_4$
0	688	723	406	796
1	294	326	421	437
5	236	201	400	387
10	196	209	391	406
30	207	183	369	342

[a] Data from Grant (*74*). S-phase nuclei were isolated 3 hours after mitosis, and G2 nuclei 2.5 hours before mitosis.

the class II enzyme has been extensively purified (*80*). The presence of an RNA polymerase III, apparently unstable, has been tentatively suggested (*81*, *82*). In crude extracts, the activities of the RNA polymerases are reduced by the presence of inhibitory compounds, and the enzymes can be assayed accurately only after these have been removed (*83*).

The relative amounts of RNA polymerases I and II in extracts prepared from plasmodia at different times during the mitotic cycle have been determined (*81*). Homogenates containing 0.5 M NH_4Cl were sonicated to allow quantitative extraction of the two enzymes. After precipitation with $(NH_4)_2SO_4$ they were separated by chromatography on DEAE-Sephadex to permit the determination of the relative amount of each type. However, no significant change in the ratio of the two enzymes during the mitotic cycle was detected; approximately equal proportions of enzymes I and II were present in each sample. These findings suggest that changes in transcriptional activity during the cycle (*74*) cannot be due to, or be accompanied by, changes in the relative amounts of the two major RNA polymerases.

C. Synthesis of Ribosomal RNA

1. General Aspects

The organization of the genes that code for the ribosomal RNAs has already been described (Section II, A). As in other organisms the 19 S, 26 S, and 5.8 S molecules are derived by posttranscriptional cleavage and modification of a single, primary transcript with a molecular weight estimated to be approximately 4×10^6 (*84*). The structure of actively transcribing genes has been examined by electron microscopy (*42a*). In the center of each nucleolar DNA molecule, there is a nontranscribed region, 6.0 to 6.5 μm in length, that exhibits a beaded nucleosome structure. The sites of initiation of rRNA synthesis lie at each end of the central region, and matrices of rRNA fibrils, associated with protein, extend from them to the ends of the DNA molecule as expected from its palindromic structure (*32*, *33*). Each matrix of growing rRNA molecules is accommodated by 4.2 μm of nucleolar chromatin. As 26 S rRNA hybridizes closer to the ends of the nucleolar DNA than the 19 S molecule (*33*), the polarity of the matrices of nascent rRNA chains establishes the order of transcription as 19 S to 26 S, in accord with experiments with other eukaryotes (*84a*).

The 44 S primary transcript is cleaved to give a 34 S molecule, which constitutes a major fraction of the RNA in isolated nucleoli (*85*), and separate cleavage products of this species ultimately give rise to the mature 19 S and 26 S rRNAs. Although the 26 S molecule can be detected in both the nucleolus and nucleoplasm (*85*), little 19 S rRNA is found in either fraction, suggesting that the small ribosomal subunit is rapidly transported to the cytoplasm (*86*). Rapid movement of the smaller rRNA molecule from the nucleus to the cytoplasm has been detected in other eukaryotes (*87*).

The accumulation of 19 S and 26 S rRNAs during the mitotic cycle has been followed using a procedure based on the technique of isotope-dilution (*20*). A plasmodium was prepared from microplasmodia that had been labeled with [^{3}H]uridine, as described in Section I, G, and rRNA, from samples taken at intervals during the nuclear division cycle, was purified by zone sedimentation in sucrose-density gradients. The curve that describes the relative synthesis of rRNA during the cycle was obtained from the specific activities of the pooled 19 S and 26 S fractions, normalized to a value of 1.0 at M-II (Fig. 11). Combined analysis of the two large rRNAs was considered to be justified, since they are derived from a common precursor (*84, 86*) and there is no evidence to suggest that they are not processed in equal amounts, as judged by their relative proportions and estimated molecular weights (*84, 86*).

On the basis of the results presented in Fig. 11, several observations may be made concerning the synthesis of rRNA. First, it is continuous throughout the mitotic cycle, although it is reduced to a low rate during the actual process of nuclear division, specifically between prophase and telophase-early reconstruction. Second, the doubling in the amount of rRNA between mitoses II and III suggests both that growth of a plasmodium can occur in a balanced manner (see also Section V) and that the turnover of rRNA during growth is low; a net doubling in the amount of rRNA with concomitant turnover and mixing of the products of breakdown with unlabeled nucleotides would produce a decrease in specific activity to less than half the initial value. Third, the rate of synthesis of rRNA clearly increases throughout interphase. These findings are discussed in more detail below.

2. Variation in the Rate of Synthesis during the Cycle

The low rate of synthesis of rRNA during mitosis reflects a general characteristic of transcription at this stage in the nuclear division cycle, and an equivalent reduction is also evident, for example, in the accumulation curve for tRNA (see Fig. 14). Kessler (*88*), using autoradiography to examine pulse-labeled cells, found a decrease in total RNA synthesis during mitosis in *Physarum*. This experiment could be criticized on the basis that the uptake of uridine might change during

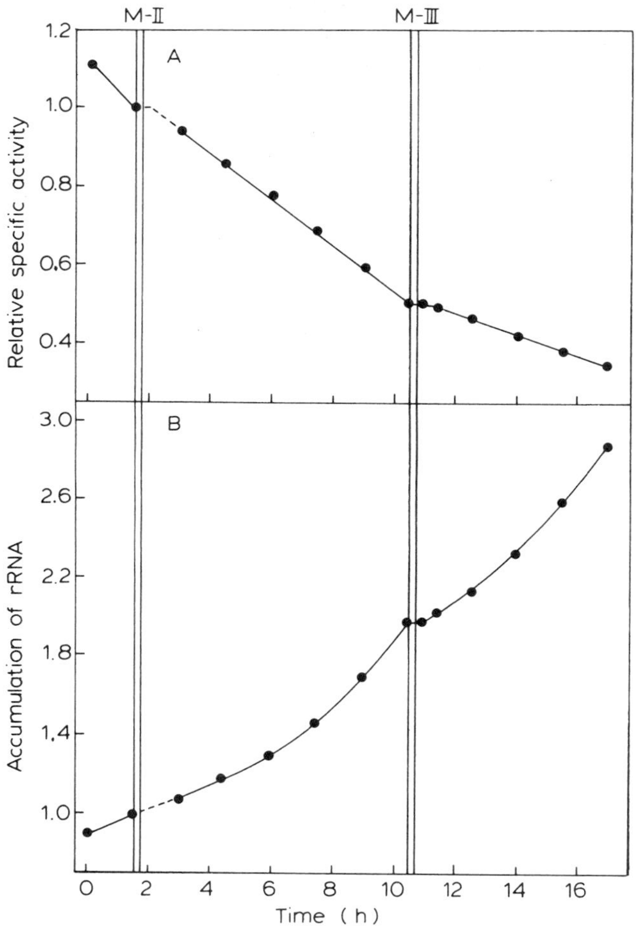

FIG. 11. Synthesis of rRNA during the mitotic cycle in strain M_3C-VIII. A large plasmodium was established from microplasmodia that had been labeled with [^{3}H]uridine, and rRNA was isolated from samples taken from just before M-II until after M-III. (A) The specific activity of rRNA as a function of time. (B) The corresponding accumulation curve for rRNA, each point being the reciprocal of the corresponding specific activity. The data are normalized to a specific activity of 1.0 at M-II. From Hall and Turnock (*20*).

mitosis. However, nuclei isolated during mitosis possess very little ability to synthesize RNA (Table III and also Fig. 13). In contrast, chromatin prepared from metaphase nuclei shows endogenous RNA polymerase activity comparable to that found with chromatin from

TABLE III
ENDOGENOUS RNA POLYMERASE ACTIVITY IN NUCLEI ISOLATED AT DIFFERENT TIMES DURING THE MITOTIC CYCLE[a]

Phase	pmol UMP incorporated/mg DNA	
	Low Salt	High salt
S	420	500
G2	450	698
Metaphase	20	22

[a] Data from Davies and Walker (*17*).

interphase nuclei (*17*). This suggests that the reduction in transcription during mitosis has its basis in structural changes associated with nuclear division, and that these changes may be abolished by disrupting the structural organization of the nucleus. The importance of structure for accurate transcription is emphasized by the observation (*17*) that, although chromatin could retain up to 60% of the capacity of nuclei to synthesize RNA, the RNA made had a very different size distribution.

The data for the synthesis of rRNA (Fig. 11) have been fitted to three possible equations to estimate the magnitude and form of the increase in the rate of synthesis during interphase (*20*). Although it was not possible to define the precise form of the differential, the overall increase in the rate of synthesis was found to be five- to sixfold between the beginning of S phase and the end of G2 phase.

The technique of isotope-dilution used to follow the synthesis of rRNA during the mitotic cycle (Fig. 11) clearly measures net synthesis. In consequence, the rate of synthesis, d[RNA]/dt, obtained as the derivative of the accumulation curve, will be a direct index of transcription of the rRNA genes *only* if the synthesis of rRNA is controlled primarily by regulation of transcription itself. Regulation of the proportion of the primary transcript processed to yield mature, functional rRNA may also occur, as has recently been shown, for example, in the biosynthesis of rRNA in *E. coli* (*47*).

The possibility of posttranscriptional control of rRNA synthesis in *Physarum* has been suggested by determination of the activity of RNA polymerase I in nuclei isolated at different times during the mitotic cycle (*75*). As in the experiments referred to earlier (*74*), α-amanitin was used to inhibit RNA polymerase II selectively, and the results for both enzymes are shown in Fig. 12. The measurements are expressed in terms of the amount of UMP incorporated per nucleus; on this basis,

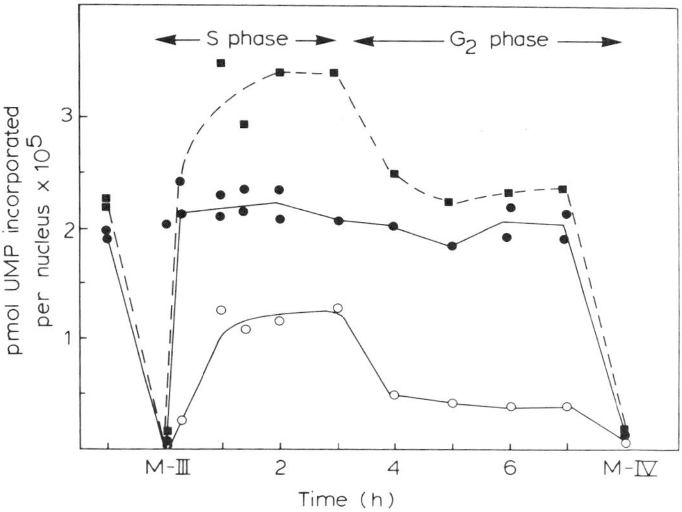

FIG. 12. Synthesis of RNA, measured by incorporation of radioactivity from [^{3}H]UTP, in nuclei isolated at different times during the mitotic cycle. ■--■, Total synthesis; ●——●, synthesis in the presence of α-amanitin, which inhibits RNA polymerase II; ○——○, the difference curve, which is an index of the activity of RNA polymerase II. From Davies and Walker (75).

the activity of RNA polymerase I in the *in vitro* assay does not vary during interphase. A similar result was obtained using isolated nucleoli (75); in this instance the inhibitor of initiation, rifamycin AF/013, was used to show that the proportion of new chains initiated in nucleoli isolated at different times throughout the cycle remained at about 50%.

During the mitotic cycle there must be a doubling in the amount of RNA polymerase, although it is not known whether the synthesis of this enzyme is continuous or occurs at a discrete period in the cycle. Equally, the genes that code for rRNA must double in number. The evidence that the replication of rDNA takes place not only during S phase but also during G2 has already been described (Section II, D). Thus, provided the measurements of the activity of RNA polymerase I in isolated nuclei (Fig. 12) are an accurate reflection of transcription *in vivo*, there must be precise control of the enzyme that maintains a constant activity per nucleus, even though the amount of the enzyme and the number of genes both double during the cycle. At the same time, the proportion of the primary transcript that is incorporated into mature, functional ribosomes must increase five- to sixfold during the cycle to give the pattern of accumulation determined by isotope dilution (Fig. 11).

D. Synthesis of Transfer RNA

The synthesis of tRNA during the mitotic cycle in *Physarum* has also been determined by the isotope-dilution method and compared to that of rRNA (89). The specific activity of tRNA was determined after electrophoresis in polyacrylamide gels and elution of the RNA into buffer. The 5 S rRNA was separated from the tRNA by electrophoresis, but unfortunately the amount obtained was too small to permit accurate determination of its specific activity.

An accumulation curve for tRNA from before M-II until M-III is given in Fig. 13. It shows that, as for rRNA, there is a break in synthesis during nuclear division. The rate of synthesis increases during S phase and then appears to remain approximately constant for the remainder

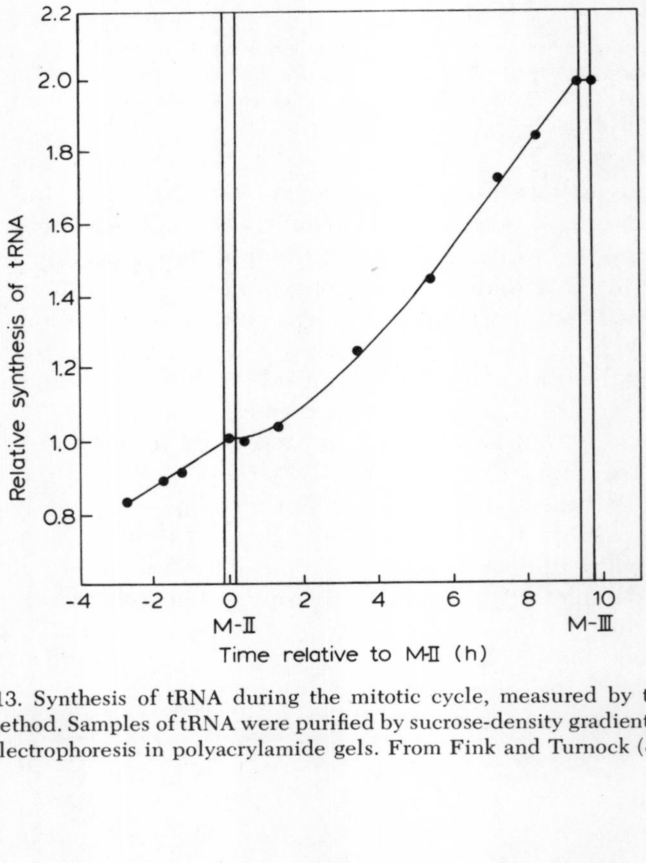

FIG. 13. Synthesis of tRNA during the mitotic cycle, measured by the isotope-dilution method. Samples of tRNA were purified by sucrose-density gradient centrifugation and electrophoresis in polyacrylamide gels. From Fink and Turnock (89).

of interphase, suggesting that there could be a simple gene-dosage effect. However, it should be emphasized that rigorous analysis of the accumulation curve in terms of its derivative would not be easy to interpret because, in contrast to the 19 S and 26 S rRNAs, which are derived from a common precursor, tRNA is composed of a large number of individual molecules each specified by a different gene.

E. Synthesis of Other Classes of RNA

The isotope-dilution method used to characterize the synthesis of rRNA and tRNA cannot be applied to the metabolically unstable nucleoplasmic RNAs and cytoplasmic mRNAs, and little is known about the detailed metabolism of these related classes of RNA. Experiments with isolated nuclei (*74, 75*) suggest a relative decline in nucleoplasmic RNA synthesis as the mitotic cycle progresses from S to G2 phase, as evidenced by the titration of the activity of RNA polymerase II with α-amanitin (Fig. 12). There have been brief studies of transcription during S phase and G2 phase using RNA · DNA hybridization techniques (*90, 91*), which also indicate changes in transcription of putative mRNAs during the cycle.

Characterization of the size distribution of nucleoplasmic RNA by gel electrophoresis indicates that the predominant species has the same mobility as 26 S rRNA with smaller amounts of 19 S and 30 S RNA. Figure 14 compares the RNA species found in the nucleolar and nucleoplasmic fractions of isolated nuclei. When nuclei were isolated from a plasmodium labeled with [^{3}H]uridine for 15 minutes, the nucleoplasmic RNA, which had a broad size distribution centered on 30 S,

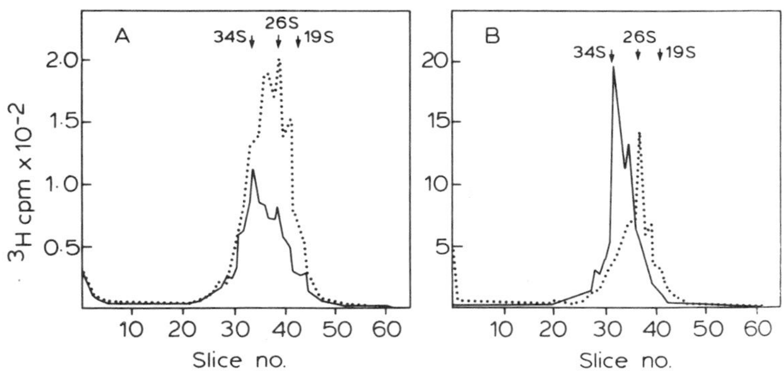

FIG. 14. Analysis of RNA, labeled *in vivo* with [^{3}H]uridine, by gel electrophoresis. (A) RNA labeled for 15 minutes in S phase. (B) RNA labeled for 30 minutes in S phase. In each panel, — denotes nucleolar RNA, and . . . . denotes nucleoplasmic RNA. From Davies and Walker (*85*).

contained more radioactivity than the nucleolar RNA (Fig. 14A). Lengthening the period of labeling from 15 to 30 minutes reversed the relative amount of radioactivity in the two nuclear fractions (Fig. 14B). The distribution of RNA in the nucleoli reflects the pathway of biosynthesis of rRNA that has already been discussed (Section III, C). Between 15 and 30 minutes of labeling, the 26 S peak in the nucleoplasmic RNA became relatively more prominent, and it has been suggested (*85*) that the broad distribution of RNA molecules on which the 26 S peak is superimposed may correspond to heterogenous nuclear RNA with a relatively short half-life. Otherwise, we still know very little about the characteristics of this class of RNA in *Physarum* and its relationship to mRNA in the cytoplasm. Even if the 30 S nucleoplasmic RNA represents an average size for the primary transcript of a structural gene, it is still considerably smaller than the hnRNA of animal cells (*92*). On the other hand, the hnRNA of the cellular slime mold *Dictyostelium discoideum* is only slightly larger than the cytoplasmic mRNA (*93*).

IV. Mitochondrial Nucleic Acids

A. DNA

Physarum is an obligate aerobe and possesses well-developed mitochondria. A prominent feature of the mitochondrial structure is a central, elongated nucleoid (*94–96*) composed principally of DNA (*96*). Measurements of the amount of DNA per mitochondrion show that mitochondria of *Physarum* contain considerably more DNA than those of many other organisms (*96*). The functional significance of this observation has not been established.

Preparations of DNA from *Physarum* exhibit two satellite bands when centrifuged to equilibrium in CsCl. One of these has a density slightly greater than that of the main band of nuclear DNA, and originates in the nucleolus (Section II, A). The other, with a density of 1.686 g cm^{-3} (*97*), is less dense than the nuclear DNA and consists of DNA from the mitochondria (*98*). It comprises about 7% of the DNA of *Physarum* (*49*).

To try to resolve differences in earlier studies (*99–101*), which agreed in finding principally linear molecules with, however, considerable variation in size in the mtDNA of *Physarum*, mitochondria were purified by differential centrifugation and treated with DNase to remove contaminating nuclear DNA prior to extraction of the mitochondrial DNA (*98*). This method produced mainly linear molecules with a maximum length of 20 μm, although 10% of the structures had an open, circular conformation with a circumference of 19.1 ± 0.5 μm corresponding to a molecular weight of 41 × 10^6. This figure agrees with the maximum molecular weight calculated from sedimentation analysis of mitochondrial DNA (*101*). By combining the molecular weight with the amount of DNA per nucleoid (*96*), the latter may be expressed as the number of DNA molecules per mitochondrion, about 40.

Lysis of mitochondria under mild conditions releases 50 to 70% of the DNA molecules complexed with protein and with no visible ends (*98*). It is possible that the small proportion of circular DNA molecules observed in conventional preparations of DNA may be maintained by DNA–protein interactions. The importance of proteins in the general organization of the DNA in the mitochondrion is demonstrated also by the observation that treatment with trypsin releases DNA from the central nucleoid (*102*). Preparations of nucleoids have been obtained by treating mitochondria successively with the detergents Triton X-100 and Nonidet P40 (*103*). The proteins of the isolated nucleoids have been analyzed by gel electrophoresis and found to include at least one basic protein.

The interest in the analysis of the replication of the mitochondrial DNA and of the division of the mitochondria lies in their relationship, if any, to the synchronous nuclear division cycle. The incorporation of radioactive thymidine into mitochondrial DNA occurs at an approximately constant rate throughout interphase (*49*). The number of mitochondria per nucleus, measured in stained smears, increases during late S phase, whereas the majority of mitochondria appear to synthesize DNA from mid-S phase until well on into G2 phase (*104*). Nevertheless, the proportion of labeled mitochondria in smears of samples exposed to [^{3}H]thymidine for three hours never fell below about 20%, so that for the population of mitochondria as a whole, the division cycle would appear to be less tightly controlled than is the case for nuclei. For an individual mitochondrion, however, it is surmised (*104*) that there are substantial "G1" and "G2" phases.

B. RNA

There is one substantial study of the RNA of mitochondria from *Physarum* and of the ability of mitochondria to synthesize RNA *in vitro* (*105*). Two putative rRNAs, with electrophoretic mobilities distinct from those of the cytoplasmic species, were isolated from purified mitochondria (Fig. 15). The quantity of mitochondrial RNA as a percentage of total RNA was such as to be barely detectable in the electrophoretic profile (Fig. 15A).

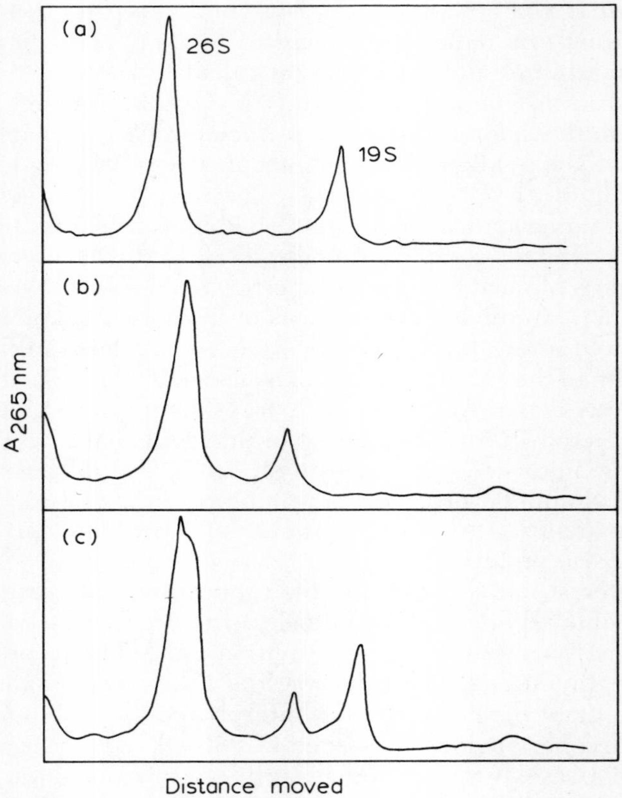

FIG. 15. Electrophoretic profiles of RNA from (a) whole plasmodia of *Physarum*, (b) isolated mitochondria of *Physarum*, and (c) coelectrophoresis of RNA extracted from plasmodia and mitochondria of *Physarum*. From Grant and Poulter (*105*).

The incorporation of [^{3}H]UTP into RNA in purified mitochondria was inhibited both by actinomycin D and by rifampicin, distinguishing the reaction from transcription in the nucleus, which is insensitive to the second inhibitor (*105*). Also in contrast to RNA synthesis in isolated nuclei, the incorporation of [^{3}H]UTP by mitochondria was not dependent on the addition of the other three substrates or on Mg^{2+}. This suggests that isolated mitochondria retain substantial concentrations of nucleoside triphosphates, a possibility that was confirmed for UTP itself by comparing total uptake of [^{3}H]UTP with its incorporation into acid-insoluble material. Radioactive RNA synthesized by mitochondria *in vitro* was characterized by gel electrophoresis; the principal component was fairly small with a mobility significantly greater than 16 S rRNA from *E. coli*. Radioactivity was not incorporated into either of the two large mitochondrial rRNAs.

Regarding the possible function of the RNA synthesized *in vitro*, there is an interesting correlation between transcription and protein synthesis in isolated mitochondria in that both processes are inhibited by rifampicin (*105*). Furthermore, addition of the inhibitor to mitochondria that had incorporated [^{3}H]UTP for 30 minutes demonstrated that the radioactive RNA has a half-life of about 5 minutes. Equally, the ability of mitochondria to incorporate leucine into protein declined at a similar rate after addition of rifampicin. These results suggest that some of the RNA synthesized by mitochondria *in vitro* is mRNA.

V. Integration of the Synthesis of Macromolecules

The determination of the composition of cells in terms of the major macromolecules—DNA, RNA, and protein—could be regarded as an unrewarding goal by comparison, for example, with the isolation of mutants defective in their ability to synthesize nucleic acids and proteins. However, it is clear from the history of the study of bacterial growth in particular (*106, 107*) that such information is an essential prerequisite for a proper understanding of the interrelationships between nucleic acid and protein synthesis; without this knowledge the aberrant behavior of mutants cannot be properly defined or exploited.

Maaløe and his colleagues (*106, 107*) emphasized the importance of defining the composition of bacteria, with the crucial proviso that measurements must be made in the early exponential phase of growth of a culture. This ensures that all the major parameters (e.g., cell number, protein, RNA, DNA) are determined under conditions in which they are all increasing with the same growth-rate constant. For an asynchronous population of cells, a balanced state of growth (*108*) is then defined as A/B = constant, where A and B can be any two components of the cell.

In the analysis of bacterial growth, important ideas about the regulation of the synthesis of macromolecules have been derived from a comparison of the composition of cultures with different growth rates, and also by studying the changes that accompany the transition from one rate of growth to another. A seminal observation is that the composition of a bacterium such as *E. coli* or *S. typhimurium* is essentially independent of the chemical constituents of the medium used to generate a particular growth rate; the simple nutritional requirements of the coliform bacteria ensure a wide choice of sources of carbon and nitrogen for the construction of different growth media.

The two principal subjects that have been stimulated by the detailed study of bacterial composition as a function of growth rate are the relationship among DNA synthesis, cell division, and cell mass on the one hand, and between ribosome production and function on the other. In the former case, a mechanism that links initiation of DNA synthesis to increase in cell mass has been identified (*109, 110*), while in the latter the most important finding is that, over quite a wide range of growth rates, the synthesis of ribosomes is controlled such that they function with a nearly constant efficiency in protein synthesis (*107*).

Analogous studies with eukaryotes have been much less extensive, although some data are available for yeast (*111–113*), a colorless alga *Prototheca zopfii* (*114*), and *Physarum* (*15*). The experiments with *Physarum* employed microplasmodia in suspension culture, and a range of growth rates was obtained using both complex and defined media and by means of a chemostat.

Figure 16 illustrates the variation in the RNA and protein contents, normalized to DNA, of microplasmodia of *Physarum* as a function of

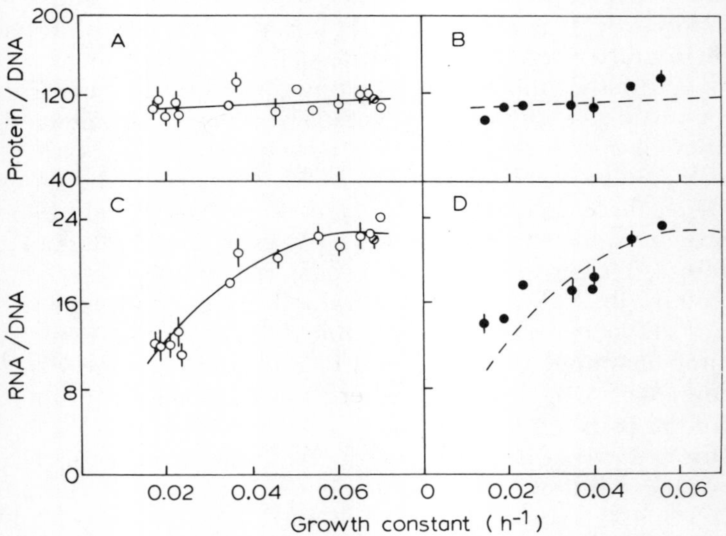

FIG. 16. Variation in the chemical composition of microplasmodia of strain CL with growth constant. (A) and (B): Protein/DNA ratio; (C) and (D): RNA/DNA ratio. ○, Batch cultures; ●, chemostat cultures; ⊘, batch culture of a diploid strain, a × i. The lines fitted to the data for batch cultures by a least-squares procedure are superimposed as dotted lines on the data obtained from chemostat cultures. From Plaut and Turnock (*15*).

growth rate. The results show that, per unit of DNA (or per nucleus), the amount of protein is almost independent of growth rate while the amount of RNA increases, approaching a constant value at fast rates of growth. A protein/DNA ratio invariant with growth rate is characteristic both of bacteria (*106, 107*) and of simple eukaryotes (*111, 115*). This suggests that, in both types of organism, replication of the DNA is regulated by a mass (of which protein comprises a major part)-titration mechanism. The change in the intermitotic time subsequent to inactivation of a proportion of the nuclei in a plasmodium of *Physarum* by UV light (*12*) is in accord with this view. Haploid and diploid strains of *Physarum* have similar protein/DNA ratios (Fig. 16), indicating that it is the number of genomes, rather than the number of nuclei, that is defined by the regulatory mechanism.

The change in the RNA/DNA ratio of microplasmodia as the growth rate increases is more complex than that observed for bacteria, which exhibit a linear relationship at fast growth rates. The rate of protein synthesis per ribosome has been calculated (*15*); the values increased from 85 to 130 amino acids per minute between the lowest and the highest growth rate in batch culture. The calculation took into account the proportion of RNA that is rRNA, and the fraction of ribosomes found in polyribosomes. Nevertheless, the values obtained for the net rate of protein synthesis per ribosome can only be taken as minimum estimates in the absence of knowledge of the rate of protein turnover, a parameter that is difficult to determine.

The studies described so far have been concerned with asynchronous populations of cells. Of course, it is equally valid to consider balanced growth in relation to the cell cycle. To do so requires the simple definition, A/B = constant, to be redefined, because the relative amounts of many cell constituents will certainly change during the progress of the cell from one division to the next. Accordingly, A/B = constant will apply only when cells in the same relative position in successive cycles are compared. However, there are many practical difficulties that hinder adequate testing of this simple statement. The most serious problem is that, even when it is possible to synchronize a population of cells, the synchrony decays at a rate that prevents measurements being made over more than two cycles. Even *Physarum*, with its natural synchrony, imposes constraints on the investigator. Ideally, for example, one would like to obtain data from a single plasmodium over several mitotic cycles; in practice this is not possible without replacing the medium and/or surgery to prevent a plasmodium becoming too large. Such manipulations may not perturb the growth of a plasmodium, but it is not easy to prove this.

Bearing in mind these cautionary remarks, what is the evidence that a plasmodium can exhibit balanced growth? Using the isotope-dilution method to follow the synthesis of nucleic acids, we have shown that nuclear DNA (*20*), rRNA (*20, 89*), and tRNA (*89*) can double in amount between postfusion mitoses M-II and M-III; some of the data are reproduced in Figs. 11 and 13. It is tempting to conclude that the results obtained for plasmodia growing between the two reference points (M-II and M-III) represent the normal patterns of accumulation, even though it is not yet technically possible to carry out the ideal experiment, namely, to follow cultures growing through three or four doublings in mass.

The synthesis of protein during the mitotic cycle in *Physarum* has also been determined (*24*) using an indirect application of the isotope-dilution method. A plasmodium was prepared from microplasmodia that had been labeled with [^{14}C]uridine, and samples were assayed for nucleic acid and protein. The accumulation curve for nucleic acid was obtained from the specific activity of the nucleic acid fraction and that for protein was derived by dividing the relative value for nucleic acid by the nucleic acid/protein ratio (Fig. 17). The results are of immediate interest because they demonstrate that for the same culture both total nucleic acid and total protein can essentially double in amount between successive nuclear divisions.

The protein measurements are less precise than those of nucleic acid because more manipulations are involved and for each sample total recovery of nucleic acid and protein is required. Nevertheless, the data (Fig. 17B) indicate that the rate of protein synthesis is lower during S phase than later on in the cycle. Labeling experiments with [^{35}S]methionine also suggest that the rate of protein synthesis increases during the mitotic cycle (*116*); the results obtained fitted an exponential pattern of synthesis. An increase in the rate of protein synthesis appears to correlate with an increase in the rate of ribosome synthesis (Fig. 11) in G2 phase, although the data are not sufficiently rigorous to confirm a causal relationship between an increased availability of ribosomes and an increase in the rate of protein synthesis.

In general, even though the DNA comprises only a minor part of the mass of the cell, it seems that the synthesis of other macromolecules is held at a relatively low level during S phase. Whether this reflects a reduction in transcription imposed by the requirements of replication of the DNA or whether, for example, there are positive controls that inhibit protein synthesis in the cytosol awaits further investigation. The differences in the characteristics of transcription in

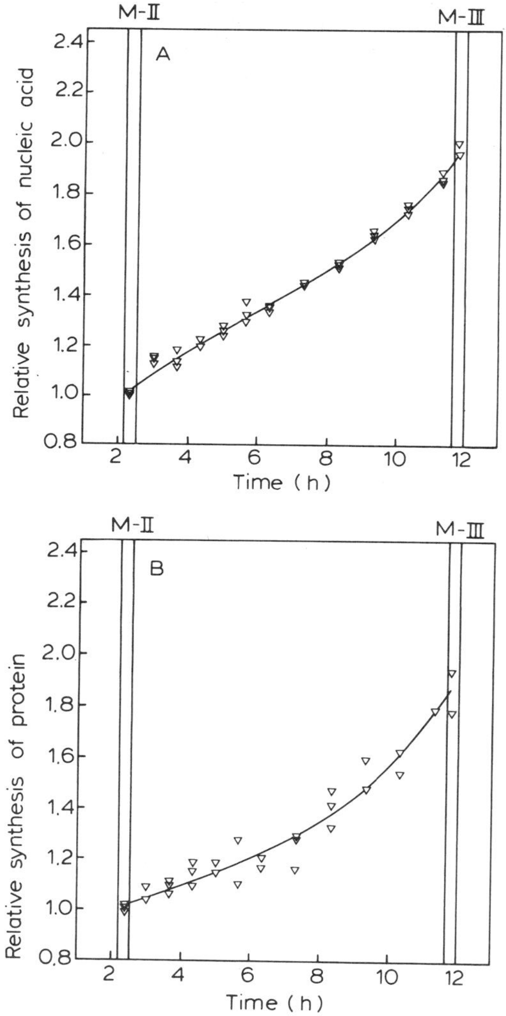

FIG. 17. Synthesis of total nucleic acid (A) and total protein (B) during the mitotic cycle of strain M_3C-VIII. A plasmodium was prepared from microplasmodia that had been labeled with [^{14}C]uridine. The accumulation curve for nucleic acid was determined from the specific activities of hot acid extracts, whereas that for protein was obtained by dividing the relative nucleic acid values by the corresponding nucleic acid/protein ratios. The lines were obtained by fitting the data to a polynomial using a computer program. From Birch and Turnock (*24*).

S phase and G2 phase, described in Section III, suggest that mRNA is a major product in early interphase but rRNA synthesis predominates in the later part of the cycle. The logic of this may be that, after nuclear division, the existing protein-synthesizing machinery (ribosomes and certainly some mRNAs) is sufficient to cope with the specialized demands of protein synthesis essential for replication of the genome, whereas the bulk synthesis of the major part of the protein population requires concomitant production of more ribosomes.

That balanced growth of plasmodia of *Physarum* is possible under the conditions of culture currently employed is demonstrated by the results already presented, although it would be much more satisfactory if reliable measurements could be obtained over several cycles. In some experiments, increases of less than a factor of two between successive nuclear divisions have been observed; for example, increments in DNA of only 75 to 85% are not uncommon. These results do not represent random variation in the measurements because increments greater than 100% are rarely if ever observed. Equally, care is taken not to include material from the initial area of the inoculum in the samples, since it has been known for some time that this part of a plasmodium does not remain in good synchrony with the remainder of the cell (*8*), and we have shown that DNA synthesis in nuclei in this region is much less than in the remainder of a plasmodium. For this reason we now remove the "inoculum ring" from plasmodia growing in the large tray apparatus (*24, 89*). This is done just before M-II to reduce the possibility of contamination of samples with inoculum material.

The phenomenon of "unbalanced growth" has been observed, under certain conditions, in a particularly acute form with the Colonia strain of *Physarum*. Figure 18 illustrates the synthesis of rRNA in a plasmodium of this strain; the experiment was carried out in a similar manner and with the same growth apparatus as that shown in Fig. 11, in which strain M_3C-VIII was used. However, with strain CL, the increase in rRNA between M-II and M-III was only 56% and an identical increment was obtained in a second experiment. Despite the lack of doubling in the amount of rRNA between mitoses, the form of the accumulation curve is very similar to that found with strain M_3C-VIII, the rate of synthesis being very low just after nuclear division and increasing to a maximum at the end of G2 phase. The increase in the rate of rRNA synthesis between the beginning and end of interphase, estimated by drawing tangents to the accumulation curve, was about fivefold, again very similar to strain M_3C-VIII. This result is encouraging because it indicates that, even when the synthesis of a particular

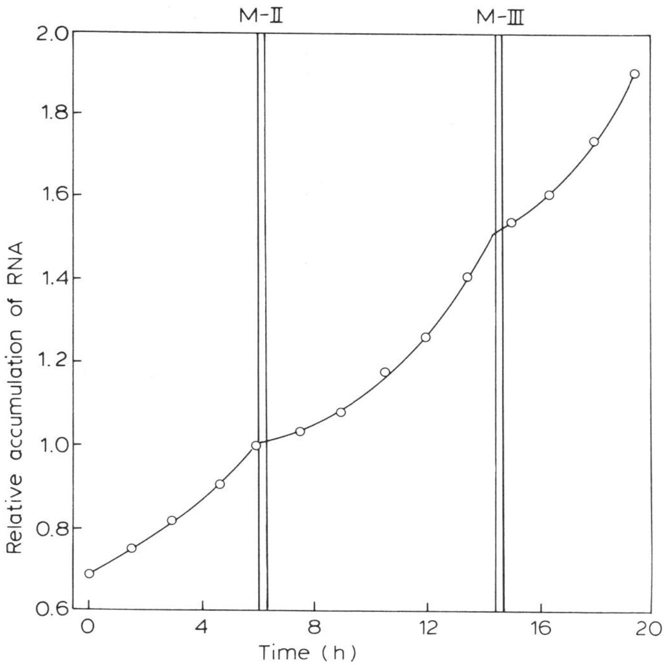

FIG. 18. Synthesis of rRNA during the mitotic cycle in strain CL, determined by the isotope-dilution method (L. Hall and G. Turnock, unpublished results).

macromolecule is not in step with the nuclear division cycle, the characteristics of the accumulation curve may still approximate that observed in true balanced growth.

The synthesis of nuclear DNA in plasmodia of strain CL has also been investigated. Table IV shows that in three large plasmodia the increase in DNA between M-II and M-III was only 30 to 40%, accentuating the problem of unbalanced growth in this strain. However, in experiment 3, in which small, Petri-dish cultures (up to 6 cm in diameter) were established from the same suspension of microplasmodia used to prepare the large, tray culture, the DNA content of the small plasmodia *did* double in amount between M-II and M-III. Hence plasmodia of strain CL can exhibit balanced growth under some conditions, as can microplasmodia in suspension culture (*15*). Furthermore, at M-II, nuclear DNA from both the large and small plasmodia had the same specific radioactivity, indicating that up to that point both cultures had had the same growth characteristics.

Two possibilities can be considered as explanations of the inade-

TABLE IV
NUCLEAR DNA SYNTHESIS IN STRAIN *CL*[a]

Experiment	Specific activity of DNA (cpm/μg DNA) M-II	M-III	Increment in DNA (%)
1. Large plasmodium	612	432	42
2. Large plasmodium	479	362	32
3(i). Large plasmodium	251	193	30
3(ii). Small plasmodia	249	125	99

[a] DNA synthesis was determined by the isotope-dilution method (Section I,G).

quate increase in DNA content between M-II and M-III in strain CL. One is that only 30 to 40% of the nuclei double their DNA content in the normal way, and the other is that some or all of the nuclei replicate their DNA only partially. To distinguish between these possibilities the DNA contents of individual nuclei, isolated at various times up to M-III, were determined, after staining, by microdensitometry and the results (Fig. 19) clearly show the first to be correct. Apart from a small number (about 10%) of nuclei that appear to be diploid (*26*), those isolated from small plasmodia had a G2 DNA content of about 0.6 pg/nucleus. A similar G2 value was also found with nuclei from large plasmodia up to early prophase of M-II. However, just before the next nuclear division, two sizes of nuclei were clearly present, the larger one being the expected 0.6 pg/nucleus and the smaller one about half of this value. Following nuclear division, a single peak was found with a mean DNA content of 0.3 pg/nucleus.

The data presented in Fig. 19 illuminate the reason for the low increment in nuclear DNA synthesis in large plasmodia of strain CL, but do not in any way provide an explanation as to why such a large proportion of the nuclei should fail to enter S phase after M-II. Given that nuclear DNA synthesis follows the same pattern in both Petri dish and tray cultures up to M-II, the most likely explanation is that some unidentified aspect of growth in the large tray apparatus is unfavorable and that strain CL is more sensitive to this than M_3C-VIII. Even though it is customary to replace the growth medium of Petri-dish cultures when they reach M-II, the ratio of medium to inoculum is still three to four times greater in the tray apparatus, so that depletion of the medium and/or accumulation of toxic metabolites are not obviously of prime importance in the phenomenon. On the other hand, as described in Section II, B, "abnormal" nuclei have previously been detected in plasmodia of another strain (*50*). The occurrence of these nuclei, which

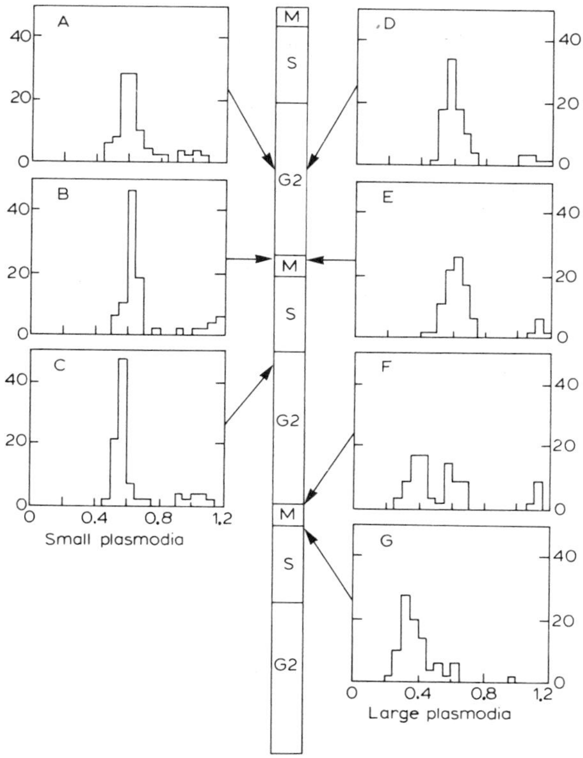

FIG. 19. DNA contents of nuclei isolated from small and large plasmodia of strain CL between postfusion mitosis M-I and M-III. Nuclei were stained by the Feulgen method, and the stain intensity is measured with a Vickers Scanning Microdensitometer. Nuclei from chicken erythrocytes were used as a standard. (A) and (D): three hours before M-II; (B) and (E): early prophase of M-II; (C): three hours after M-II; (F): early prophase of M-III; (G): early reconstruction of M-III. Abscissas: picograms of DNA. Ordinates: nuclei (% of total) (L. Hall and G. Turnock, unpublished results).

may parallel the production of diploid nuclei in CL (Fig. 19), and the cessation of DNA synthesis in a significant proportion of nuclei in large plasmodia may reflect unrecognized aspects of physiology and regulation in a multinucleate cell.

VI. Conclusion

The particular attraction of *Physarum* is the high degree of natural synchrony of nuclear division in the plasmodial phase of the life cycle, and this review has surveyed some of the ways in which this has been exploited in the context of nucleic acid synthesis and function. We now have a reasonable picture of the patterns of synthesis of the major classes of nucleic acids and total protein, and considerable detail has been obtained about the organization of the replication of DNA during the S phase.

For the future, an obvious development will almost certainly be the use of DNA recombination techniques to provide DNA probes for the analysis both of the replication of particular segments of the genome and of the products of transcription. The restriction enzyme map of nucleolar DNA has already been determined, and phage and plasmid vectors carrying different segments of this defined component of the genome should be invaluable for investigating variations in the primary rate of transcription of rRNA during the mitotic cycle.

Acknowledgments

I wish to thank Barbara Birch, Martin Cunningham, and Stephen Morris for patient and constructive criticism during the preparation of this article.

References

1. J. M. Ashworth and J. Dee, "The Biology of Slime Moulds." Arnold, London, 1975.
2. J. Dee, *J. Protozool.* **13,** 610 (1966).
3. R. W. Anderson, D. J. Cooke, and J. Dee, *Protoplasma* **89,** 29 (1976).
4. C. H. R. McCullough and J. Dee, *J. Gen. Microbiol.* **95,** 151 (1976).
5. J. W. Daniel and H. H. Baldwin, *Methods Cell Physiol.* **1,** 9 (1964).
6. R. T. M. Poulter, Ph.D. Thesis, University of Leicester (1969).
7. C. H. R. McCullough, D. J. Cooke, J. L. Foxon, P. E. Sudbery, and W. D. Grant, *Nature NB* **245,** 263 (1973).
8. J. Mohberg and H. P. Rusch, *J. Bact.* **97,** 1411 (1969).
9. M. M. Laane and F. B. Haugli, *Norw. J. Bot.* **21,** 309 (1974).
10. J. Mohberg, *in* "The Cell Nucleus" (H. Busch, ed.), Vol. 1, p. 187. Academic Press, New York, 1974.
11. A. Lord, L. Nicole, and J. G. Lafontaine, *J. Cell Sci.* **23,** 25 (1977).
12. P. E. Sudbery and W. D. Grant, *Exp. Cell Res.* **95,** 405 (1975).
13. J. J. Wille and S. Kauffman, *BBA* **407,** 158 (1975).
14. J. Mohberg and H. P. Rusch, *Exp. Cell Res.* **66,** 305 (1971).
15. B. S. Plaut and G. Turnock, *Mol. Gen. Genet.* **137,** 211 (1975).
16. L. Hall and R. Braun, *EJB* **76,** 165 (1977).
17. K. E. Davies and I. O. Walker, *J. Cell Sci.* **26,** 267 (1977).
18. L. Hall, G. Turnock, and B. J. Cox, *EJB* **51,** 459 (1975).
19. B. Birch and G. Turnock, *EJB* **69,** 257 (1976).
20. L. Hall and G. Turnock, *EJB* **62,** 471 (1976).
20a. K. Fink and P. Nygaard, *EJB* **89,** 417 (1978).
21. L. Lindahl, *JMB* **92,** 15 (1975).
22. P. G. W. Plagemann, *J. Cell. Physiol.* **77,** 241 (1971).
23. K. Fink, *BBA* **414,** 85 (1975).
24. B. Birch and G. Turnock, *FEBS Lett.* **84,** 317 (1977).
25. J. Mohberg, *J. Cell Sci.* **24,** 95 (1977).
26. T. G. Laffler and W. F. Dove, *J. Bact.* **131,** 473 (1977).

27. H. Fouquet, B. Bierweiler, and H. W. Sauer, *EJB* **44**, 407 (1974).
28. N. Hardman and P. L. Jack, *EJB* **74**, 275 (1977).
29. N. Hardman and P. L. Jack, *NARes* **5**, 2405 (1978).
30. W. Jelinek, G. Molloy, M. Solditt, R. Wall, D. Sheiness, and J. E. Darnell, *CSHSQB* **38**, 891 (1973).
31. R. Williamson, *Nature* **270**, 295 (1977).
32. V. M. Vogt and R. Braun, *JMB* **106**, 567 (1976).
33. H. V. Molgaard, H. R. Matthews, and E. M. Bradbury, *EJB* **68**, 541 (1976).
34. A. Zellweger, U. Ryser, and R. Braun, *JMB* **64**, 681 (1972).
35. U. Ryser and R. Braun, *BBA* **361**, 33 (1974).
36. H-U. Affolter and R. Braun, *BBA* **519**, 118 (1978).
37. L. Hall, U. Gubler, and R. Braun, *EJB* **86**, 45 (1978).
38. S. Guttes, E. Guttes, and R. A. Ellis, *J. Ultrastruct. Res.* **22**, 509 (1968).
39. J. Engberg, P. Andersson, V. Leick, and J. Collins, *JMB* **104**, 455 (1976).
40. M-C. Yao and J. G. Gall, *Cell* **12**, 121 (1977).
41. C. M. Johnson, V. C. Littau, V. G. Allfrey, E. M. Bradbury, and H. R. Matthews, *NARes* **3**, 3313 (1976).
42. J. L. Compton, R. Hancock, P. Oudet, and P. Chambon, *PNAS* **73**, 4382 (1976).
42a. R. M. Grainger and R. C. Ogle, *Chromosoma* **65**, 115 (1978).
42b. M. J. Butler, K. E. Davies, and I. O. Walker, *NARes* **5**, 667 (1978).
43. J. M. Mitchison, "The Biology of the Cell Cycle." Cambridge Univ. Press, London and New York, 1971.
44. O. F. Nygaard, S. Guttes, and H. P. Rusch, *BBA* **38**, 298 (1960).
45. R. Braun, C. Mittermayer, and H. P. Rusch, *PNAS* **53**, 924 (1965).
46. H. H. Evans, S. R. Littman, T. E. Evans, and E. N. Brewer, *JMB* **101**, 169 (1976).
47. K. Gausing, *JMB* **115**, 335 (1977).
47a. F. Bovey and F. Ruch, *Histochemie* **32**, 153 (1972).
48. C. S. Newlon, G. E. Sonenshein, and C. E. Holt, *Bchem* **12**, 2338 (1973).
49. R. Braun and T. E. Evans, *BBA* **182**, 511 (1969).
50. E. Guttes and S. Guttes, *J. Cell Biol.* **43**, 229 (1969).
50a. S. Funderud, R. Andreasson, and F. Haugli, *Cell* **15**, 1519 (1978).
51. R. Braun and H. Wili, *BBA* **174**, 246 (1969).
52. J. A. Huberman and A. D. Riggs, *JMB* **32**, 327 (1968).
53. F. W. Studier, *JMB* **11**, 373 (1965).
54. E. N. Brewer, *JMB* **68**, 401 (1972).
55. E. N. Brewer, T. E. Evans, and H. H. Evans, *JMB* **90**, 335 (1974).
56. S. Funderud and F. Haugli, *NARes* **2**, 1381 (1975).
57. S. Funderud, R. Andreasson, and F. Haugli, *NARes* **5**, 713 (1978).
58. H. Weintraub, *Nature NB* **236**, 195 (1972).
59. E. N. Brewer and P. Ting, *J. Cell. Physiol.* **86**, 459 (1975).
60. E. N. Brewer, *BBA* **402**, 363 (1975).
61. S. Funderud and F. Haugli, *BBRC* **74**, 941 (1977).
62. W. Schiebel and U. Schneck, *Zp Chem* **355**, 1515 (1974).
63. V. M. Vogt and R. Braun, *EJB* **80**, 557 (1977).
64. E. Stimac, D. Housman, and J. A. Huberman, *JMB* **115**, 485 (1977).
65. J. E. Cummins and H. P. Rusch, *J. Cell Biol.* **31**, 577 (1966).
66. J. J. Muldoon, T. E. Evans, O. F. Nygaard, and H. E. Evans, *BBA* **247**, 310 (1971).
67. S. Funderud and F. Haugli, *NARes* **4**, 405 (1977).
68. F. B. Haugli, W. F. Dove, and A. Jimenez, *Mol. Gen. Genet.* **118**, 97 (1972).
69. A. Travers, *Nature* **263**, 641 (1976).
70. C. Mittermayer, R. Braun, and H. P. Rusch, *BBA* **91**, 399 (1964).

71. R. Braun, C. Mittermayer, and H. P. Rusch, *BBA* **114**, 527 (1966).
72. U. Wiegers, G. Kramer, K. Klapproth, and H. Hilz, *EJB* **64**, 535 (1976).
73. C. Mittermayer, R. Braun, and H. P. Rusch, *BBA* **114**, 536 (1966).
74. W. D. Grant, *EJB* **29**, 94 (1972).
75. K. E. Davies and I. O. Walker, *FEBS Lett.* **86**, 303 (1978).
76. P. Chambon, *ARB* **44**, 613 (1975).
77. A. Hildebrandt and H. W. Sauer, *FEBS Lett.* **35**, 41 (1973).
78. S. Z. Gornicki, S. B. Vuturo, T. V. West, and R. F. Weaver, *JBC* **249**, 1792 (1974).
79. A. B. Burgess and R. R. Burgess, *PNAS* **71**, 1174 (1974).
80. S. S. Smith and R. Braun, *EJB* **82**, 309 (1978).
81. A. Hildebrandt and H. W. Sauer, *BBA* **425**, 316 (1976).
82. G. H. Ernst and H. W. Sauer, *EJB* **74**, 253 (1977).
83. S. S. Smith and R. Braun, *Experientia* **32**, 813 (1976).
84. P. Melera and H. P. Rusch, *Exp. Cell Res.* **82**, 197 (1973).
84a. I. B. Dawid and P. K. Wellauer, *Cell* **8**, 443 (1976).
85. K. E. Davies and I. O. Walker, *J. Cell Sci.* **29**, 93 (1978).
86. D. N. Jacobson and C. E. Holt, *ABB* **159**, 342 (1973).
87. R. A. Weinberg, *ARB* **42**, 329 (1973).
88. D. Kessler, *Exp. Cell Res.* **45**, 676 (1967).
89. K. Fink and G. Turnock, *EJB* **80**, 93 (1977).
90. H. Fouquet and R. Braun, *FEBS Lett.* **38**, 184 (1974).
91. H. Fouquet and H. W. Sauer, *Nature* **255**, 253 (1975).
92. R. P. Perry, E. Bard, B. D. Hames, D. E. Kelly, and U. Schibler, This Series **19**, 275 (1976).
93. R. A. Firtel and H. F. Lodish, *JMB* **79**, 295 (1973).
94. W. Stockem, *Histochemie* **15**, 163 (1968).
95. T. J. Nicholls, *J. Cell Sci.* **10**, 1 (1972).
96. T. Kuroiwa, *Exp. Cell Res.* **78**, 351 (1973).
97. E. Guttes, P. C. Hanawalt, and S. Guttes, *BBA* **142**, 181 (1967).
98. H. J. Bohnert, *Exp. Cell Res.* **106**, 426 (1977).
99. T. E. Evans and D. Suskind, *BBA* **228**, 350 (1971).
100. D. Kessler, *J. Cell Biol.* **43**, 68A (1969).
101. G. E. Sonenshein and C. E. Holt, *BBRC* **33**, 361 (1968).
102. T. Kuroiwa, *J. Cell Biol.* **63**, 299 (1974).
103. T. Kuroiwa, S. Kawano, and M. Hizume, *Exp. Cell Res.* **97**, 435 (1976).
104. T. Kuroiwa, M. Hizume, and S. Kawano, *Cytologia* **43**, 119 (1978).
105. W. D. Grant and R. T. M. Poulter, *JMB* **73**, 439 (1973).
106. O. Maaløe and N. O. Kjeldgaard, "Control of Macromolecular Synthesis." Benjamin, New York, 1966.
107. O. Maaløe, *Dev. Biol., Suppl.* **3**, 33 (1969).
108. A. Campbell, *Bacteriol. Rev.* **21**, 263 (1957).
109. R. H. Pritchard, *Philos. Trans. R. Soc. London, Ser B* **267**, 303 (1974).
110. R. H. Pritchard, *in* "DNA Synthesis: Present and Future," p. 1. Plenum, New York, 1978.
111. I. McMurrough and A. H. Rose, *BJ* **105**, 189 (1967).
112. C. T. Wehr and L. W. Parks, *J. Bact.* **98**, 458 (1969).
113. C. Waldron and F. Lacroute, *J. Bact.* **122**, 855 (1975).
114. R. O. Poyton, *J. Bact.* **113**, 203 (1973).
115. V. Leick, *C.R. Trav. Lab. Carlesberg* **36**, 113 (1967).
116. M. Schwärzler, B. M. Jockusch, L. Hall, and R. Braun, *EJB* **80**, 43 (1977).

Biochemical Effects of the Modification of Nucleic Acids by Certain Polycyclic Aromatic Carcinogens[1]

DEZIDER GRUNBERGER AND
I. BERNARD WEINSTEIN

Cancer Center/Institute of Cancer Research, Departments of Biochemistry and Medicine and Division of Environmental Sciences,
Columbia University College of Physicians and Surgeons
New York, New York

[1] This work was supported by Grants No. CA 21111 and CA 13696 awarded by the National Cancer Institute, DHEW.

Progress in Nucleic Acid Research
and Molecular Biology, Vol. 23

ISBN 0-12-540023-3

I. Introduction[2]

A fundamental principle underlying the success of modern molecular biology is that elucidation of the structure of biological macromolecules at the primary, secondary, and tertiary levels is essential to an understanding of biological function. There is increasing evidence that covalent binding of carcinogens to cellular macromolecules, particularly nucleic acids, is the initial critical event in the encounter between an environmental carcinogen and target cells in the exposed host (*1, 2*). Therefore, to understand carcinogenesis at a molecular level, it becomes necessary to understand the complete chemical structure and stereochemistry of the carcinogen-macromolecular adducts, to then determine the associated conformational changes in the target macromolecules, and, finally to relate these chemical and physical findings to possible aberrations in the functioning of the chemically modified macromolecules. We have pursued these long-term goals with the aromatic carcinogens 2-acetamidofluorene (*3, 4*) and benzo[*a*]pyrene (*5–7*), concentrating on the latter because it is a ubiquitous carcinogen and is representative of polycyclic aromatic hydrocarbons, which are of increasing concern as environmental carcinogens.

The chemical structures of carcinogen-nucleic acid adducts and the associated conformational changes in nucleic acids have been reviewed (*8, 9*). The present purpose is to summarize what is known about the effects of such modifications on the functioning of nucleic acids. Possible implications with respect to the carcinogenic process are also discussed. We concentrate on three polycyclic aromatic carcinogens, the two mentioned above (AAF and B[*a*]P, the latter frequently generated in the pyrolysis of fossil fuels) and the highly potent mycotoxin aflatoxin B_1. Observations on a few related compounds are also included. We do not discuss the simpler alkylating agents, such as the nitrosamides and nitrosureas, which methylate or ethylate nucleic

[2] Abbreviations used:
AAF, 2-acetamidofluorene
HO-AAF, 2-(*N*-hydroxy)acetamidofluorene
AcO-AAF, 2-(*N*-acetoxy)acetamidofluorene
B[*a*]P, benzo[*a*]pyrene
B[*e*]P, benzo[*e*]pyrene
B[*a*]PDE, (±)7,8-dihydrobenzo[*a*]pyrene-7,8-diol 9,10-epoxide
B[*a*]PDE-I, B[*a*]P-7β,8α-diol 9α,10α-epoxide
B[*a*]PDE-II, B[*a*]P-7β,8α-diol 9β,10β-epoxide
Afl-B_1, aflatoxin B_1

acids, since this subject is reviewed in detail by Singer and Kröger in this volume.[3]

II. Structural Considerations

A. Structure of Nucleoside Adducts

The chemical structures of the major deoxynucleoside adducts formed from activated derivatives of AAF, B[*a*]P, and Afl-B_1 are shown in Fig. 1. The corresponding adducts are also seen in RNA. A complicating feature in attempting to relate chemical structure to functional effects is that each of these agents forms more than one type of nucleoside adduct. This is also the case with the simpler alkylating agents, where almost every nitrogen and oxygen residue of all the nucleic acid bases can be modified (*10*).[3]

In addition to the structures depicted in Fig. 1, some studies suggest that AAF may also modify adenine residues in nucleic acids (*11–13*), although such an adduct has not actually been isolated and characterized. The fluorene adducts shown in Fig. 1 can also exist in their deacetylated forms (*14*). The situation is even more complicated with benzo[*a*]pyrene. Multiple isomers of its activated form, the 7,8-dihydrodiol 9,10-epoxide (B[*a*]PDE), are known, and isomers of the adducts depicted in Fig. 1 have also been identified (*8*, *15*). A minor reaction product probably resulting from the reaction of the epoxide with the N^4 of cytosine has also been detected (*16*). Some studies (*17*) suggest that the epoxide can also modify the N-7 of guanine residues in DNA, and this could lead to depurination. There is also indirect evidence that it can modify phosphoric residues of the nucleic acid backbone (*18*); this might enhance strand breakage. The possibility that other oxides, or radical ion derivatives of B[*a*]P play a role in carcinogenesis has also not been excluded, although there is little evidence to support such. Indeed, several studies indicate that the guanosine adduct depicted in Fig. 1c is the major RNA and DNA adduct formed when human or rodent tissues or cell cultures are exposed to B[*a*]P (*19*). It is also the major adduct formed when DNA is modified with the epoxide *in vitro* (*20*). Therefore, we emphasize the effects of this adduct, although the possible significance in carcinogenesis of other types of nucleic acid modification by activated forms of B[*a*]P has not been excluded. In addition to the guanine

[3] See the article by Singer and Kröger on p. 151.

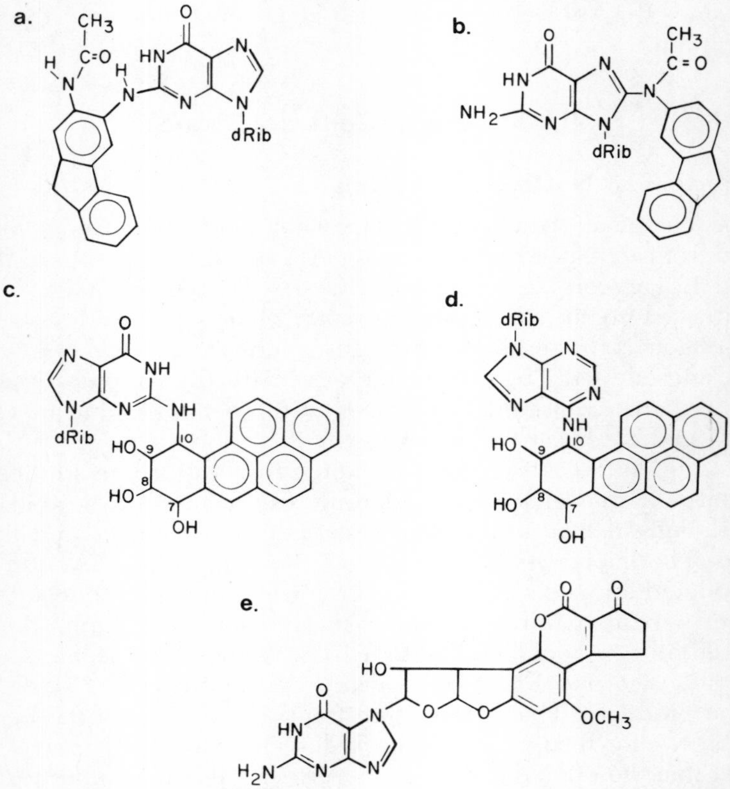

FIG. 1. The structures of carcinogens and nucleoside adducts. (a) 2-Acetamido-3-(2′-deoxy-N^2-guanosyl)fluorene. (b) 3-[*N*-(2′-Deoxy-8-guanosyl)acetamido]fluorene. (c) 2′-Deoxy-N^2-(7,8,9,10-tetrahydro-7β,8α,9α-trihydroxybenzo[*a*]pyren-10-yl) guanosine. (d) 2′-Deoxy-N^6-(7,8,9,10-tetrahydro-7β,8α,9α-trihydroxybenzo[*a*]pyren-10-yl)adenosine. (e) 7-Guanyl-dihydrohydroxyaflatoxin B_1.

adduct depicted in Fig. 1, recent studies suggest that aflatoxin B_1 may also modify the N-7 of adenine residues (*21*), although there is no definitive evidence for this derivative.

B. Conformational Aspects

Fundamental to an understanding of the functional consequences of nucleic acid modification by these bulky carcinogens is information on two points: (*a*) the orientation of the covalently bound carcinogen residues within a single- or double-stranded nucleic acid; and (*b*) possible alterations in the native conformation of the nucleic acid resulting from this modification. These aspects have been reviewed in detail elsewhere (*3*, *4*, *9*). In the case of the fluorene modification of the C-8 of guanine residues in double-stranded DNA, considerable evidence has been obtained for a conformational distortion termed "base displacement." To accommodate the bulky residue, modification of the C-8 of guanine requires rotation of the base about the glycosyl bond of deoxyguanosine from the usual *anti* to the *syn* position. In addition, the planar fluorene ring system is inserted into the helix, occupying the position vacated by the affected guanine residue. These conformational changes cause a marked destabilization of the double-stranded helix at the sites of this modification. The effects with respect to base-pairing and template function are discussed below. On the other hand, AAF modification of the N^2 of guanine, which occurs to about one-fifth the extent of the C-8 modification, does not appear to produce major destabilization of the DNA helix (*22*).

Studies on double-stranded linear DNA modified *in vitro* by the B[*a*]P diol-epoxide suggest that the covalently bound carcinogen residue lies in the minor groove of the DNA helix and that there is much less destabilization of the double-stranded helix than in the case of the AAF–C-8 guanine adducts (*6*, *23*, *24*). It is not apparent whether this structure in itself would directly impair the base-pairing capacity of the modified guanine residue. In contrast to the latter results, another study (*25*) suggests that when supercoiled circular SV40 DNA is modified by B[*a*]PDE, the carcinogen residue is actually intercalated into the DNA helix. Model-building studies indicate that this would require displacement from the helix of the modified guanine residue, which would interfere with its availability for Watson–Crick base-pairing. Further studies are required to determine which of these structures pertain *in vivo* during different functional states of nucleic acids modified by the epoxide. The problem is further complicated by results (*26*) suggesting that sites of B[*a*]PDE modification of adenine residues produce a greater destabilization of the DNA helix than occurs at sites of guanine modification.

At present, there is little information available on the conformation of nucleic acids modified by aflatoxin B_1. Modification of the N-7 of guanine, which lies in the major groove of the helix, would not directly interfere with base-pairing.

III. Alterations in DNA Repair

Perhaps the earliest response of the cell to damage of its DNA by chemical carcinogens is a DNA excision-repair process and the associated "unscheduled" DNA synthesis. The precise enzymic mechanisms by which the chemically modified bases are recognized and excised are not known at the present time. Specific endonucleases that act at sites of X-ray or UV-induced lesions in DNA, and N-glycosylases that remove uracil residues in DNA have been identified in bacterial and mammalian systems [for review, see Hanawalt (27)].[4] O^6-Methylguanine lesions may be repaired in rat liver by a demethylase activity (*28*). The enzymology responsible for the removal of bases in DNA modified by the polycyclics is not known. The fluorene and benzo[*a*]pyrene adducts depicted in Fig. 1 are chemically very stable, but the aflatoxin adduct, like all N-7 adducts, is unstable under slightly alkaline conditions. Therefore, spontaneous depurination of the latter could occur to some extent *in vivo*, but enzymic excision appears to be the only biological mechanism for the removal of the fluorene and benzo[*a*]pyrene adducts.

It is of interest that patients with xeroderma pigmentosum belonging to certain complementation groups (notably A and C) are deficient in the repair of DNA lesions induced by UV, AAF, B[*a*]P, Afl-B_1, and certain other bulky carcinogens, whereas they are not impaired with respect to the repair of lesions induced by X-ray or methylating agents (for reviews, see *27, 29, 30*). This suggests that one or more steps are common to the repair of the former types of lesions. These cells are more sensitive to killing and to mutagenesis by UV and polycyclic aromatic carcinogens (*27*), presumably because of deficient excision of these lesions. However, it has not yet been demonstrated that these cells are more sensitive to *in vitro* transformation. If mouse 3T3 cells are exposed either to 4-nitroquinoline 1-oxide or to 3-methylcholanthrene and held for a period of time at confluence so that they do not divide, transformation is markedly reduced compared to that seen when they are allowed to divide (*31*). It is concluded that the nondividing cultures had time to excise the damaged regions before replication began.

Differential rates of excision of carcinogen adducts may play an important role in carcinogenicity. This aspect has been reviewed with respect to methylating and ethylating agents.[3] The C-8 guanine adduct of AAF is excised from rat liver DNA *in vivo* much more rapidly than the N^2 adduct (*32*). This may relate to the different conformational states of the two types of AAF-DNA adducts [see the Introduction and Yamasaki *et al.* (*22*)]. Preliminary evidence suggests that the multiple deoxynucleoside adducts formed by the B[*a*]P-diol-epoxide also

[4] See article by Lindahl in Vol. 22 of this series.

undergo differential rates of excision during DNA repair (*33, 34*). Further studies are required to determine to what extent the ability of specific DNA adducts to elude excision-repair mechaisms plays a role in organ specificity and carcinogen potency.

The evidence suggests that the DNA excision-repair process operates with high fidelity and, therefore, is a host-protective mechanism (*27*). Presumably, carcinogen damage to DNA would be functionally significant in terms of mutagenesis only if the DNA replication mechanism reaches the modified region of DNA before excision repair of that region had been completed. On the other hand, it is possible that carcinogens do not act simply by producing errors in DNA replication at the sites at which they damage DNA. In bacteria, physical and chemical agents that damage DNA, including chemical carcinogens or other factors that interfere with DNA replication, induce a highly pleiotropic response called "SOS functions" (*35, 36*). These functions include induction of lysogenic phage, filamentous growth, and mutagenesis. Activated forms of Afl-B^1, B[*a*]P, and other polycyclic hydrocarbons cause induction and mutagenesis of prophage λ in *Escherichia coli* K12; this is referred to as the "induct test," and it is proposed that it may be a useful screening test for potential carcinogens (*37*). The latter type of mutagenesis appears to result from the induction of an error-prone DNA replication mechanism that results in mutagenesis of even undamaged DNAs. It is thought to have a positive function because it allows the replication mechanism to read through a damaged region of DNA and thus permit cell survival. It is not yet clear that similar responses to DNA damage occur in eukaryotic cells, although recent experiments support this possibility (*38–40*). In *E. coli,* the induction of phage synthesis and mutagenesis following damage to DNA appears to involve the action of a protease that destroys repressor protein(s) (*35, 36*). The carcinogenic process also appears to be frequently associated with increased synthesis of the protease plasminogen activator, and perhaps other proteases [for a review of this subject, see Reich *et al.* (*41*)]. Further studies are required to determine the possible role in the carcinogenic process of inducible DNA repair and error-prone DNA synthesis mechanisms, and the significance of protease induction.

Ether-permeabilized (nucleotide-permeable) *E. coli* cells have been used to study the details of DNA repair in response to a variety of DNA damaging agents (*42, 43*). A variety of "ultimate" carcinogens (i.e., compounds that do not require metabolic activation) induce DNA excision-repair in wild-type *E. coli,* as measured by [^{3}H]dTMP incorporation. These compounds simultaneously inhibit replicative DNA synthesis (see Section IV). Studies on the *E. coli* mutants UVrA and UVrB, which are defective in the pyrimidine-dimer-specific UV-endonuclease, indicate that this enzyme is involved in the removal of agents, such as AcO-AAF, 7-bromomethylbenz[*a*]anthracene, mitomycin C, and some very reactive alkylating or carboxyethylating agents, that create steric distortions or localized denaturation in duplex DNA. During this process, ATP is required as a cofactor for the incision step. A mutant (KMBL 1789:$polA_{107}$) lacking the 5′-3′ exonucleolytic activity of DNA polymerase I (exonuclease VI) was unable to carry out DNA repair polymerization, thus demonstrating the role of this enzyme in the DNA repair process. Other studies also show that exonuclease VI excises thymine dimers and carcinogen-nucleotide adducts from damaged double-stranded DNA *in vitro* (*44, 45*). A detailed study (*46*) of permeabilized *E. coli* also indicates that the DNA excision-repair induced by the "K-region epoxides" (i.e., 7-methyl- and 7,12-dimethylbenz[*a*]anthracene 5,6-epoxide, benzo[*a*]pyrene 4,5-epoxide, and chrysene 5,6-epoxide) requires the UVrA and UVrB functions as well as DNA polymerase I. In view of the recent evidence that "Bay-region diol-epoxides" (i.e., the 3,4-dihydrodiol 1,2-oxide of the benz[*a*]anthracenes, the 7,8-dihydrodiol 9,10-oxide of benzo[*a*]pyrene, and the 9,10-dihydrodiol 7,8-oxide of 3-methylcholanthrene) (*47*) of the polycyclics, rather than the above-mentioned K-region epoxides, are involved in their carcinogenicity, it would be important to study these compounds in this system.

Compounds that do not bind covalently to DNA in the above *E. coli* system, such as the intercalating agents quinacrine, ethidium bromide, adriamycin, and *cis*-Pt(II) diammine,[5] or AAF, dimethylnitrosamine, aflatoxins B_1, B_2, and G_1, and 4-nitroquinoline 1-oxide (which are not converted to activated forms by *E. coli*), do not induce DNA excision, even though some of them inhibit replicative DNA synthesis. Presumably the proximate carcinogens would induce DNA repair if an activation system (microsomes and cofactors) were added to the *E. coli* system. Also, the tumor promoters 12-*O*-tetradecanoyl phorbol-13-acetate (TPA) and phorbol 12,13-didecanoate do not induce excision repair and do not inhibit repair synthesis induced by AcO-AAF or *N*-methyl-*N*-nitrosourea, a result consistent with other evidence that these agents are not mutagens (*48*).

Although the above studies indicate that the UV-endonuclease in *E. coli* may be involved in the excision of bulky covalently bound carcinogens, other mechanisms also appear to be involved in the DNA

[5] See article by Roberts and Thomson in Vol. 22 of this series.

repair induced by these agents. If they give rise to spontaneous depurination (which may be the case with aflatoxin B_1 adducts), or if they lead to enzymically catalyzed depurination, via an N-glycosylase mechanism, then the apurinic sites would be substrates for the apurinic endonucleases (*46, 49, 50*). N-Glycosylases that recognize nucleoside adducts formed from activated forms of AAF, B[*a*]P, Afl-B_1, or other bulky carcinogens, have not yet been identified, and it is possible that they are excised only at the nucleotide level. Interestingly, the thymine-dimer excision enzyme purified by Sekiguchi *et al.* (*51*) does not recognize these chemical adducts, and the uracil-DNA-glycosylase (*52*),[4] the 3-methyladenine-DNA glycosylase (*53*), and the hypoxanthine-DNA-glycosylase (*53a*) enzymes are specific for their respective substrates. Thus, the enzymes involved in the excision of certain carcinogen-DNA adducts could be quite specific with respect to their substrates. It also remains to be determined whether the excision-repair system for carcinogen-induced lesions is constitutive or inducible in mammalian systems.

Several distinct endonucleases that recognize and cleave different types of damage in DNA have been purified from *Micrococcus luteus*. These include: (*a*) endonucleases that recognize pyrimidine dimers in UV-irradiated DNA (*54, 55*); (*b*) endonucleases that selectively cleave irradiated DNA (*56, 57*); and (*c*) endonucleases for apurinic sites in DNA (*53, 58*). An endonuclease for apurinic sites has been separated from a DNA glycosylase that excises 3-methyladenine from Me_2SO_4-treated DNA (*53*). Recently, another exonuclease that recognizes apurinic sites has been purified (*59*). This enzyme also appears to recognize DNA damaged by a variety of alkylating, arylating, and arylamidating carcinogens. This preparation may be a mixture of both an apurinic endonuclease and a DNA glycosylase that recognizes a variety of chemically modified bases, but this remains to be established. Thus, as stressed above, N-glycosylases or endonucleases that specifically recognize polycyclic aromatic adducts have not yet been identified with certainty. Linn *et al.* (*60*) have recently described an "insertase" reaction[6] in which a base in DNA appears to be excised and replaced at the glycoside level without breakage of the phosphodiester backbone. The possible significance of this rather direct mechanism of DNA repair remains to be determined.

A puzzling aspect of the DNA excision process is the large size of the DNA patch removed during the excision of the bulky lesions in

[6] The action of this enzyme identifies it as a transferase or nucleosidase. See also J. R. Katze and W. R. Farkas, *PNAS* **76**, 3271 (1979). [Ed.]

DNA. In normal mammalian cells exposed to UV, the polycyclics under discussion, or 4-nitroquinoline 1-oxide, this "patch" (number of nucleotides excised per modified residue excised) has been estimated to be in the range of 30 to 100 nucleotides, whereas with X-ray or methylating agents it is much smaller (*29, 45*). A large patch size (30 to 120 nucleotides) in the permeabilized *E. coli* system has also been seen with 7-bromomethylbenz[*a*]anthracene and the fluorene lesions (*43*). Although it is tempting to correlate the large patch sizes with the DNA-denaturing effects of these chemically induced lesions, as measured by susceptibility to single-strand-specific nucleases and other *in vitro* devices (*32, 61–63*), in several cases the patch size appears to be larger than the length of the denatured region estimated from the *in vitro* data. Presumably the patch size reflects not only the conformational changes in the damaged region of the DNA, but also the functional properties of the excision enzymes.

The phenomenon of "postreplication repair," a process by which gaps in newly replicated strands of DNA resulting from lesions in the parental strand are eventually filled, is even less well understood in mammalian systems than is the excision-repair process. It has been described with activated derivatives of the three polycyclics (*27, 64*), and a recent study suggests that the process may be inducible in mammalian cells (*65*). There is evidence that in bacteria these gaps may be filled by a process of recombination, but no such evidence has emerged in studies of mammalian cells (*27*). It seems likely that the formation of these gaps reflects the inability of the DNA replication mechanism to "read through" lesions in the parental DNA strand produced by these bulky agents and not removed by the excision-repair process. The arrest in chain elongation during *in vitro* replication or transcription of carcinogen-modified DNA templates is discussed in Sections IV and VII.

IV. Effects on DNA Synthesis

A. *In Vivo* Effects

In contrast to its stimulation of unscheduled or excision-repair DNA synthesis, carcinogen modification of DNA generally inhibits replicative DNA synthesis in both bacterial and mammalian systems (*46, 66, 67*). In fact, it has been proposed that this inhibition can be used as a simple screening test for potential mutagens and carcinogens (*67, 68*). Several DNA damaging agents, including aflatoxin B_1, cause a rapid inhibition of DNA synthesis in HeLa cell cultures. With

4-nitroquinoline 1-oxide and X-ray, there is evidence that the initial and major effect is due to inhibition of replicon initiation, with a later effect on inhibition of chain elongation, although with certain agents (i.e., adriamycin) the inhibitory effect was primarily on chain elongation (*67, 68*). These results, as well as evidence that supercoiling of the DNA is required for the replication of phage or plasmid DNA (*69, 70*), have led to the suggestion that a supercoiled subunit of DNA within mammalian chromatin serves as a replicon cluster on which DNA synthesis is initiated, and that certain damaging agents may alter the conformation of these supercoiled regions, thereby blocking DNA replication (*67, 68*). With regard to this hypothesis, there is evidence that superhelicity of DNA can also influence transcription. Increasing the superhelical density of phage DNA templates enhances the rate of transcription initiation by a bacterial polymerase, presumably by influencing the interaction between the promoter region of the DNA and the polymerase enzyme. The state of superhelicity may in turn be influenced by an enzyme, "DNA gyrase," which is capable of inserting negative superhelical turns into closed circular DNA [for review of these aspects, see Trevers (*71*)]. Presumably, carcinogen modification of DNA conformation could also influence these aspects of DNA replication and transcription.

Dimethylbenzanthracene induces tumors in regenerating rat or mouse liver, although it is not tumorigenic in the normal adult rat liver. It appears (*72*) that its binding to DNA is greater in regenerating liver than in the intact liver, and that this is associated with a more marked inhibition of DNA synthesis in the regenerating liver. It was suggested (*72*) therefore, that the "binding" may be causally involved in this induced inhibition of DNA synthesis. A curious aspect is that although the inhibition of DNA synthesis was transient in the treated animals, with recovery to near control values by 72 hours, the binding to DNA persisted at a high level for at least four weeks. The mechanism by which the tissue overcomes this inhibition is not known.

When isolated rat mammary epithelial cells are exposed in culture to low doses of Me_2-benzanthracene (0.0005 to 0.05 μg/ml), there is an inhibition of DNA synthesis that persists for at least three days. By the sixth day, however, DNA synthesis in the treated cultures increases over that in control cultures (*73*).

B. *In Vitro* Effects

Rat liver DNA modified by *in vivo* administration of AAF has a decrease in template activity with respect to both RNA and DNA synthesis when assayed *in vitro* (*74*). The inhibition of transcription is greater than the inhibition of replication. Treatment of rat liver DNA with AcO-AAF results in an extensively modified DNA. This material has no priming activity for DNA polymerase. It competitively inhibits the priming of DNA polymerases by untreated DNA at concentration ratios of 1 : 100.

Studies on DNA synthesis by isolated nuclei from regenerating rat liver reveal that the addition of exogenous nuclease-activated DNA stimulates [^{3}H]dT incorporation from [^{3}H]dTTP (75). Activated DNA previously treated with AcO-AAF is less effective in this system, and the modification diminished, in a dose-dependent fashion, the incorporation to below basal levels. Methylnitronitrosoguanidine does not have this inhibitory effect. These results led to the suggestion (75) that the localized regions of denaturation in the AAF-DNA prevent chain elongation and thus yield nonfunctional enzyme–DNA complexes.

The effects of modification of single-stranded ϕX174 DNA by (±)B[*a*]PDE-I on its *in vitro* template activity during DNA synthesis have been studied (76). DNA, containing various amounts of the covalently bound adduct, was added to a partially purified extract from *E. coli* H560 in the presence of all four deoxynucleotides and appropriate cofactors. In this system, DNA replication usually results in conversion of the single-stranded template to the double-stranded replicative intermediate RF2. When the template contained three residues of B[*a*]P per ϕX174 DNA molecule, the incorporation of [^{3}H]dT from [^{3}H]TTP into acid-insoluble product occurred at a reduced rate, and the amount of label incorporated was less than with unmodified template. In neutral sucrose gradients, with both unmodified and modified templates, the newly synthesized DNA sedimented like the replicative intermediate RF2. However, when the products were examined by sedimentation under alkaline conditions to dissociate the complementary strands, a large fraction of the DNA product made with the diol-epoxide-modified DNA as template sedimented more slowly and was more heterogeneous than that obtained with the unmodified template. These results led the authors to suggest (76) that "benzo[*a*]pyrene groups on DNA templates block the propagation of new chains to full-sized molecules and imply that polymerization through the alkylated template site does not occur." This conclusion is similar to that reached when DNAs modified with AcO-AAF or B[*a*]PDE were used as templates for DNA-directed RNA synthesis *in vitro* (see Section VII).

In the above study (76), evidence was adduced that, under the conditions used, with up to eight covalently bound B[*a*]P residues per ϕX DNA, there was no evidence of DNA strand breakage, or intermolecular cross-linking. Another study (*77*) indicates that when single-stranded ϕX174 DNA is used as a substrate for (±)B[*a*]PDE-I, the adducts formed are more heterogeneous than those obtained with a double-stranded DNA. In the latter case they obtained almost pure (+)B[*a*]PDE-dG adducts, whereas, in the former case they obtained, in addition, the (−) adduct and lesser amounts of dA adducts. One cannot ascribe with certainty, therefore, the effects on ϕX174 template activity described by Hsu *et al.* (*76*) to a specific B[*a*]PDE-deoxynucleoside adduct. However, in view of its general predominance, particularly under conditions of low extents of modification, it seems likely that the effects were mainly due to the B[*a*]PDE-dG adduct.

Berthold *et al.* (*78*) examined the effects of AcO-AAF modification of DNA on template activity in a subcellular system utilizing DNA-primed single-stranded phage fd with *E. coli* polymerase III holoenzyme or DNA polymerase I as enzymes. A concentration range of 4 to 40 μM AcO-AAF caused appreciable inhibition of both types of reaction, although the actual extent of DNA modification and the mechanism of inhibition were not examined. In parallel studies, a 1000-fold higher concentration of methylnitrosourea was required to produce equivalent inhibition.

When the fdDNA was complexed with DNA-binding-protein I prior to the addition of the AcO-AAF, there was a marked protection of the DNA template activity, whereas the binding protein did not protect the DNA from the inhibition exerted by the methylnitrosourea. This may relate to the steric aspects associated with modification by the bulky AAF residue (see Section II).

Although there is evidence that aflatoxin B_1 inhibits DNA synthesis *in vivo* (*79*), we are not aware of any studies on the effects of DNA so modified on subcellular DNA synthesis.

Loeb *et al.* (*80*) studied the effects of carcinogens on the *in vitro* fidelity of DNA synthesis. Utilizing poly (dA-dT) and similar synthetic templates, and the DNA polymerase (EC 2.7.7.7) from avian myeloblastosis virus (AMV), they measured the extent of misincorporation of noncomplementary nucleotides into the polynucleotide products. Some of these studies have also employed *E. coli* Pol I, or the mammalian DNA polymerases α and β. A variety of metal salts, known to be mutagens and carcinogens, decrease the fidelity in these systems, particularly with the AMV polymerase; the effect of the metal appears to be on the enzyme rather than the template (*81*). The *in vitro* effects of metals on fidelity of replication of a natural DNA template were extended in a system in which errors in DNA replication are scored in terms of reversion of an amber mutation in ϕX174 DNA, utilizing DNA polymerase I as enzyme and assaying biologic activity by transfection into *E. coli* spheroplasts (*82, 83*). Modification of synthetic templates with the alkylating agent β-propiolactone (*81*) and the presence of regions of depurination in the template (*84*) also decreased the fidelity of nucleotide incorporation. At present, there is no evidence that the polycyclic aromatic carcinogens have a similar miscoding influence in these systems. It is suggested that depurinated sites occurring in DNA *in vivo* might also cause errors in DNA replication (*83*). Hence, it is conceivable that chemical depurination due to instability of the guanine N-7 adducts with Afl B_1 and B[*a*]PDE, or the action of an N-glycosylase on such adducts, could lead to these effects, but this remains to be demonstrated.

V. Mutagenesis and Viral Effects

A. Mutagenesis

The subject of mutagenic activity by chemical carcinogens has been reviewed in detail elsewhere (*85, 86*). Therefore, we review here only the major features of mutagenesis by the polycyclic compounds, emphasizing those aspects that relate to mechanism. Whereas, in early studies, there was no strong correlation between the mutagenic and carcinogenic activity of many chemicals, the finding that many of these chemicals are metabolized *in vivo* to highly reactive intermediates, and the application of this principle to mutagenesis assays, has considerably strengthened this correlation (*1, 85, 86*).

The exposure of *Bacillus subtilis* DNA to the highly reactive 2-(*N*-acetoxy)acetamidofluorene (AcO-AAF) results in induction of mutations, as well as inactivation, when the modified DNA is assayed for transforming activity (*87*). Activated forms of acetamidobiphenyl, acetamidostilbene, acetamidophenanthrene, and *N*-methyl-4-aminoazobenzene have similar effects, whereas, the parent compounds are inactive (*87, 88*).

The most extensive studies on mutagenesis have been done in the *Salmonella typhimurium* strains developed by Ames *et al.*, which are assayed in the presence or absence of rat liver microsomal fraction, termed "S-9," and appropriate cofactors to provide metabolic activation of the test compound. The assay of appropriate tester "his$^-$" mutants for induced reversion can distinguish between frame-shift and base-pair substitution-type mutations. Activated derivatives of the polycyclics produce mainly frame-shift mutations in this system; these occur in (G · C)-rich sequences of the "his" gene (*85*). These results are consistent with the chemistry of DNA binding of these agents (see Sections I and II). With a series of polycyclic hydrocarbons, potency with respect to mutagenesis does not correlate with potency in terms of carcinogenicity; this correlation does obtain, however, if the "Bay-region dihydrodiols" (defined on page 112) are used in the assay, rather than the parent compound (*89*). The latter results presumably reflect the complexities of metabolic activation of these compounds. Aflatoxin B_1 is a highly potent carcinogen and one of the most potent frame-shift mutagens in the *S. typhimurium* system (*85*).

Utilizing the Chinese hamster ovary cell line, and a microsomal activation system, Hsie *et al.* (*90*) have demonstrated the mutagenicity of several types of carcinogens. Similar results have also been obtained with the V79 cell line when combined with metabolizer cells (*91*). These compounds are also mutagenic in *Neurospora crassa* and in *Drosophila* (*85*). Thus far, demonstrations that the activated forms of the polycyclic aromatic carcinogens can produce frame-shift mutations have been possible only in the bacterial systems.

B. Viral Effects

Hsu *et al.* (*92*) utilized *E. coli* spheroplasts infected with nucleic acid prepared from either the RNA phage MS2 or the DNA phage ϕX174 to assay for the extent of inhibition of viral replication after addition of polycyclic hydrocarbons. When the phage nucleic acid was first treated with the hydrocarbon, purified, and then used to infect spheroplasts, the most active compound with respect to inhibition was (±)B[*a*]PDE-I. (±)B[*a*]PDE-II was also active but a lesser extent, thus indicating that the stereochemistry of the compound is an important feature of its biologic activity. The finding that I was much more active than II is consistent with a variety of other biochemical and biologic effects, including carcinogenicity (*93*).

Other compounds that showed strong activity in the above assay system were the synthetic compounds 7-bromomethyl-12-methylbenzanthracene, 7-oxiranyl-12-methylbenzanthracene, dimethylbenzanthracene 5,6-oxide, and the "anti"-isomer of (±)-*trans*-10,11-epoxy-7,8,9,10-tetrahydrobenzo[*a*]pyrene-8,9-diol. The "*syn*" isomer of the latter was much less active. Similar results with the same compounds were obtained with QβRNA or ϕX174 DNA. Evidence was also obtained that *E. coli* can metabolize certain polycyclic hydrocarbons.

A 50% inhibition of the biologic activity of Qβ RNA was obtained when the molar ratio of RNA nucleotide to B[*a*]PDE-I during the incubation was about 300 : 1, corresponding to about 14 molecules of B[*a*]PDE per molecule of Qβ RNA (*92*). The actual extent of modification of the RNA was not determined in this study, but the same group subsequently indicated that one bound molecule is sufficient to inhibit the replication of a single molecule of ϕX174 DNA (*92*).

The ability of activated forms of these polycyclics and certain other carcinogens to induce prophage λ in *E. coli* is discussed above (Section III). Carcinogen treatment of certain polyoma-virus-infected mammalian cells has also been found to induce the synthesis of viral DNA and infectious virus (*94, 95*).

VI. Effects on Chromatin

Other sections of this review are largely concerned with the effects of nucleic acid modifications by carcinogens in terms of base-pairing interactions[3] during replication, transcription, and translation, and generally emphasize the properties of the free nucleic acids. It is likely, however, that *in vivo* the quaternary structure of nucleic acids, i.e., their association with proteins, confers additional specificity with respect to both susceptibility to carcinogen modification and the effects on function resulting from this modification. These aspects may be particularly important in chromatin, where the DNA is intimately associated with proteins that appear to play an important role in the control of both replication and transcription.

Chromatin appears to consist of a series of nucleoprotein subunits (nucleosomes), each containing about 200 base-pairs of DNA and two molecules each of the four major classes of histones. Digestion of nuclei with staphylococcal nuclease first releases multimers and monomers of nucleosomes, indicating digestion of the DNA between the nucleosomes. As digestion proceeds, the DNA of the nucleosome is reduced in size to a core containing 140 base-pairs. These fragments presumably reflect the organization of DNA and histones within the subnucleosome structure. On the other hand, DNase I digests DNA within, as well as between, nucleosomes and evidence has been obtained that it initially preferentially digests the transcriptionally active regions of chromatin (for reviews of these aspects, see *96–98*).

Two types of studies related to carcinogens and chromatin have been performed. The first is concerned with whether carcinogens preferentially modify specific regions of DNA within the chromatin structure. The second asks whether carcinogen modification of DNA perturbs the association of the DNA with histones in the formation of the nucleosome structure.

In a study with [^{3}H]benzo[*a*]pyrene, the carcinogen was incubated with calf thymus nuclei in the presence of NADPH and rat liver microsomes, and the time-course of digestion of the carcinogen-modified DNA within the nuclei, by staphylococcal nuclease or DNase I, was determined (*99*). Both enzymes preferentially digested the modified DNA, suggesting that the carcinogen is bound more extensively to the nuclease-susceptible regions of DNA in chromatin. When isolated duck reticulocyte chromatin was treated with [^{14}C]B[*a*]PDE-I and the time course of digestion of the modified chromatin with staphylococcal nuclease was followed, it was observed that the carcinogen preferentially modified the "open" regions of DNA, i.e., regions susceptible to nuclease digestion (*100*). It was estimated that 65% of the carcinogen was bound to the "open" regions and 35% to the closed regions. Studies (*101*) with AcO-AAF-modified chromatin also indicated that the staphylococcal-nuclease-susceptible regions of DNA had about twice the specific activity of the nuclease-resistant regions. Thus it appears that activated derivatives of both AAF and B[*a*]P preferentially react with the internucleosomal DNA. At the same time, there is appreciable modification of regions of DNA within the nucleosome structure, even though these regions are protected from nuclease digestion by their association with histones.

These conclusions from *in vitro* studies appear to apply to the *in vivo* situation as well. When either hamster or human cell cultures are incubated with [^{3}H]B[*a*]P, there is covalent binding of radioactivity to both "open" and "closed" regions of nuclear DNA, as defined by susceptibility to staphylococcal nuclease digestion (*100*). Similar results were obtained with rat liver following *in vivo* administration of [^{14}C]AAF (*102*). There is evidence that the internucleosomal regions of DNA not only are more susceptible to damage, but they are also more susceptible to excision repair (*29*, *103*).

The results obtained with staphylococcal nuclease digestion do not indicate whether or not carcinogens preferentially modify transcriptionally active regions of chromatin, since this nuclease does not discriminate between these and transcriptionally inactive regions. The results obtained with DNase I digestion of [^{3}H]B[*a*]P-modified calf thymus nuclei (*100*), on the differential binding of polycyclic hydrocarbons to eu- and heterochromatin (*104*), and on the differential binding of HO-AAF to eu- and heterochromatin (*105*), suggest that these carcinogens may initially bind preferentially to the transcriptionally active regions of chromatin, but this requires further study.

Studies have also been performed to determine whether carcinogen modification of DNA impairs its ability to associate with histones to form nucleosome structures. Double-stranded duck-reticulocyte DNA was modified with either AcO-AAF or B[*a*]PDE-I so that about one to two percent of the nucleotides contained the covalently bound carcinogen adducts. These DNAs were then mixed *in vitro* with histones under conditions used for chromatin reconstitution. The carcinogen-modified DNAs formed apparently normal nucleosome structures, as judged by sedimentation profiles on sucrose gradients, digestion kinetics with *Staphylococcal* nuclease, and gel electrophoresis patterns of the DNA fragments after nuclease digestion (*105, 106*). On the other hand, completely denatured DNA did not function in the reconstitution of chromatin. These results suggest that although the DNA must have a structure that is predominantly double-stranded, localized distortions in base-pairing and the presence of fairly bulky substituents do not interfere with the DNA-histone associations involved in nucleosome formation. This is consistent with other evidence indicating that in the formation of nucleosomes, the histone molecules do not appear to recognize specific base sequences in the DNA (*96*).

B[*a*]PDE reacts predominantly with the 2-amino group of guanine, which lies in the minor groove of the helix, and AcO-AAF reacts predominantly with the C-8 position of guanine, which lies in the major groove. The evidence indicates that, in nucleosomes, the DNA is wrapped around the outside of a histone core (*96*). Apparently neither the major nor the minor grooves of the double helix are completely shielded in terms of susceptibility to chemical modification.

Further studies are required to determine the possible biological significance of the findings in these model systems. The results predict that all regions of chromatin-associated DNA are susceptible to attack by activated derivatives of these chemical carcinogens, although the attack is not totally random. In addition, DNA modification by carcinogens is unlikely to disturb the gross structure of cellular nucleosomes.

It is, of course, still possible that carcinogens act by disturbing more subtle aspects of protein–nucleic acid recognition, chromatin structure, and the control of gene expression. The effects of DNA modification in the binding of specific nonhistone proteins to DNA and chromatin has not been studied in detail. There is evidence that administration of aflatoxin B_1 impairs the binding of the glucocorticoid and receptor complex in rat liver nuclei (*107*).

Since several carcinogens bind covalently not only to DNA but also to protein, it is important to determine to what extent the covalent binding of carcinogens to chromatin-associated proteins disrupts chromatin structure and function. There is evidence from both *in vivo* and *in vitro* studies that radiation and some bulky chemical carcinogens induce DNA–DNA and DNA–protein cross-links (*108, 109*). The effects of such cross-links on chromatin function are not known. Since it has been postulated that DNA methylation plays an important role in differentiation and the control of gene expression (*110*), it is of interest that *in vitro* modification of DNA by AcO-AAF impairs its capacity to be enzymically methylated (*111*). L-Ethionine, which is also carcinogenic, enhances differentiation of Friend erythroleukemia cells, and this appears to be associated with inhibition of DNA methylation (*112*).

VII. Effects on Transcription

A number of studies indicate that a variety of carcinogens lead to a rapid inhibition of RNA synthesis *in vivo* (*16, 43*). In several cases this is paralleled by effects of carcinogens in RNA synthesis *in vitro*. The results of several of these studies are described in detail below.

A. 2-Acetamidofluorene

When the liver DNA of rats fed a diet containing 2-acetamidofluorene for periods of 2 to 32 weeks was used as template in an RNA polymerase system, the template capacity of this DNA, measured by the incorporation of label from [^{3}H]UTP and [^{3}H]ATP into RNA, showed a considerable decrease. The RNA polymerase enzyme used was prepared from *Micrococcus lysodeikticus* cells. Similarly, modification of rat-liver DNA *in vitro* with AcO-AAF resulted in an extensively modified DNA that had no activity as a template for RNA polymerase (*113*). Since metabolites of AAF are known to bind not only to DNA (*114–119*), but also to RNA (*120, 121*) and proteins (*122, 123*), there is the question of whether the inhibition of RNA synthesis by AAF *in vivo* is due to alterations of the DNA template, effects on the RNA polymerase, or other mechanisms. Zieve (*124*) found that a single intraperitoneal injection of HO-AAF to rats inhibited the activity in isolated liver nuclei of both RNA polymerase I and II by as much as 80%. Surprisingly, there was no effect on the *in vitro* template capacity of chromatin or DNA isolated from the livers of the treated rats when assayed with *E. coli* RNA polymerase. From these observations, it was suggested that the *in vivo* inhibition of RNA synthesis of HO-AAF is due to inactivation of the RNA polymerases, not to decreased activity of DNA for transcription.

To resolve the divergent conclusions concerning the mechanism of HO-AAF inhibition of RNA synthesis *in vivo*, Grunberger *et al.* (*125*) applied a different technique. Two hours after HO-AAF injection, RNA synthesis was measured in separated nucleolar and nucleoplasmic fractions from hepatic nuclei of the carcinogen-treated rats. With endogenous template, the inhibitory effect was extensive in the nucleolar fraction, which is the site of ribosomal RNA synthesis. Much less inhibition occurred in the nucleoplasmic fraction, where mRNA is formed. When the endogenous template function of the nucleoli was abolished by actinomycin D, and the RNA polymerase activity was determined in the presence of the exogenous synthetic template, poly(dA-dT), the polymerase activity of the nucleolar fraction from the control and the carcinogen-treated rats became identical. Under similar conditions, the activity of the nucleoplasmic fraction was slightly inhibited. These results suggest that the *in vivo* inhibition of RNA synthesis, after acute HO-AAF treatment, could reflect both inhibition of ribosomal RNA synthesis due to effects on the DNA template, and impaired RNA synthesis due to inhibition of nucleoplasmic RNA polymerase. The latter enzyme is presumably involved in the synthesis of messenger RNA, or at least a DNA-like RNA product (*126, 127*).

Glazer *et al.* (*128*) examined the template activity of DNA and the activities of crude and partially purified RNA polymerases from hepatic nuclei isolated from partially hepatectomized rats treated with HO-AAF. DEAE-Sephadex chromatography of nuclear RNA polymerases from treated rats revealed stimulation of the activity of RNA polymerase I (nucleolar) and inhibition of RNA polymerase II (nucleoplasmic). However, no alterations were observed in the template activity of rat liver DNA from similarly treated animal. They concluded that, whereas the inhibition of nucleoplasmic RNA polymerase can account for the impairment of synthesis of extranucleolar species of RNA, the inhibition of ribosomal RNA synthesis cannot be a direct result of either reduction of nucleolar RNA polymerase or DNA template activity.

Differences in the above studies could be due to the use of different experimental techniques. For example, Glazer *et al.* (*128*) found that poly(dA-dT) is a poor template for liver RNA polymerases (*128*). However, this contrasts with other reports that it is a good template (*130–132*). This discrepancy is probably due to the fact that poly(dA-dT) cannot be used at high salt concentration [e.g., over 0.2 M $(NH_4)_2SO_4$] to assay RNA polymerase activity (*141*). In addition, they used partially hepatectomized rats, and it is well known that partial hepatectomy alone causes dramatic alterations in gene expression (*129*). Glazer *et al.* (*133, 143*) evaluated the effects of AAF and HO-AAF on the incorporation of [^{3}H]orotic acid into nuclear ribosomal RNA and heterogeneous RNA in normal and regenerating liver. The distribution of radioactivity in rRNA and hnRNA between normal and regenerating liver was not markedly different. Both types of RNA were inhibited to the same extent by treatment of hepatectomized rats with HO-AAF.

In a study of the action of HO-AAF on *in vivo* synthesis of ribosomal and poly(A)-RNA, the two types of RNA were prepared from polyribosomal RNA by chromatography on poly(U)-Sepharose (*142*). Treatment of normal and partially hepatectomized rats with HO-AAF showed a differential dose-dependent effect on rRNA and poly(A)-RNA synthesis. The synthesis of rRNA seemed to be more sensitive to the inhibitory effect of the carcinogen. On the other hand, no change in the content of poly(A)-RNA was apparent as a result of HO-AAF treatment.

When cultured rat liver epithelial cells were exposed for 10 minutes to increasing concentrations (0.2 to 1.5 μml) of AcO-AAF and, after removal of the carcinogen, the cells received [^{3}H]uridine for 10 minutes, there was a progressive inhibition of incorporation of label into RNA (*144*). Since only the 45 S ribosomal RNA and heterogenous RNA (hnRNA 30 to 100 S) incorporate significant amounts of labeled precursor during this time period (*145*), they determined the extent of inhibition of both 45 S and hnRNA by Aco-AAF. Utilizing a low concentration of actinomycin D, which selectively inhibits the synthesis of 45 S RNA (*146*), they obtained evidence that under the conditions employed AcO-AAF produced a 66% inhibition of 45 S RNA synthesis and only a 6% inhibition of hnRNA synthesis (Table I). The actual amount of 45 S and hnRNA made in both control and carcinogen-treated cultures was also directly measured after separation by dodecyl-sulfate/polyacrylamide-gel electrophoresis. The results confirmed a preferential inhibition of 45 S RNA synthesis. These results indicate that the inhibition is at the level of transcription of the 45 S

TABLE I
DISTRIBUTION OF [^{3}H]URIDINE IN DIFFERENT RNA FRACTIONS AFTER PULSE LABELING OF CONTROL AND 2-(*N*-ACETOXY)ACETAMIDOFLUORENE-TREATED CULTURES

	RNA fraction (cpm × $10^{-3}/A_{260}$				
	>45 S	45 S	<45–28 S	28–18 S	% in 45 S RNA[a]
Control	36.6	31.4	27.9	7.6	30
Treated	35.7	18.0	23.6	8.7	21

[a] Total cpm in 45 S region × 100 divided by total counts per minute recovered from gel (*144*).

ribosomal RNA precursor. It could be that AAF is either preferentially bound to the ribosomal cistrons of mammalian cells because of their relatively high guanine content (*147*) or because of aspects related to their secondary or tertiary structure.

Very high levels of HO-AAF (80 mg per kilogram of body weight) given to rats causes an inhibition of liver nuclear RNA synthesis (*132*). This effect was reflected in an *in vitro* inhibition of RNA synthesis by isolated whole nuclei. The reduced RNA synthesis was considered to be due to decreased activity of the RNA polymerases, and not to a decreased availability of the chromatin template. This conclusion was based on the results of three kinds of experiments: (a) solubilization and assay of the polymerases; (b) measurement of the polymerase activity in intact nuclei under conditions independent of the endogenous template; and (c) determination of the template capacity of chromatin. After solubilization and ion-exchange chromatography, a decreased activity of both the nucleoplasmic and nucleolar RNA polymerases was found (*132*).

Because of these conflicting results, Yu and Grunberger (*138*) reinvestigated this problem. They separated RNA polymerases according to the functional states of the enzymes. One state, referred to as the "engaged" enzyme, is active in transcribing the endogenous chromatin template, while the other, referred to as the "free" enzyme, transcribes the exogenous template and is functionally inactive toward the endogenous chromatin template (*139*). In these experiments, male rats were injected with HO-AAF (30 mg per kilogram body weight). Two hours after the injection, nuclei were isolated in hypotonic buffer, and nuclear and nucleolar RNA synthesis were measured using endogenous DNA or exogenous poly(dI-dC) as templates. With the endogenous template, there was a 57% inhibition of RNA synthesis by nuclei from treated animals. However, when poly(dI-dC) was used as a template and the endogenous template function was blocked by actinomycin D, the total "free" RNA polymerase activity was slightly but consistently increased above that of the untreated control. When nucleolar fractions were isolated, HO-AAF treatment inhibited nucleolar RNA synthesis by 80%. On the other hand, the nucleolar "free" RNA polymerase activity was decreased only by 20% when measured with the exogenous poly(dI-dC) template. The level of "free" RNA polymerase in soluble extracts of nuclei was measured also in the presence and in the absence of α-amanitin, which selectively inhibits RNA polymerase II activity. The results indicate that the residual RNA polymerase activity from the HO-AAF-treated group was almost totally insensitive to α-amanitin (3.2 μg/ml) inhibition, providing evidence that the HO-AAF treatment inhibits only the RNA polymerase II activity.

The above conclusions were confirmed by studies using DEAE-Sephadex chromatography to separate the total nuclear RNA polymerases (*138*). The elution profiles of the nuclear RNA polymerases of the control and HO-AAF-treated animals showed that RNA polymerase II activity was markedly reduced, whereas RNA polymerase I activity was not appreciably affected by the treatment. Figure 2 shows the elution profiles of the nuclear-engaged RNA polymerases from control (A) and from the treated (B) groups. Here, too, there was a marked reduction of RNA polymerase II, but not RNA polymerase I activity, as a result of the treatment.

A conclusion suggested by these results is that, since HO-AAF had little effect on RNA polymerase I activity, the inhibition of nucleolar RNA synthesis is probably due to impairment in function of the nucleolar DNA template; on the other hand, the decrease in nucleoplasmic RNA synthesis may be due to direct inhibition of the enzyme activity of RNA polymerase II.

Support for the conclusion that AAF inhibits nucleolar DNA template function comes from observations of soluble RNA polymerase I activity, and also the capacity for RNA synthesis, under conditions optimized for RNA polymerase I activity, by intact hepatic nuclei isolated from rats fed a diet containing AAF (*140*). Increasing the duration of carcinogen exposure (4, 7, 14 days) resulted in a progressive increase in soluble RNA polymerase I activity, which reached 230% of the control value after 14 days of carcinogen feeding. However, the capacity for RNA synthesis by intact nuclei remained quite constant during the treatment. Since the synthesis of RNA in intact nuclei is a function of both the level of RNA polymerase activity and the capacity of chromatin to serve as a template, these results suggest that AAF leads to an inhibition of DNA template function.

A detailed examination of the transcription of modified DNA utilized an *in vitro* transcription system containing DNA-dependent RNA polymerase from *E. coli* and AcO-AAF-modified bacteriophage T7 DNA as template (*134*). The T7 DNA was chosen because its *in vitro* transcription product is predominantly a single RNA species (2.2×10^6 daltons) corresponding to the early transcription region (*135*), easily characterized by its mobility in polyacrylamide gel electrophoresis. The major effect of AAF modification of T7 DNA was an inhibition of the rate of RNA synthesis. At very high levels of modification (>0.72% of the bases modified), transcription was completely

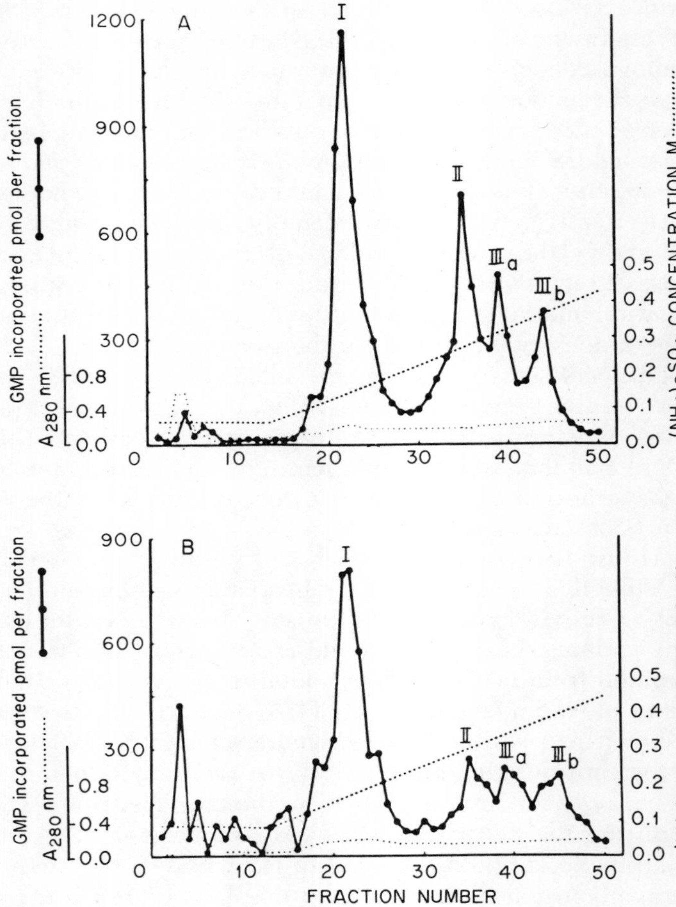

FIG. 2. DEAE-Sephadex (A-25) column chromatography of "nuclear-engaged" RNA polymerase. "Nuclear-engaged" enzyme was prepared by gentle homogenization of isolated liver nuclei in 0.34 M sucrose from control rats (A), and from rats injected with 2-(*N*-hydroxy)acetamidofluorene (3 mg/100 g body weight) (B). The homogenate was centrifuged for 10 minutes at 700 *g*. The supernatant contained the free enzyme, and the pellet retained the "engaged" RNA polymerase, which was solubilized by sonication. Separation was performed on a DEAE-Sephadex column (15 × 0.3 cm) equilibrated with buffer [0.05 M TrisCl, pH 7.9; 25% (v/v) glycerol; 5 mM $MgCl_2$; 0.1 mM EDTA; and 20 mM 2-mercaptoethanol containing 50 mM $(NH_4)_2SO_4$]. (A) Control: 1.5 ml of the RNA polymerase sample (6.1 g liver equivalent) was applied to the column. (B) 2-(*N*-Hydroxy)acetamidofluorene; 1.5 ml of the enzyme (6.3 g liver equivalent) was used [details are described in Yu and Grunberger (*138*)].

blocked. Analyses of the sizes of the RNA products indicated that, with increasing extents of AAF modification of the template DNA, there was a progressive decrease in the size of the RNA products. The kinetics showed that the modification had very little effect on the initiation of transcription. These results suggest that fluorene residues on the coding strand of the DNA cause premature termination of transcription at or near the sites of modification, with release of the RNA polymerase. As a result, shorter RNA chains are formed with increasing amounts of bound carcinogen. In experiments with RNA polymerases I or II from normal and HO-AAF-treated rat liver, it was also found that impairment of RNA synthesis with an AAF-modified DNA template was due predominantly to inhibition of elongation (*136*). Additional studies indicate that increasing modification of DNA leads to decreasing template activity for *in vitro* transcription with *E. coli* RNA polymerase either when the DNA is used as such or after reconstitution into chromatin (*130*).

Possible effects on the transcriptional specificity of polymerases obtained from AAF-treated rats were examined by nearest-neighbor analysis (*136*). The RNA products showed an altered distribution of [α-^{32}P]CMP and [α-^{32}]GMP. These results imply modification in the specificity of the polymerases from the treated animals. It is not known by what mechanism this effect occurs, i.e., indirectly or via acetylation or arylamidation of the enzymes. Arylamidation of lysine and arginine residues in nuclear proteins by HO-AAF (*123*), as well as acetylation *in vitro* of lysine residues in ribonuclease by AcO-AAF (*137*), suggest that either mechanism may be operative.

Since there were changes in the nearest-neighbor frequencies in transcripts from native DNA obtained with polymerase I or II from treated animals, Glazer also examined the possibility of misincorporation by using synthetic DNA templates (*148*). The degree of misincorporation of noncomplementary nucleotides was assessed with the double-stranded poly(dG-dC) or poly(dA-dT) using hepatic RNA polymerases I and II from either normal or HO-AAF-treated rats or *E. coli* RNA polymerase. With poly(dG-dC), there was misincorporation of AMP, which occurred to about equivalent extents with RNA polymerase I from control or HO-AAF-treated animals, or with *E. coli* RNA polymerase. The nearest-neighbor analysis indicated that this was due to misreading of the guanine residues in the template strand. However, transcription of poly(dG-dC) by RNA polymerase II from carcinogen-treated animals, in the presence of Mg^{2+}, showed a statistically greater misincorporation of UMP than with RNA polymerase II from control animals, and this was also due to misreading of guanine residues in the template strand, i.e., a dG · rC to dG · rU transition. These results suggest that RNA polymerases from AAF-treated animals can miscopy synthetic DNA templates.

What conclusions can be drawn from these studies? For one, increasing modification of DNA by AcO-AAF *in vitro* leads to increasing impairment of its template capacity. Concerning the mechanism of impairment, initiation of transcription does not appear to be affected, but there is premature termination of chain elongation, presumably because of the distortions in DNA conformation of sites of AAF modification (see the Introduction). The mechanism responsible for initiation of RNA synthesis after *in vivo* treatment of the animal with HO-AAF is less clear. There is a consensus that the activity of RNA polymerase II, the enzyme responsible for the synthesis of mRNA, is inhibited by this treatment. It is not clear, however, whether this inhibition is due to direct binding of AAF to amino-acid residues of the enzyme, or whether the effect is indirect. Recent studies suggest that nuclear RNA synthesis catalyzed by RNA polymerase II involves a regulatory protein, called S-II (*149*). Presumably, S-II stimulates transcription by interaction with RNA polymerase II. This raises the possibility that either synthesis of regulatory protein(s) is inhibited by AAF, or AAF blocks the interaction of the stimulatory factor with RNA polymerase II.

Although the synthesis of ribosomal RNA is more profoundly inhibited than that of mRNA by HO-AAF administration, the mechanism is even more obscure. A few authors find direct effects of HO-AAF on the RNA polymerase I enzyme (*128, 132*). Other investigators cannot confirm these results, and suggest a direct inhibitory effect of the carcinogen on DNA template function (*138, 140*). These discrepancies may reflect differences in experimental design, the use of different strains and sexes of rats, and the fourfold differences in doses of carcinogens used in the various studies. Since a number of structurally diverse agents preferentially inhibit ribosomal RNA synthesis, the nucleolar structure may be particularly susceptible.

B. Aflatoxin B_1

Aflatoxin B_1, a difuranocoumarin metabolite of certain strains of *Aspergillus flavus* (see Fig. 1), is hepatotoxic and hepatocarcinogenic in several animal species (*150*). It is also a potent mutagen in mammalian and bacterial cells (*151, 152*). Injected into rats, it inhibits DNA (*153*), RNA (*154*), and protein synthesis (*155*). Further, soon after administration, there is a marked inhibition of RNA synthesis in isolated rat liver nuclei (*156*). Similarly, in liver slices, it inhibits incorporation of [^{14}C]orotic acid into RNA (*157*). On the other hand, calf thymus DNA treated *in vitro* directly with the compound and transcribed by bacterial (*158, 159*) or by rat testicular RNA polymerases (*160*), shows no decrease in template activity. These results suggest that a metabolite, rather than the aflatoxin itself, is the inhibitor of DNA-dependent RNA synthesis. Other evidence indicates that it requires metabolism to the 2,3-oxide to exert its carcinogenic or mutagenic effect (*150, 161*). Evidence for the formation of the 2,3-oxide as a major activated metabolite came from the isolation of the 2,3-dihydro-2,3-dihydroxy derivative from acid hydrolyzates of nucleic acid adducts formed *in vitro* with rat and human liver microsomes, or adducts obtained from the liver of rats (*162*). Two groups of investigators independently established that the principal adduct formed between aflatoxin B_1 and DNA, both *in vitro* in the presence of liver microsomes and *in vivo*, is 2,3-dihydro-2-(7-guanyl)-3-hydroxy-aflatoxin B_1 (*163–165*) (see Fig. 1).

As with AAF, the mechanism responsible for *in vivo* inhibition of RNA synthesis by aflatoxin B_1 could be the result of its interaction with the DNA template, the polymerase enzyme, or other mechanisms. Crossover experiments have been performed in which *in vitro* RNA synthesis was determined when a solubilized enzyme preparation from aflatoxin-treated rat liver was incubated with the chromatin from control rat liver, or the reverse (*166*). When the chromatin from untreated rats was used as template, there was no inhibition of RNA synthesis. In contrast, chromatin from treated rats was 28 to 46% less effective as a template for RNA synthesis than chromatin from control rats. These results emphasize an inhibitory effect on the chromatin template rather than on RNA polymerases.

The effect on aflatoxin B_1 on the activity of the RNA polymerases was studied in greater detail by measuring the nucleoplasmic and nucleolar RNA polymerase activity in intact rat liver nuclei and in crude solubilized or fractionated preparations, with denatured DNA as template (*167*). Two and 24 hours after aflatoxin administration to rats, nucleolar polymerase activity was not significantly different from control values. On the other hand, the fractionated nucleoplasmic enzyme activity was reduced to 40% of the control values, suggesting that aflatoxin B_1 inhibits the activity of RNA polymerase II *in vivo*. The endogenous template activity was not measured in these studies. Addition of aflatoxin to incubations of whole nuclei, crude solubilized enzymes, or separated enzymes failed to inhibit RNA synthesis, confirming the requirement for metabolic activation. With a more vigorous technique of ultrasonication in a high-salt medium for the solubilization of both the nucleolar and nucleoplasmic enzymes from nuclei of aflatoxin-treated rat liver, the nucleolar enzyme retained full activity but there was a decrease in the activity of the nucleoplasmic enzyme (*168, 169*).

To determine the effect of aflatoxin on DNA template activity, DNA was incubated with microsomes, cofactors, and the carcinogen before transcription was measured with control nucleoplasmic enzyme. This led to about 75% inhibition of RNA synthesis. On the other hand, when the enzyme was first incubated with the aflatoxin and microsomes, its activity was not inhibited. The latter result is in contrast to the finding that a metabolite of AFB_1 can inhibit the activity of bacterial RNA polymerase *in vitro* (*170*).

Yu (*171*) attempted to resolve the above differences by applying the concept of an "engaged" polymerase enzyme, active in transcribing the endogenous template, and a "free" enzyme, one not active but available to an exogenous template. A similar approach had been used previously to determine the mechanism of RNA synthesis by AAF (*138;* see also above section on AAF). The results are very similar to those obtained with AAF. Two hours after aflatoxin injection, rat hepatic nuclear and nucleolar *in vitro* RNA synthesis were inhibited 70 and 90%, respectively. When total nuclear free and engaged RNA polymerases were solubilized and assayed in the presence of α-amanitin (which selectively inhibits RNA polymerase II), only the α-amanitin-sensitive activity was reduced (50 to 70%) by the aflatoxin treatment. DEAE-Sephadex chromatography provided further evidence that RNA polymerase II was selectively inhibited. Since the aflatoxin markedly inhibited nucleolar RNA synthesis, but had little effect on RNA polymerase I and III activity per se, it was concluded that in addition to its direct inhibitory effect on the enzymic function of RNA polymerase II, it probably impairs nucleolar DNA function.

It is not clear why both HO-AAF and aflatoxin B_1 administration cause selective inhibition of RNA polymerase II. It may be that this enzyme is directly inactivated by both (*172*), or that it is metabolically unstable. It is interesting that the nucleolar DNA template appears to be much more susceptible to direct attacks by a variety of chemical agents than the nucleoplasmic chromatin counterparts. It may be that the nucleolar DNA is more exposed to reactions with drugs and carcinogens than the rest of the nuclear DNA.

C. Polycyclic Hydrocarbons

1. *In Vivo* STUDIES

The effects of various polycyclic aromatic hydrocarbons on RNA synthesis have been less well studied than those of 2-acetamidofluorene or aflatoxin B_1: Most of the information has been derived from *in vivo* experiments.

Alexandrov *et al.* (*173*) compared the effect of B[*a*]P, Me_2-benzanthracene, and B[*e*]P on epidermal RNA synthesis after application to mouse skin. Within the first two hours after the application of the first two, there was a decrease in the rate of RNA synthesis, followed by an abrupt increase that doubled above the control level at 24 hours. No inhibition was found with the noncarcinogenic B[*e*]P. The electrophoretic patterns of Sephadex-agarose gels of skin RNA (4 S, 5 S, 18 S, 28 S) at different intervals of time after application of the first two did not differ from the RNA pattern of control mouse skin.

The effects of Me_2 benzanthracene on RNA synthesis in mammary epithelial cells have been compared in rat strains of high and low mammary tumor incidence (*174*). After one to four days of feeding it to rats with high tumor susceptibility, the incorporation of [^{3}H]uridine into rapidly labeled nucleolar and cytoplasmic RNA extracted from mammary gland epithelial cells was depressed to 60% of the control value; this was followed by a 40 to 100% stimulation of RNA specific activity throughout the 28-day test period. RNA synthesis in the treated, low-tumor-susceptibility rats did not differ from that in untreated animals. The alterations in RNA synthetic patterns were specific to the mammary epithelium and could not be demonstrated in other tissues. The exposure of mammary epithelial cells to Me_2-benzanthracene for 20 hours *in vitro* showed an early decrease of RNA synthesis that paralleled that seen in mouse epidermis after a single initiating dose of the substance.

We are not aware of any detailed subcellular studies on RNA synthesis after *in vitro* administration of polycyclic hydrocarbons.

2. Template Activity of DNA Modified with the Diol-Epoxide of Benzo[*a*]pyrene

After the exposure of rodent tissues or the exposure of either rodent, bovine, or human tissue culture cells to B[*a*]P, the major DNA adduct is derived from its diol-epoxide, which is linked from its 10-position to the N^2 of guanine (Fig. 1c). The stereochemistry of this adduct and the presence of lesser amounts of other adducts are discussed in Section II.

The effect of the major adducts (see Fig. 1) on the template activity of DNA was studied by *in vitro* transcription with *E. coli* DNA-dependent RNA polymerase under two different conditions (*5*). In one case, the RNA polymerase was allowed to recycle continuously on the DNA template, while in the other case, only a single round of initiation was permitted (*175*). In this way, the initiation and elongation phases of transcription were distinguishable.

When RNA synthesis was studied under nonreinitiating conditions, benzo[*a*]pyrene-modified DNA showed a progressive inhibition of incorporation with increasing extents of modification (Table II). A DNA sample with 1.5% modified bases showed approximately 55% inhibition of RNA synthesis compared to native unmodified DNA. When the assays were performed under conditions that permitted multiple rounds of initiation, the modified samples were much more inhibited in their template capacities than in the nonreinitiating assay. Thus, the sample in which there was only 0.2% modification showed about 40% inhibition. These results are consistent with a model in which benzo[*a*]pyrene modification of DNA inhibits chain elongation but not chain initiation during transcription. Presumably, movement of the polymerase along the DNA template is blocked when it encounters sites of modification, and the enzyme cannot recycle.

To monitor the effects of benzo[*a*]pyrene modification of DNA on chain elongation directly during transcription, the transcripts were subjected to sucrose density gradient centrifugation (Fig. 3). Transcripts of native DNA consisted of heterogeneous high-molecular-weight products ranging in size from about 30 S to 7 S with a broad peak at about 20 S. On the other hand, transcripts obtained from DNA containing a 0.5% modification showed a lower size distribution ranging from about 25 to 7 S with a peak at approximately 14 S; those

TABLE II

In Vitro TEMPLATE ACTIVITY OF CALF THYMUS DNA MODIFIED WITH BENZO[*a*]PYRENE-7,8-DIOL 9,10-EPOXIDE[a]

DNA	Modified bases (%)	No reinitiation incorporation relative to native DNA (%)	With reinitiation incorporation relative to native DNA (%)
Native	0	100	100
Mock modified	0	93	115
Modified	0.2	75	59
	0.5	56	47
	1.5	45	27
	2.2	26	15
	4.5	11	3

[a] Incorporation of ^3H from [^3H]UTP with native DNA under nonreinitiating conditions was 300 pmol and under reinitiation conditions 760 pmol. For details, see Leffler *et al.* (*5*).

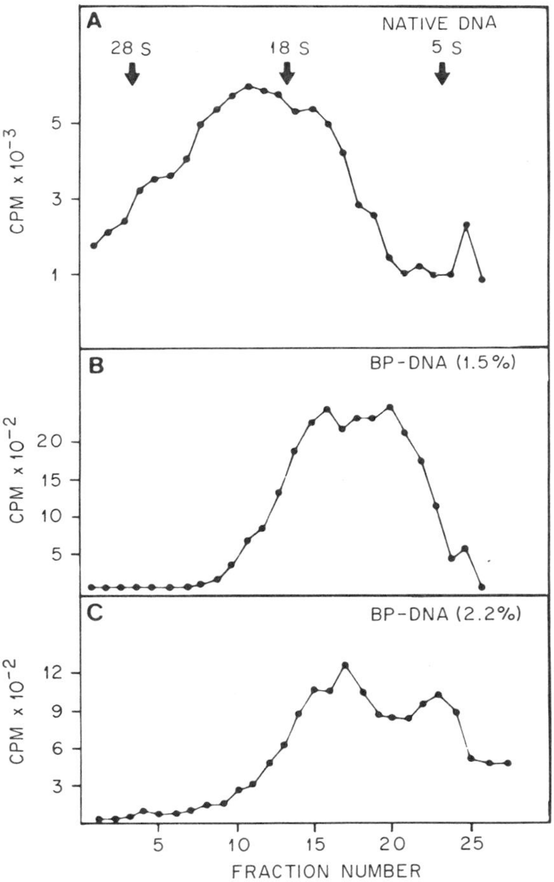

FIG. 3. Sucrose density gradient centrifugation of RNA transcripts from benzo-[*a*]pyrene-DNA templates. The reaction mixture contained 10 mM TrisCl (pH 7.9), 1 mM $MnCl_2$, 0.08 mM each of ATP, GTP, CTP, and [^{3}H]UTP (specific activity 27 Ci/mmol), 3 units of *E. coli* DNA-dependent RNA-polymerase, and 1 μg of DNA or modified DNA. After incubation at 37°C for 20 minutes, the samples were layered on a 5 to 20% linear sucrose gradient and centrifuged for 3 hours at 50,000 rpm in a Spinco VW65 rotor at 25°C. The acid-precipitable counts in each fraction were assayed by a liquid scintillation spectrometer. Eukaryotic ribosomal RNA was used as a sedimentation marker. For details, see Leffler *et al.* (*5*) and Weinstein *et al.* (*8*).

obtained from DNA containing either 1.5 or 2.2% modification tended to be even smaller. These results are consistent with the interpretation that B[*a*]P modification tends to interrupt chain elongation during transcription.

When the sedimentation values were converted to chain length(s), the native DNA sample yielded products with an average nucleotide chain length of 1400 to 1450 (*175*). The corresponding value for the 0.5% B[*a*]P-DNA sample was 1260; for the 1.5% sample, 1020; and for the 2.2% sample, 750. Thus, although increasing modification of the template led to decreasing average chain-lengths of the products, the average chain-lengths were greater than one would expect if the derivatives were randomly distributed along the DNA template and if there was complete interference of chain elongation at every site of modification. Under the latter conditions, a 1% modification would yield transcripts with a maximum chain-length of less than 100 nucleotides, which would correspond to about 5 S. A possible explanation for these differences could be either that the B[*a*]P derivatives are clustered rather than scattered randomly on the DNA template, or, if the distribution is random, the polymerase may "bypass" some, but not all, of the sites of modification during the process of chain elongation. Other studies indicate that at low levels of modification there is no clustering of the B[*a*]P derivatives on the DNA (*176*).

Since the extent of covalent binding of B[*a*]P to cellular DNA is much lower *in vivo*, one residue per 10^4 to 10^5 nucleotides versus one per 10^2 to 10^3 nucleotides *in vitro*, it is difficult to extrapolate from *in vitro* results to the *in vivo* situation. If, however, qualitatively similar, but quantitatively less extensive, changes occur *in vivo* when cells are exposed to B[*a*]P, then the carcinogen could produce aberrations in the process of transcription. A more complete understanding of the effects of this carcinogen on transcription requires further *in vivo* studies, as well as *in vitro* studies using isolated RNA polymerase enzymes and defined templates.

VIII. Functional Changes in the Translational System

A. Effect of Aflatoxin B_1 on Protein Synthesis

A single injection of aflatoxin B_1 to rats inhibits protein synthesis *in vivo* within 30 minutes; reversibility depends on the dose (*177*). In similarly treated rats, a marked but reversible disaggregation of liver polysomes occurs (*178*). The disaggregation is evident at 3 hours and persists for 36 hours. Partial reaggregation begins at 36 hours, and the process is essentially complete five days after dosing. Inhibition of incorporation of amino acids into proteins of liver slices and cell-free preparations is also associated with alterations in ribosomal aggregation (*179*).

The time course of the effects of aflatoxin B_1 on protein synthesis *in vivo* does not correlate with the disaggregation of polysomes. Incorporation of amino acids is only transiently inhibited and, by 12 hours after injection, it has returned to control levels. Induction of the microsomal enzyme zoxazolamine hydroxylase is completely inhibited when aflatoxin B_1 is administered after the application of the inducer benzo[*a*]pyrene (*180*).

At least two steps may be seen in the complex alteration of translation following aflatoxin administration: after 7 hours, protein synthesis inhibition appears chiefly as a consequence of transcription impairment that decreases both ribosomal and messenger RNAs; by 24 hours, total cellular RNA has decreased 50% (*181*). A concomitant disaggregation of polysomes reaches about 55% 20 hours after dosing. Up to 5 hours, however, the amount of polysomes in the cell does not decrease significantly, even though a very strong inhibition of translation culminates 2 hours after treatment. This early translational alteration is probably due to a direct and specific action of the carcinogen on the polysomes.

In an attempt to localize the translational step inhibited *in vivo*, Sarasin and Moule (*182*) used the simulation study developed by Li *et al.* (*183*), which is based on two parameters: the kinetics of polysome labeling to follow the nascent peptide synthesis, and the kinetics of supernatant labeling to follow the completed protein synthesis.

Up to 5 hours after injection, aflatoxin specifically inhibits the elongation and/or termination steps during protein synthesis; after longer periods of time, inhibition occurs essentially at the initiation step. When the intracellular concentration of aflatoxin is too high, particularly two hours after dosing, both steps of protein synthesis are blocked. Polypeptide synthesis by the postmitochondrial supernatants isolated from aflatoxin-treated animals is impaired in the same proportion as protein synthesis *in vivo*. The damage caused by aflatoxin is mostly observed on microsomes. However, purified polysomes isolated from aflatoxin-treated rats synthesize proteins *in vitro* to the same extent as those from controls. These results suggest that aflatoxin metabolite(s) are bound to polysomes with noncovalent bonds. These active metabolites are probably lost during polysome isolation procedures.

B. Coding Properties of Oligo- and Polynucleotides Modified with 2-Acetamidofluorene

Although changes in the structure and function of DNA have constituted the focal point for many studies on the molecular basis of carcinogenesis, RNA is also modified *in vivo* by several carcinogens

(*184–187*). Hydrolysis of liver RNA obtained from rats given AAF, and of RNA treated with AcO-AAF *in vitro*, yield the same major nucleoside derivative, 2-[*N*-(8-guanosinyl)]acetamidofluorene (*188*). 2-(*N*-Acetoxy)acetamidofluorene, therefore, provides a convenient reagent with which chemically defined oligo- and polynucleotides can be treated to produce a chemical alteration similar to what occurs upon administration of 2-acetamidofluorene *in vivo*.

Binding of AAF to C-8 of G in the GUU codon for valine completely inactivates this triplet in ribosomal binding of valyl-tRNA (*189*). Similarly, modification with AAF of the G in an AAG codon for lysine abolishes the template activity of this triplet. These effects would be explained if (a) the modified oligonucleotides cannot bind to ribosomes, or (b) guanosine residues containing AAF cannot base-pair properly with the complementary base in the anticodon regions of related tRNAs. Since poly(U_3G), modified with AAF, impairs only the binding of valyl-tRNA, not phenylalanyl-tRNA (codon UUU) to ribosomes, it seems that the modified polymer is still capable of binding to ribosomes. It appears, therefore that AAF modification of synthetic mRNA impairs translation primarily by inhibiting codon–anticodon interaction. The inhibition is limited to those codons in which the G residue is modified and is independent of its position in the codon. Sequences that do not contain modified G residues, such as UUU in a modified polymer, still function normally with respect to binding of the corresponding tRNA to ribosomes.

On the other hand, poly(U_3G) preparations modified by 2-acetamidofluorene are impaired in directing incorporation of both phenylalanine and valine into polypeptides. This can be explained by assuming that protein synthesis is blocked each time the translation arrives at a modified G residue. Direct evidence for this was obtained by finding that the modified poly(U_3G) stimulates the synthesis of di- and tripeptides, but not of larger products. The lack of base-pairing between the modified G residues in mRNA and complementary bases in the anticodon region of tRNA is considered to be the result of the conformational changes that occur at the sites of the modification (*190, 191*). This type of change, designated as the "base displacement model," includes rotation of the modified G residue from the *anti* to *syn* conformation, and insertion of the AAF residue parallel to the neighboring bases and perpendicular to the phosphodiester backbone. If an A residue is adjacent to either the 5′ or 3′ side of the modified G, as in A_3G or GA_3, there is partial inhibition of Lys-tRNA (Lys codon is AAA) binding to ribosomes (*191*). This inhibitory effect extends only to the function of the immediately adjacent A, since modification of G residues in A_4G, or GA_4 does not impair the coding for lysine. In contrast, a U residue adjacent to the modified G, as in U_3G or GU_3, has no effect on Phe-tRNA binding to ribosomes. The different behavior between A and U residues in this system is probably due to the stronger stacking interaction that occurs between a modified G and a neighboring A than between a modified G and an adjacent U.

Although the above results were obtained in a highly artificial model system, they suggest that AAF might block translation *in vivo* by a similar mechanism, i.e., by binding to G residues in exposed regions in mRNA and blocking the base-pairing capacity of the modified residues with anticodons in tRNA.

The *in vitro* modification of rabbit globin mRNA with the diolepoxide inactivates its capacity to direct globin synthesis in a subcellular system, apparently due to inhibition of the initiation of translation (D. Grunberger *et al.*, *JBC*, in press). Sagher *et al.* (*192*) correlated the degree of inhibition of MS2 RNA infectivity, at various levels of alkylation by B[*a*]PDE, with the translation efficiency *in vitro* of the same alkylated RNA for the synthesis of viral synthetase and of maturation and coat proteins. The results indicate that this modification of MS2 RNA impairs its template capacity for the synthesis of phage-specific proteins. For the three viral proteins synthesized *in vitro*, the translation of RNA synthetase is much more sensitive to MS2 RNA modification than either coat or maturation protein synthesis. However, the inhibition is insufficient to account for the loss of RNA infectivity at lower molar ratios of alkylation. These results also indicate that the loss of viral RNA infectivity follows a single-hit inactivation mechanism, whereas several alkylation events in the viral RNA synthetase cistron may be necessary to block translation of this gene product.

C. Modification of tRNA *in Vivo* and *in Vitro*

It has been established that tRNA, in particular, may be heavily modified *in vivo* by various chemical carcinogens (*193*, *194*). Although its role in protein synthesis is usually emphasized, it is important to recognize that tRNA is involved in many other biological functions (*195*). In bacteria, specific tRNAs play a role in regulating the transcription of RNAs for enzymes associated with biosynthesis of the related amino acids. This was first discovered of the operon for histidine biosynthesis (*196*). It is possible, therefore, that modification of tRNA by chemical carcinogens could affect functions associated with metabolic regulation, cellular differentiation, or neoplastic transformation (*197*).

The interaction of tRNA with chemical carcinogens also provides a tool for studies of the three-dimensional structures of tRNA. Certain chemical carcinogens, for example AAF, behave as selective reagents and fulfill the requirements discussed by Cramer for tertiary structural studies (*198*). The detailed three-dimensional structure of $tRNA^{Phe}$ obtained by X-ray crystallography (*199, 200*)[7] can be compared with data obtained from chemical modification experiments to see if they agree and whether the proposed $tRNA^{Phe}$ structure applies to other tRNAs.

1. 2-Acetamidofluorene

Agarwall and Weinstein (*120*) found that 18 hours after an intraperitoneal injection of radioactive 2-(*N*-hydroxy)acetamidofluorene AAF to rats, the tRNA fraction had a specific activity two to three times higher than the 5 S, 18 S, and 28 S RNA fractions. Chromatography of the modified tRNA suggested that AAF reacts with several types of tRNA. Cochromatography on a BD-cellulose column of tRNAs obtained from rats fed either normal or AAF-containing diets revealed no gross changes in the profile of newly synthesized liver tRNAs. The attachment of AAF to tRNAs impairs the capacity of certain tRNAs to accept amino acids (e.g., lysine, arginine), and also inhibits the function of specific tRNAs during ribosomal binding and codon recognition (*201*). Since total tRNA was used, no correlation between the site of modification and function could be made. This aspect was examined in further studies with purified *E. coli* formylmethionine tRNA for *in vitro* modification with AcO-AAF (*202*). After separation of modified from unmodified tRNA, the modified molecules were enzymically cleaved and the site(s) of AAF-guanosine residues located within the nucleotide sequence. It was found that the primary target of AAF modification in $tRNA^{fMet}$ is the C-8 of the G_{20} residue in the dihydrouridine loop.[8] This position is analogous to G_{19} or to G_{20} of yeast $tRNA^{Phe}$. In the latter tRNA, G_{20} is in a single-strand conformation and G_{19}, although it base-pairs with C_{56} in the three-dimensional structure of $tRNA^{Phe}$, would also be exposed to chemical modification in the tertiary structure (*201a*). Thus, the reactivity of G_{20} in $tRNA^{fMet}$ toward AAF is consistent with the X-ray structure of $tRNA^{Phe}$.

The above-described AAF modification of the G_{20} residue of $tRNA^{fMet}$ causes about a 60% decrease in methionine acceptor activity (*202*). The partial rather than complete decrease in acceptor activity may reflect the existence of two different conformations of the modified tRNA; one of these may be completely inactive and the other normally

[7] See article by Kim in Vol. 17 of this series.

[8] See the article by Singhal and Fallis on p. 227.

active in methionine acceptor activity. The reduction of acceptor activity could also be due to steric hindrance by the bulky fluorene residue, which might prevent appropriate orientation of the tRNA during its interaction with the cognate synthetase. In accord with this interpretation is the higher K_m value found with the modified tRNA (*202*). With other types of modification of $tRNA^{fMet}$ that result in decreased acceptor activity, there are also changes in kinetic parameters (*203*). It is unlikely that the decrease of acceptor activity after modification of G_{20} indicates that this region of the tRNA is normally involved in synthetase recognition.

Other studies indicate that complete removal of nucleotides from the D loop by nucleolytic excision does not destroy synthetase recognition (*204, 205*). In addition, photooxidation of G_{20} does not have an appreciable effect on the methionine acceptor activity of $tRNA^{fMet}$ (*206*). As expected, AUG codon recognition in a ribosomal binding assay is not affected by the attachment of AAF to G_{20} in $tRNA^{fMet}$ (*202*).

In solutions containing 3 mM Mg^{2+} and 0.1 M K^+, conditions likely to preserve the three-dimensional structure of tRNA, modification of yeast tyrosine tRNA by AcO-AAF is less extensive and more specific than in the absence of these ions. (*207*). On the average, one residue of carcinogen is bound to one tRNA molecule. Partial modification occurred at three different sites, at the 2′-*O*-MeG_{18} and G_{19} in the D loop, and at G_{36} in the anticodon. In the absence of Mg^{2+} and K^+, and with 3 mM EDTA, the binding of AAF increased to 1.6 residues per tRNA. Under these conditions, in addition to modification of G_{18}, G_{19}, and G_{36} there was also modification of G_{15} in the D loop, presumably reflecting the less ordered structure of the tRNA. The tyrosine acceptor activity of $tRNA^{Tyr}$ modified in the presence of Mg^{2+} and K^+ ions was decreased by about 18%; whereas without these ions and with EDTA, it was decreased to 50% that of unmodified molecules. The UAU codon-dependent binding of the modified $tRNA^{Tyr}$ to ribosomes decreased by approximately the fraction of tRNA molecules modified in the anticodon.

These results with $tRNA^{fMet}$ and $tRNA^{Tyr}$ indicate that modification by AcO-AAF is quite specific and is related to the three-dimensional structure of the tRNA molecule. Ionic environments that alter the conformation of tRNA affect the extent and specific sites of tRNA modification. The modification can also inhibit the functions of tRNA with respect to both aminoacyl-tRNA synthetase and codon recognition.

2. Binding of Other Carcinogens to tRNA

The alkylation of a number of purified tRNA preparations by reaction with the carcinogens *N*-methyl-*N*-nitrosourea and *N*-

ethyl-*N*-nitrosourea was studied in order to investigate the role of nucleic acid structure on the distribution of alkylation products within the nucleotide sequence (*208*). The rate of alkylation was greatly increased by increasing the pH over the range six to eight, and the degree of alkylation was directly proportional to the concentration of the nitrosamide added and independent of the amount of tRNA present. There was no significant difference in the degree of alkylation of any of the tRNA preparations tested. Reaction with the ethyl derivative resulted in a degree of alkylation 1/13th that achieved with a similar concentration of the methyl compound. The major product of the reaction was 7-alkylguanine, amounting to about 80% of the total, but 3-methyladenine was also identified as a product of the reaction of $tRNA^{fMet}$ with *N*-methyl-*N*-nitrosourea.

A highly purified $tRNA^{Phe}$ was isolated from livers of rats fed a diet containing 3′-methyl-4-dimethylaminoazobenzene for three weeks (*194*). No structural characterization of modified nucleotides or assays for possible functional changes of the modified $tRNA^{Phe}$ molecules were reported. To our knowledge, there are no detailed published studies on tRNA modification by aflatoxin or benzo[*a*]pyrene.

IX. Summary

From the above review, certain general principles emerge. The acute effects of exposure of tissues or cells in culture to the polycyclic aromatic carcinogens 2-acetamidofluorene, benzo[*a*]pyrene, and aflatoxin B_1 include an inhibition of replicative DNA synthesis, DNA transcription, and protein synthesis. An almost immediate response of the cell to DNA damage by these agents is excision of the modified adducts and repair of the DNA. *In vitro* studies reveal that the function of DNA during replication and transcription, and the function of tRNAs and mRNAs during translation, are inhibited by the covalent binding of activated forms of these carcinogens. All the evidence points to inhibition of base-pairing rather than to errors in base-pairing with these bulky carcinogens, whereas the simpler methylating and ethylating agents often induce mispairing.[3] In bacteria, the polycyclic aromatic carcinogens produce mainly frame-shift mutations, but the specific types of mutations induced in mammalian cells are not known. For reasons that are not apparent, there is *in vivo* a preferential inhibition of nucleolar synthesis as well as an apparent inhibition of the activity of RNA polymerase II. At present, very little is known about the effects of these agents in chromatin structure and the control of specific gene expression.

Extrapolations from results obtained with carcinogens and nucleic acids in subcellular systems to the *in vivo* situation are complicated by several factors. The first is that, in the *in vitro* studies, the extent of nucleic acid modification is often much greater than occurs after *in vivo* administration of the carcinogen. *In vivo* nucleic acids are covalently modified to an extent of one residue per about 10^4 to 10^5 nucleotide residues, whereas, for detectable effects *in vitro*, this ratio is usually in the range of one per 10^2. A second complication, stressed in this review, is that many of the carcinogens give more than one type of nucleic acid adduct, both *in vivo* and *in vitro*, and it is often difficult to relate the functional effects to a specific adduct. A third complication is that the *in vitro* synthesis of RNA and DNA is usually assayed with highly artificial systems that may not resemble the *in vivo* situation. For example, RNA synthesis is often accomplished with an *E. coli* polymerase and mammalian templates. The results may not accurately reflect the *in vivo* process with respect to the role of promoters and other *in vivo* determinants of initiation of RNA synthesis. Excepting the ϕX174 phage system, most subcellular DNA-synthesizing systems do not fulfill the natural conditions of DNA replication. Mammalian cells contain multiple DNA polymerases, and the actual mechanism of DNA replication in these systems is poorly understood.

A final reservation is that the effects of carcinogens on macromolecular synthesis *in vivo* may occur via more indirect mechanisms than are apparent in subcellular systems. This is well illustrated with respect to RNA synthesis. We have reviewed the evidence that, although the modification of DNA by carcinogens impairs their template activity for RNA synthesis, the inhibition of RNA synthesis *in vivo* may be due, at least in part, to effects on the polymerase enzyme. It is essential, therefore, to consider the possibility that certain effects of carcinogens *in vivo* may be due to alterations in membrane function, effects on energy metabolism, modifications of polymerases, chromatin-associated proteins, and other cellular proteins, RNA processing, etc.; rather than to covalent binding to DNA or RNA.

We may contrast the effects of modification of nucleic acids by methylating or ethylating agents with those obtained with polycyclic aromatic compounds like AAF or B[*a*]P. There is considerable evidence that the former types of modification produce errors in base-pairing at the sites of modification presumably by interfering with hydrogen bonding and the formation of Watson–Crick base-pairs.[3] On the other hand, the evidence we have cited indicates that the latter type of modification blocks base-pairing and chain elongation during DNA, RNA, and protein synthesis. The latter effects presumably relate to the conformational changes in nucleic acids produced by these bulky agents, yet both classes of agents are carcinogenic. Thus, the full significance of these structural and functional effects with respect to carcinogenesis is not apparent at the present time.

Acknowledgment

The authors wish to thank Ms. Sue A. Allen for valuable assistance in the preparation of this article.

References

1. J. A. Miller, *Cancer Res.* **30**, 559 (1970).
2. I. B. Weinstein, *Colloq. Int. C.N.R.S. Cancerog. Chimi.* **256**, 2 (1977).
3. I. B. Weinstein and D. Grunberger, *in* "Model Studies in Chemical Carcinogenesis, Part 2" (P.O.P. Ts's and J. DiPaolo, eds.), p. 217. Dekker, New York, 1974.
4. D. Grunberger and I. B. Weinstein, *in* "Biology of Radiation Carcinogenesis" (J. M. Yuhas, R. W. Tennant, and J. D. Regan, eds.), p. 175. Raven, New York, 1976.
5. S. Leffler, P. Pulkrabek, D. Grunberger, and I. B. Weinstein, *Bchem* **16**, 3133 (1977).
6. P. Pulkrabek, S. Leffler, I. B. Weinstein, and D. Grunberger, *Bchem* **16**, 3127 (1977).
7. I. B. Weinstein, A. M., Jeffrey, K. W., Jennette, S. H. Blobstein, R. G. Harvey, C. Harris, H. Autrup, H. Kasai, and K. Nakanishi, *Science* **193**, 592 (1976).
8. I. B. Weinstein, A. M. Jeffrey, S. Leffler, P. Pulkrabek, H. Yamasaki, and D. Grunberger, *in* "Polycyclic Hydrocarbons and Cancer" (H. Gelboin and P. O. P. Ts'o, eds.), Vol. 2, p. 3. Academic Press, New York, 1978.
9. D. Grunberger and I. B. Weinstein, *in* "Chemical Carcinogens and DNA" (P. L. Grover, ed.), p. 59. CRC Press, Boca Raton, Fl., 1979.
10. B. Singer, *Nature* **264**, 333 (1976).
11. A. F. Levine, L. M. Fink, I. B. Weinstein, and D. Grunberger, *Cancer Res.* **34**, 319 (1974).
12. A. M. Kapuler and A. M. Michelson, *BBA* **232**, 436 (1971).
13. E. Kriek and J. Reitsema, *Chem.-Biol. Interact.* **3**, 397 (1971).
14. E. Kriek, *BBA* **355**, 177 (1974).
15. V. Ivanovic, N. E. Geacintov, H. Yamasaki, and I. B. Weinstein, *Bchem* **17**, 1597 (1978).
16. K. M. Straub, T. Meehan, A. L. Burlingame, and M. Calvin, *PNAS* **74**, 5285 (1977).
17. M. R. Osborne, R. G. Harvey, and P. Brookes, *Chem. Biol. Interact.* **20**, 123 (1978).
18. M. Koreeda, P. D. Moore, H. Yagi, J. C. Yeh, and D. M. Jerina, *JACS* **98**, 6721 (1976).
19. A. M. Jeffrey, I. B. Weinstein, K. W. Jennette, K. Grzeskowiak, K. Nakanishi, R. G. Harvey, H. Autrup, and C. Harris, *Nature* **269**, 348 (1977).
20. Jeffrey, A. M., K. W., Jennette, S. H. Blobstein, I. B. Weinstein, F. A. Beland, R. G. Harvey, H. Kasai, I. Miura, and K. Nakanishi, *JACS* **98**, 5714 (1976).
21. A. D. D'Andrea and W. A. Haseltine, *PNAS* **75**, 4120 (1978).
22. H. Yamasaki, P. Pulkrabek, D. Grunberger, and I. B. Weinstein, *Cancer Res.* **37**, 3756 (1977).

23. N. E. Geacintov, A. Gagliano, V. Ivanovic, and I. B. Weinstein, *Bchem* **17**, 5256 (1978).
24. T. Prusik, N. E. Geacintov, C. Tobiasz, V. Ivanovic, and I. B. Weinstein, *Photochem. Photobiol.* **29**, 223 (1979).
25. N. R. Drinkwater, J. A. Miller, E. C. Miller, and N. C. Yang, *Cancer Res.* **38**, 3247 (1978).
26. T. Kakefude and H. Yamamoto, *PNAS* **75**, 3247 (1978).
27. P. C. Hanawalt, E. C. Friedberg, and C. F. Fot, "DNA Repair Mechanisms." Academic Press, New York, 1978.
28. A. E. Pegg, *BBRC* **84**, 166 (1978).
29. J. E. Cleaver, *in* "Birth Defects" (J. W. Littlefield and J. de Grouchy eds.), p. 85, Excerpta Medica, Amsterdam–Oxford, 1978.
30. J. D. Regan and R. B. Setlow, *Cancer Res.* **34**, 3318 (1974).
31. M. Ikenaga and T. Kakunaga, *Cancer Res.* **37**, 3672 (1977).
32. E. Kriek, *Cancer Res.* **32**, 2042 (1972).
33. P. A. Cerutti, F. Sessions, P. V. Hariharan, and A. Lusby, *Cancer Res.* **38**, 2118 (1978).
34. V. Ivanovic, N. E. Geacintov, H. Yamasaki, and I. B. Weinstein, *Bchem* **17**, 1597 (1978).
35. E. M. Witkin, *Bacteriol Rev.* **40**, 869 (1976).
36. M. Radman, G. Villani, S. Boiteux, M. Defais, P. Caillet-Fauquet, and S. Spardari, *in* "Origins of Human Cancer" (H. H. Hiatt, J. D. Watson, and J. A. Winstein, eds.), p. 903. Cold Spring Harbor Lab., Cold Spring Harbor, New York, 1977.
37. P. Moreau and R. Devoret, *in* "Origins of Human Cancer" (H. H. Hiatt, J. D. Watson, and J. A. Winstein, eds.), p. 1451. Cold Spring Harbor Lab., Cold Spring Harbor, New York, 1977.
38. A. R. Sarasin and P. C. Hanawalt, *PNAS* **75**, 346 (1978).
39. R. S. Day, III, D. Scudiero, and M. Dimattina, *Mutat. Res.* **50**, 383 (1978).
40. U. B. DasGupta and W. C. Summers, *PNAS* **75**, 2378 (1978).
41. E. Reich, D. B. Rifkind, and E. Shaw (eds.) *in* "Proteases and Biological Control" Cold Spring Harbor Conf. on Cell Proliferation, Vol. 2, Sect. 9, Cold Spring Harbor, New York, 1975.
42. H. W. Thielmann, *EJB* **61**, 501 (1976).
43. H. W. Thielmann and H. Gersbach, *Z. Krebsforsch.* **92**, 177 (1978).
44. R. B. Kelly, M. R., Atkinson, J. A. Huberman, and A. Kornberg, *Nature* **224**, 495 (1969).
45. S. Yashida, M. Tada, and M. Tada, *NARes* **3**, 3227 (1976).
46. H. W. Thielmann and H. Gersbach, *Z. Krebsforsch.* **92**, 157 (1978).
47. D. M. Jerina, R. Lehr, M. Schaefer-Ridder, H. Yagi, J. M. Karle, D. R. Thakker, A. W. Wood, A. Y. H. Lu, D. Ryan, S. West, W. Levin, and A. H. Conney, *in* "Origins of Human Cancer" (H. H. Hiatt, J. D. Watson, and J. A. Winstein, eds.), p. 639. Cold Spring Harbor Lab., Cold Spring Harbor, New York, 1977.
48. I. B. Weinstein, H. Yamasaki, M. Wigler, L.-S. Lee, P. B. Fisher, A. Jeffrey, and D. Grunberger, *in* "Carcinogens: Identification and Mechanisms of Action" (A. C. Griffin and C. R. Shaw, eds.), p. 399. Raven, New York, 1979.
49. F. Gossard and W. G. Verly, *EJB* **82**, 321 (1978).
50. S. Ljungquist, *JBC* **252**, 2808 (1977).
51. M. Sekiguchi, H. Hayakawa, F. Makino, K. Tanaka, and Y. Okada, *BBRC* **73**, 293 (1976).
52. T. Lindahl, *Nature* **259**, 64 (1976).

53. J. Laval, *Nature* **269**, 829 (1977).
53a. P. Karran and T. Lindahl, *JBC* **253**, 5877 (1978).
54. N. V. Tomlin, E. B. Paveltchuk, and T. V. Mosevitskaya, *EJB* **69**, 265 (1976).
55. S. R. Kushner, J. C. Kaplan, H. Ono, and L. Grossman, *Bchem* **10**, 3325 (1971).
56. A. Schön-Bopp, G. Schäfer, and U. Hagen, *Int. J. Radiat. Biol.* **31**, 227 (1977).
57. R. B. Setlow and W. L. Carrier, *Nature NB* **241**, 170 (1973).
58. R. Hecht and H. W. Thielmann, *NARes* **4**, 4235 (1977).
59. R. Hecht and H. W. Thielmann, *EJB* **89**, 607 (1978).
60. W. A. Deutsch and S. Linn, *PNAS* **76**, 141 (1979).
61. R. H. Heflich, D. J. Dorney, V. M. Maher, and J. J. McCormick, *BBRC* **77**, 634 (1977).
62. M. Woodworth-Gutai, J. Lebowitz, A. C. Kato, and D. T. Denhardt, *NARes* **4**, 1243 (1977).
63. N. Camerman and A. Camerman, *Science* **160**, 1451 (1968).
64. A. R. Sarasin, C. A. Smith, and P. C. Hanawalt, *Cancer Res.* **37**, 1786 (1977).
65. S. M. D'Ambrosio and R. B. Setlow, *PNAS* **73**, 2393 (1976).
66. T. J. Slaga, G. T. Bowden, B. G. Shapas, and R. K. Boutwell, *Cancer Res.* **34**, 771 (1974).
67. R. B. Painter, *J. Environ. Pathol. Toxicol.* **2**, 65 (1978).
68. R. B. Painter, *Cancer Res.* **38**, 4445 (1978).
69. M. Gellert, M. H. O'Dea, T. Itoh, and J. Tomizawa, *PNAS* **73**, 4474 (1976).
70. M. A. Lovett, R. B. Sparks, and D. R. Helinski, *PNAS* **72**, 2905 (1975).
71. A. Trevers, *Nature* **275**, 89 (1978).
72. H. Marquardt, F. S. Philips, and A. Bendich, *Cancer Res.* **32**, 1810 (1972).
73. D. H. Janss, L. B. Malan, and S. P. Kelley, *PAACR* **17**, 145 (1976).
74. W. Troll, S. Belman, E. Berkowitz, Z. F. Chmielewicz, F. L. Ambrus, and T. J. Bardos, *BBA* **157**, 16 (1968).
75. G. J. Guzzo and R. I. Glazer, *Cancer Res.* **36**, 1041 (1976).
76. W. T. Hsu, E. J. S., Lin, R. G. Harvey, and S. B. Weiss, *PNAS* **74**, 3335 (1977).
77. T. Meehan and K. Straub, *Nature* **277**, 410 (1979).
78. V. Berthold, H. W. Thielmann, and K. Geider, *FEBS Lett.* **86**, 81 (1978).
79. C. Frayssinet, C. LaFarge, A. M. DeRecondo, and E. LeBretan, *C.R. Hebd. Seances Acad. Sci.* **259**, 143 (1967).
80. L. A. Loeb, L. A. Weymouth, T. A. Kunkel, K. P. Gopinathan, R. A. Beckman, and D. K. Dube, *CSHSQB* **43 Pt 2**, 921 (1978).
81. M. A. Sirover and L. A. Loeb, *Science* **194**, 1434 (1976).
82. L. A. Weymouth and L. A. Loeb, *PNAS* **75**, 1924 (1978).
83. M. A. Sirover and L. A. Loeb, *Cancer Res.* **36**, 516 (1976).
84. C. W. Shearman and L. A. Loeb, *Nature* **270**, 537 (1977).
85. J. McAnn and B. N. Ames, *in* "The Origin of Human Cancer" (H. H. Hiatt, J. D. Watson, and J. A. Winstein, eds.), p. 1431. Cold Spring Harbor Lab., Cold Spring Harbor, New York, 1977.
86. P. Fisher and I. B. Weinstein, *in* "Carcinogens in Industry and the Environment" (J. M. Sontag, ed.). Dekker, New York (in press).
87. V. M. Maher, E. C. Miller, J. A. Miller, and W. Szybalski, *Mol. Pharmacol.* **4**, 411 (1968).
88. V. M. Maher, J. A. Miller, E. C. Miller, and W. C. Summers, *Cancer Res.* **30**, 1473 (1970).
89. H. Bartsch, C. Malaveille, B. Tierney, P. L. Grover, and P. Sims, *Chem.-Biol. Interact.* **26**, 185 (1979).

90. A. W. Hsie, P. A. Brimer, T. J. Mitchell, and D. G. Gosslee, *Somatic Cell Genet.* **1**, 247 (1975).
91. E. Huberman, *Mutat. Res.* **29**, 285 (1975).
92. W. T. Hsu, R. G. Harvey, E. J. S. Lin, and S. B. Weiss, *PNAS* **74**, 1378 (1977).
93. A. W. Wood, R. L. Chang, W. Levin, H. Yagi, D. R. Thakker, D. M. Jerina, and A. H. Conney, *BBRC* **77**, 1389 (1977).
94. M. Fogel and L. Sachs, *Virology* **40**, 174 (1970).
95. D. Zouzia, I. Prasad, and C. Basilico, *J. Virol.* **24**, 142 (1977).
96. R. D. Kornberg, *Science* **184**, 868 (1977).
97. H. Weintraub and M. Goudine, *Science* **193**, 848 (1976).
98. A. Garel and R. Axel, *PNAS* **73**, 3966 (1976).
99. C. L. Jahn and G. W. Litman, *BBRC* **76**, 534 (1977).
100. H. Yamasaki, T. W. Roush, and I. B. Weinstein, *Chem.-Biol. Interact.* **23**, 201 (1978).
101. G. Metzger, F. X. Wilheim, and M. L. Wilhelm, *Chem.-Biol. Interact.* **14**, 257 (1976).
102. G. Metzger, F. X. Wilhelm, and M. L. Wilhelm, *BBRC* **75**, 703 (1977).
103. J. E. Cleaver, *Nature* **270**, 541 (1977).
104. H. L. Moses, R. A. Webster, G. D. Martin, and T. C. Spelsberg, *Cancer Res.* **36**, 2905 (1976).
105. E. L. Schwartz and J. I. Goodman, *Chem.-Biol. Interact.* **26**, 287 (1979).
106. H. Yamasaki, S. Leffler, and I. B. Weinstein, *Cancer Res.* **37**, 684 (1977).
107. T. W. Kensler, W. F. Busby, Jr., N. E. Davidson, and G. N. Wogan, *Cancer Res.* **36**, 4647 (1976).
108. A. J. Fornace and J. B. Little, *Cancer Res.* **39**, 704 (1979).
109. G. Metzger and H. Werbin, *Bchem* **18**, 655 (1979).
110. R. Halliday and J. E. Pugh, *Science* **187**, 226 (1975).
111. C. E. Salas, A. Pfohl-Leszkowicz, M. C. Lang, and G. Dirheimer, *Nature* **278**, 71 (1979).
112. J. K. Christman, P. Price, L. Pedrinan, and G. Acs, *EJB* **81**, 53 (1977).
113. W. Troll, E. Rinde, and P. Day, *BBA* **174**, 211 (1969).
114. E. Kriek, *BBA* **161**, 273 (1968).
115. E. Kriek, *Chem.-Biol. Interact.* **1**, 3 (1969–1970).
116. D. Szafarz and J. H. Weisburger, *Cancer Res.* **29**, 962 (1969).
117. F. Marroquin and E. Farber, *Cancer Res.* **25**, 1262 (1965).
118. C. C. Irving and R. A. Veazey, *Cancer Res.* **29**, 1799 (1969).
119. T. Matsushima and J. H. Weisburger, *Chem.-Biol. Interact.* **1**, 211 (1969).
120. S. S. Agarwal and I. B. Weinstein, *Bchem* **9**, 503 (1970).
121. C. C. Irving and R. A. Veazey, *BBRC* **47**, 1159 (1972).
122. E. J. Barry, D. Malejka-Giganti, and H. R. Gutmann, *Chem.-Biol. Interact.* **1**, 139 (1970).
123. E. J. Barry, C. A. Ovechka, and H. R. Gutmann, *JBC* **243**, 51 (1968).
124. F. J. Zieve, *JB* **247**, 5987 (1972).
125. G. Grunberger, F. L. Yu, D. Grunberger, and P. Feigelson, *JBC* **248**, 6278 (1973).
126. J. R. Tata, M. J. Hamilton, and D. Shields, *Nature* **238**, 161 (1972).
127. E. A. Smuckler and J. R. Tata, *BBRC* **49**, 16 (1972).
128. R. I. Glazer, L. E. Glass, and F. M. Menger, *Mol. Pharmacol.* **11**, 36 (1975).
129. N. L. R. Bucher, *Int. Rev. Cytol.* **15**, 245 (1963).
130. F. L. Yu and P. Feigelson, *PNAS* **69**, 2833 (1972).

131. F. L. Yu and P. Feigelson, *PNAS* **68,** 2177 (1971).
132. J. Herzog, A. Serroni, B. A. Briesmeister, and J. L. Farber, *Cancer Res.* **35,** 2138 (1975).
133. R. I. Glazer, R. C. Nutter, L. E. Glass, and F. M. Menger, *Cancer Res.* **34,** 2451 (1974).
134. R. L. Millette and L. M. Fink, *Bchem* **14,** 1426 (1975).
135. R. L. Millette and C. D. Trotter, *PNAS* **66,** 701 (1970).
136. R. I. Glazer, *Cancer Res.* **36,** 2282 (1976).
137. E. J. Barry and H. R. Gutmann, *JBC* **248,** 2730 (1973).
138. F. L. Yu and D. Grunberger, *Cancer Res.* **36,** 3629 (1976).
139. F. L. Yu, *Nature* **251,** 344 (1974).
140. M. P. Adams and J. I. Goodman, *BBRC* **68,** 850 (1976).
141. F. L. Yu and P. Feigelson, *BBA* **272,** 119 (1972).
142. R. I. Glazer, *BBA* **475,** 492 (1977).
143. R. I. Glazer, *BBA* **361,** 361 (1974).
144. L. A. Kaplan and I. B. Weinstein, *Chem.-Biol. Interact.* **12,** 99 (1976).
145. R. Weinberg, W. Loening, M. Williams, and S. Penman, *PNAS* **58,** 1088 (1967).
146. W. K. Roberts and J. F. E. Newman, *JMB* **20,** 63 (1966).
147. H. Busch, J. Hodnett, H. P. Morris, Nagy, R., K. Smetana, and T. Unuma, *Cancer Res.* **28,** 672 (1968).
148. R. I. Glazer, *NARes* **5,** 2607 (1978).
149. K. Ueno, K. Wekimizu, D. Mizuno, and S. Natori, *Nature* **277,** 145 (1979).
150. G. N. Wogan, *Methods Cancer Res.* **7,** 309 (1973).
151. D. F. Krahm and C. Heidelberger, *Mutat. Res.* **46,** 27 (1977).
152. J. J. Wong and D. P. Hsieh, *PNAS* **73,** 2241 (1976).
153. A. E. Rogers and P. M. Newberne, *Cancer Res.* **27,** 855 (1967).
154. M. B. Sporn, C. W. Dingman, H. L. Phelps, and G. N. Wogan, *Science* **151,** 1539 (1966).
155. A. Sarasin and Y. Moulè, *FEBS Lett.* **32,** 347 (1973).
156. H. V. Gelboin, J. S. Worthan, R. G. Wilson, M. Friedman, and G. N. Wogan, *Science* **154,** 1205 (1966).
157. J. I. Clifford and K. R. Rees, *Nature* **209,** 312 (1966).
158. Y. Moulè, *Nature* **218,** 93 (1968).
159. H. M. Q. King and B. H. Nicholson, *BJ* **114,** 779 (1969).
160. A. K. Roy, *BBA* **169,** 206 (1968).
161. R. C. Garner, E. C. Miller, and J. A. Miller, *Cancer Res.* **24,** 1553 (1972).
162. D. H. Svenson, J. K. Lin, E. C. Miller, and J. A. Miller, *Cancer Res.* **37,** 172 (1977).
163. J. M. Essignmann, R. G. Croy, A. M. Nadzan, W. F. Busby, Jr., V. N. Reinhold, G. Buchi, and G. N. Wogan, *PNAS* **74,** 1870 (1977).
164. R. G. Croy, J. M. Essignmann, V. N. Reinhold, and G. N. Wogan, *PNAS* **75,** 1745 (1978).
165. J. K. Lin, J. A. Miller, and E. C. Miller, *Cancer Res.* **37,** 4430 (1977).
166. G. S. Edwards and G. N. Wogan, *BBA* **224,** 597 (1970).
167. F. C. Saunders, E. A. Barker, and E. Smuckler, *Cancer Res.* **32,** 2487 (1972).
168. G. E. Neal, *Nature* **244,** 432 (1973).
169. G. E. Neal, *BJ* **130,** 619 (1972).
170. Y. Moulè and C. Frayssinet, *FEBS Lett.* **25,** 52 (1972).
171. F. L. Yu, *JBC* **252,** 3245 (1977).
172. E. O. Akinrimisi, B. J. Benecke, and K. H. Seifert, *EJB* **42,** 333 (1974).
173. K. Alexandrov, C. Vendrely, and R. Vendrely, *Cancer Res.* **30,** 1192 (1970).

174. A. B. deAngelo, W. R. Hudgins, and S. B. Kerby, *Cancer Res.* **38**, 384 (1978).
175. H. Cedar and G. Felsenfeld, *JMB* **77**, 237 (1973).
176. P. Pulkrabek, S. Leffler, I. B. Weinstein, and D. Grunberger, *176th Natl. Meet., Am. Chem. Soc.*, Abstract No. 101 (1978).
177. R. C. Shank and G. N. Wogan, *Toxicol. Appl. Pharmacol.* **9**, 468 (1966).
178. R. S. Pong and G. N. Wogan, *Biochem. Pharmacol.* **18**, 2357 (1969).
179. S. Villa-Trevino and D. D. Leaver, *BJ* **109**, 87 (1968).
180. R. S. Pong and G. N. Wogan, *Biochem. Pharmacol.* **19**, 2808 (1970).
181. A. Sarasin and Y. Moulè, *FEBS Lett.* **29**, 329 (1973).
182. A. Sarasin and Y. Moulè, *EJB* **34**, 329 (1975).
183. K. Li, R. Kisilevsky, M. T. Wasan, and G. Hammond, *BBA* **272**, 451 (1972).
184. J. A. Miller, *Cancer Res.* **30**, 553 (1970).
185. E. Farber, J. McConomy, B. Franzen, F. Marroquin, G. A. Steward, and D. N. Magee, *Cancer Res.* **27**, 1761 (1967).
186. E. Farber, *Cancer Res.* **28**, 1859 (1968).
187. C. C. Irving, R. A. Veazey, and R. F. Williard, *Cancer Res.* **27**, 270 (1967).
188. E. Kriek, J. A. Miller, U. Juhl, and E. C. Miller, *Bchem* **6**, 177 (1967).
189. D. Grunberger and I. B. Weinstein, *JBC* **246**, 1123 (1970).
190. J. H. Nelson, D. Grunberger, C. R. Cantor, and I. B. Weinstein, *JMB* **62**, 331 (1971).
191. D. Grunberger, S. H. Blobstein, and I. B. Weinstein, *JMB* **82**, 459 (1974).
192. E. Sagher, R. G. Harvey, W. T. Hsu, and S. B. Weiss, *PNAS* **76**, 620 (1979).
193. R. Axel, I. B. Weinstein, and E. Farber, *PNAS* **58**, 1255 (1967).
194. A. H. Daoud and A. C. Griffin, *Cancer Res.* **36**, 2885 (1976).
195. W. Z. Littauer and H. Inouye, *ARB* **47**, 439 (1973).
196. C. E. Singer, G. R. Smith, R. Cortese, and B. N. Ames, *Nature* **238**, 72 (1972).
197. I. B. Weinstein, D. Grunberger, S. Fujimura, and L. M. Fink, *Cancer Res.* **31**, 651 (1971).
198. F. Cramer, This Series, **11**, 391 (1971).
199. S. H. Kim, F. L. Suddeth, R. J. Quigley, A. McPherson, J. L. Sussman, A. H. J. Wang, N. C. Seeman, and A. Rich, *Science* **185**, 435 (1974).
200. J. D. Robertus, J. E. Ladner, J. T. Fink, D. Rhodes, R. S. Brown, B. F. C. Clark, and A. Klug, *Nature* **250**, 546 (1974).
201. L. M. Fink, S. Nishimura, and I. B. Weinstein, *Bchem* **9**, 496 (1970).
201a. A. Rich and U. L. RajBhandary, *ARB* **45**, 805 (1976).
202. S. Fujimura, D. Grunberger, G. Carvajal, and I. B. Weinstein, *Bchem* **11**, 3629 (1972).
203. L. H. Schulman and H. Pelka, *Bchem* **16**, 4256 (1977).
204. T. Seno, I. Kobayashi, M. Fukuhara, and S. Nishimura, *FEBS Lett.* **7**, 343 (1970).
205. T. Seno and K. Seno, *FEBS Lett.* **16**, 180 (1971).
206. L. H. Schulman, *PNAS* **69**, 3594 (1972).
207. P. Pulkrabek, D. Grunberger, and I. B. Weinstein, *Bchem* **13**, 2414 (1974).
208. A. E. Pegg, *Chem.-Biol. Interact.* **6**, 393 (1973).

Participation of Modified Nucleosides in Translation and Transcription

B. SINGER AND M. KRÖGER*

Department of Molecular Biology and Virus Laboratory
University of California
Berkeley, California

I. Introduction

Transcription and translation are fundamental processes in biology. Both involve complex interactions of enzymes with polynucleotides.

* Present address: Institut für Biologie III der Universität, Schänzlestrasse 1, D7800 Freiburg, West Germany.

Progress in Nucleic Acid Research and Molecular Biology, Vol. 23

ISBN 0-12-540023-3

However, central to this process is complex formation resulting from interactions between bases.

Most of our thinking on the parameters of base-pairing comes from experiments, primarily in the 1960s and early 1970s, on complexing of homopolymers of modified bases with various unmodified polynucleotides. In numerous papers published from 1961 to 1972, Szer and Shugar (*1–7*), Michelson (*8–16*), Ikehara and Hattori (*17–22*), Scheit (*23–25*), Reese (*26, 27*) and their collaborators concluded that in all cases where there was any change in the base, even on the non-Watson–Crick sites, there was either no complementary base-pairing, or non-Watson–Crick base-pairing, or that the thermal stability of the double strand was very different from that of the analogous unmodified double-strand. A portion of this early work was summarized in Volume 6 of this series by Michelson, Massoulié, and Guschlbauer (*28*). As more and more polynucleotides containing modified bases have been prepared, it continues, as of 1979, to be true that even the simplest modification of a base or nucleoside leads to detectable changes in the properties of the polymer containing it.

In a few cases where a modified homopolymer cannot form a complex with another [e.g., poly(3-methyluridylate)], heteropolymers containing relatively large amounts of the modified base (> 15%) were studied. In these, any apparent base-pairing was attributed only to the unmodified base, and "looping out" of the modified base in the polymer was proposed as a mechanism for polymers of N^4-methyl- and of 3-methylcytidine (*27*), for those of 8-bromoguanosine [**33**] (*17*), for those of 1-methylguanosine (*14*), and for those of 3-methyluridine (*1*). Another school of thought attributed apparent base-pairing to stacking interaction. Such stacking forces contributed by the unmodified base were believed to explain the observations on polymers containing N^2-methyl- and N^2-dimethylguanosine, ribosyl-2,6-bis(methylthio)-purine [**30**] (*18, 20*), and 6-thioguanosine [**24**] (*29*). Both "looping out" or extra-helical conformation and stacking interaction or intrahelical conformation were discussed by Lomant and Fresco in Volume 15 of this series (*30*).

The second line of research on base-pairing was the study of codon–anticodon interaction; here the requirements for binding appear to be somewhat different. For example, either 3-deaza- [**8**] or 4-deoxyuridine [**18**] in the second or third position of UUU binds a significant amount of $tRNA^{Phe}$ (*31*), and although 6-azauridine [**9**] cannot substitute for uridine (*32*), 8-azaguanosine [**7**] can substitute for guanosine (*33*).

Neither of these lines of investigation gives us a clear understanding or common hypothesis relating to the number and type of hydrogen bonds necessary for recognition or conservative transcription.

Now, with renewed interest in the effects of carcinogens or therapeutic agents on mutation and cell growth, we and others are examining the more realistic problem: What happens when a polynucleotide containing only a *few* modified bases is transcribed?[1] This should eliminate the major effects on secondary structure contributed by the modification.

II. Types of Modifications of Polynucleotides and Codons

A. Base Ring Analogs

This group of modifications involves removal, addition, or exchange of only the ring nitrogens and carbons. Laurusin[2] [**1**], formycin[2] [**2**], and pseudouridine [**3**] are "C-nucleosides," the ribose being attached to a carbon rather than nitrogen. The other derivatives in this class are 1-deaza- [**4**], 3-deaza- [**5**], and 7-deazaadenosine (tubercidin)[2] [**6**], 8-azaguanosine [**7**], and 3-deaza- [**8**] and 6-azauridine [**9**]. Laurusin can also be considered as a C-nucleoside analog of inosine (or of deamino-Guo). An example of a ring modification of a nucleoside analog is 7-deazanebularin [**10**]. No base-substituted cytidines seem to have been used in polynucleotides.

B. Exocyclic Rearrangements

Pseudouridine [**3**] is an example of this class, but is unique since it also belongs to the class of C-nucleosides. "Iso-adenosine" [**11**] also has a rearranged ribose, the attachment changed from the N-9 to the N-3, so that a pyrimidine function may result. "Iso-guanosine" (2-hydroxyadenosine) [**12**] and "iso-cytidine" (2-amino-2-deoxyuridine) [**13**] have exchanged the N^2-amino and O^6, and the N^4-amino and O^2, respectively. The nucleoside of 2-aminopurine [**14**] may be viewed as adenosine with the N^6-amino group becoming an N^2-amino group, or as a guanosine minus its 6-oxygen atom (see Section II, C).

C. Removal of Exocyclic Groups

Xanthosine [**15**] and inosine [**16**] are deaminated purines derived from guanosine and adenosine. Nebularin (6-deaminoadenosine) [**17**]

[1] For technical reasons, most work on the function of modified nucleosides has been with the ribo derivatives, polynucleotide phosphorylase being the tool for preparing homopolymers and heteropolymers. Chemical modification of a preformed polynucleotide does not generally lead to a single site of modification, and thus, with a few exceptions, studies utilizing alkylated polynucleotides are not included.

[2] See article by Suhadolnik in Vol. 22 of this series.

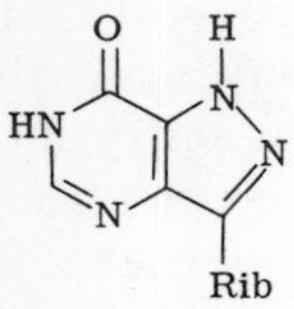

Laurusin

1

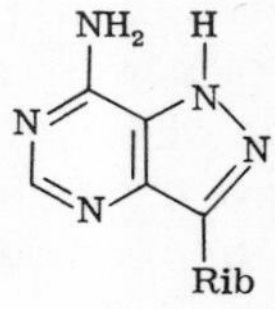

Formycin

2

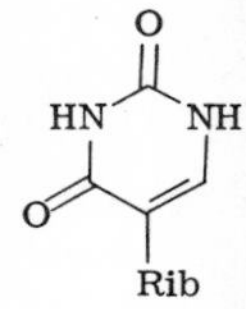

Pseudouridine

3

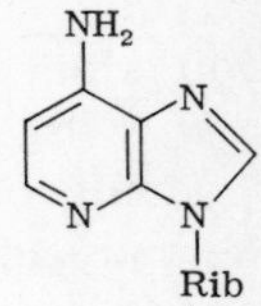

1-Deaza-
adenosine

4

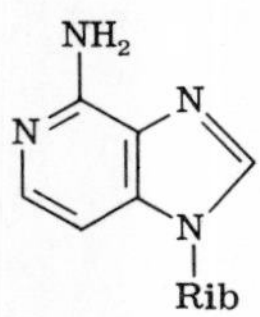

3-Deaza-
adenosine

5

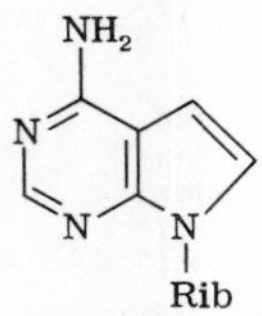

Tubercidin
7-deaza-
adenosine

6

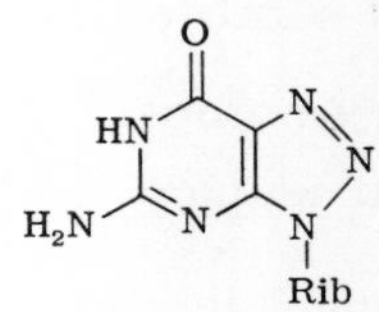

8-Azaguanosine

7

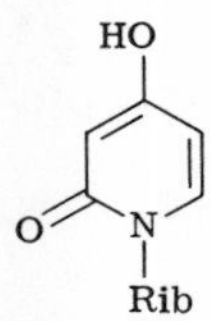

3-Deazauridine

8

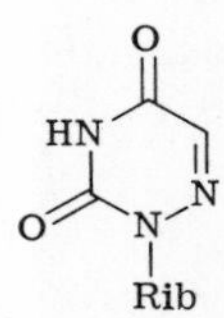

6-Azauridine

9

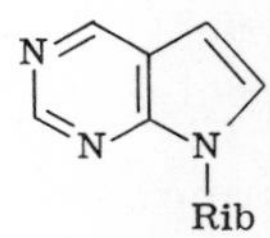

7-Deazanebularin
6-deamino-
7-deaza adenosine

10

and ribosylpyrimidone [**18**], which is 4-deaminocytidine, both lack the exocyclic amino group. Ribosylpyrimidone can also be viewed as 4-deoxyuridine, while ribosyl-2-aminopurine [**14**] is 6-deoxyguanosine. 3-Deaza-4-deoxyuridine (ribosylpyridone) has also been prepared. Nebularin (or ribosylpurine) has no exocyclic oxygen or amino group. No 2-deoxypyrimidines appear to have been studied.

D. Replacement Reactions

1. Replacement of Hydrogen by Amino or Hydroxyl Groups

The following compounds [**19–23**], with the exception of ribosyl-2,6-diaminopurine [**19**], are all modified on a part of the base not involved in Watson–Crick base-pairing. They include 5-hydroxycytidine [**20**], 5-hydroxyuridine [**21**], 8-oxyadenosine [**22**],

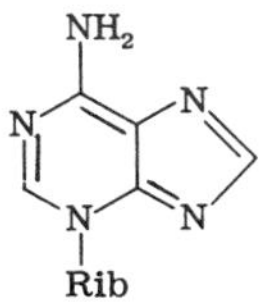

"Iso-adenosine"
3-ribosyladenine

11

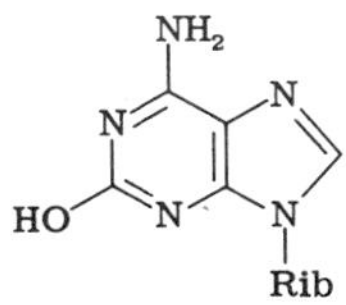

"Iso-guanosine"
2-hydroxyadenosine

12

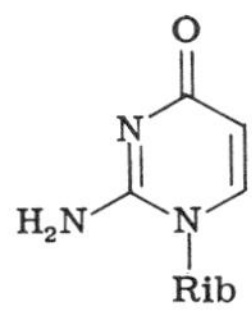

"Iso-cytidine"
2-amino-2-deoxyuridine

13

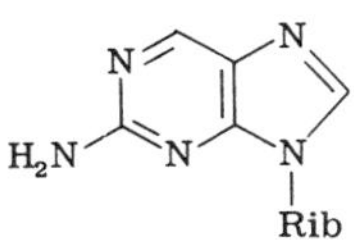

Ribosyl-2-aminopurine
6-deoxyguanosine

14

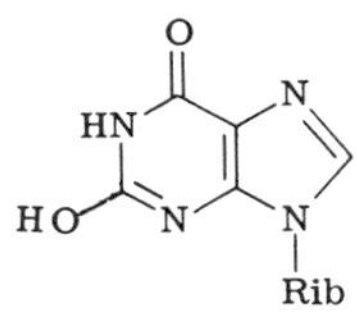

Xanthosine

15

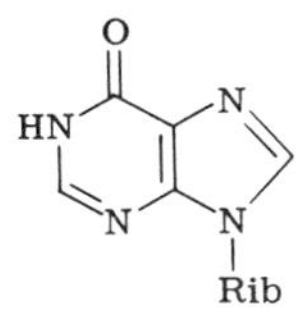

Inosine
2-deaminoguanosine

16

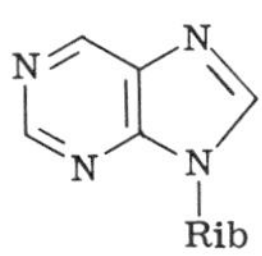

Nebularin
6-deaminoadenosine
ribosylpurine

17

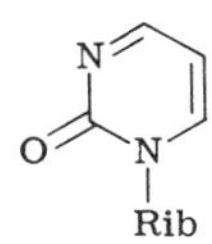

Ribosylpyrimidone
4-deoxyuridine
4-deaminocytidine

18

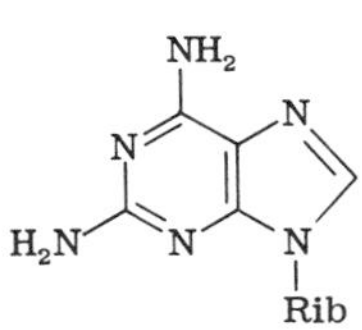

2, 6-Diaminopurine
2-aminoadenosine

19

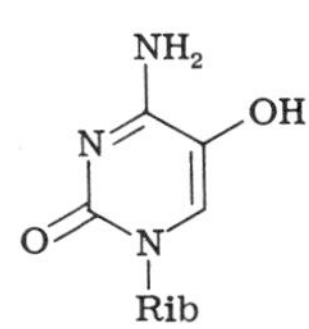

5-Hydroxycytidine

20

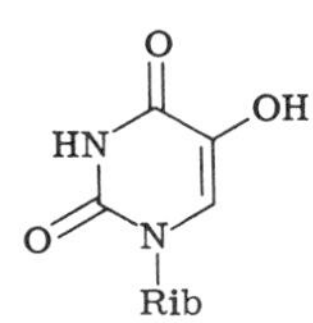

5-Hydroxyuridine

21

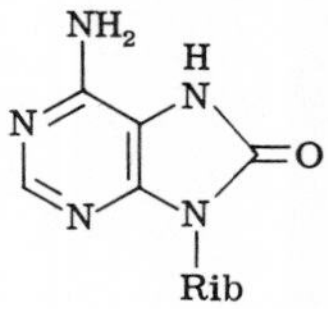

8-Oxyadenosine

22

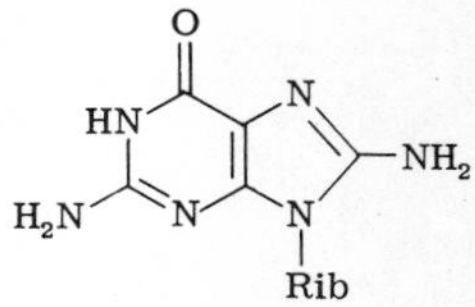

8-Aminoguanosine

23

and 8-aminoguanosine [**23**]. Ribosyl-2,6-diaminopurine [**19**] is 2-aminoadenosine.

2. Replacement of Hydrogen or Oxygen by Thio, Alkyl, Halide, Hydroxy, Methoxy, or Acetyl Groups

Sulfur has been substituted for oxygen in guanosine (6-thioguanosine) [**24**], inosine (6-thioinosine) [**25**], cytidine (2-thiocytidine) [**26**], and uridine (2-thiouridine, 4-thiouridine, 2,4-dithiouridine) [**27–29**]. Other thio derivatives are ribosyl-6-methylthiopurine, 2-methylthioinosine, and ribosyl-2,6-bis-(methylthio)purine [**30**]. Fluorine, chlorine, iodine, or bromine have been substituted at the 5 position of both cytidine [**31**] and uridine [**32**]. The 8 position of the purines has been similarly substituted by bromine [**33, 34**]. There are numerous alkyl derivatives, primarily modified by methylation or ethylation (Fig. 1). In addition to substituents on the nitrogens or oxygens of the four major bases [reviewed by Singer *(34)*], codons or polynucleotides have

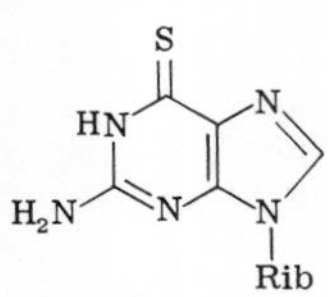

6-Thioguanosine

24

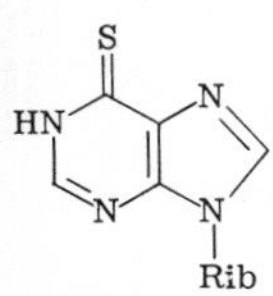

6-Thioinosine

25

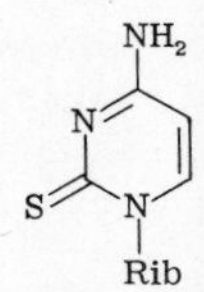

2-Thiocytidine

26

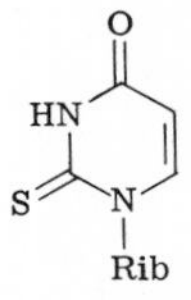

2-Thiouridine

27

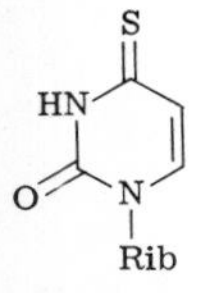

4-Thiouridine

28

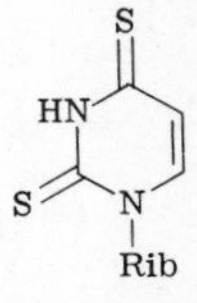

2,4-Dithiouridine

29

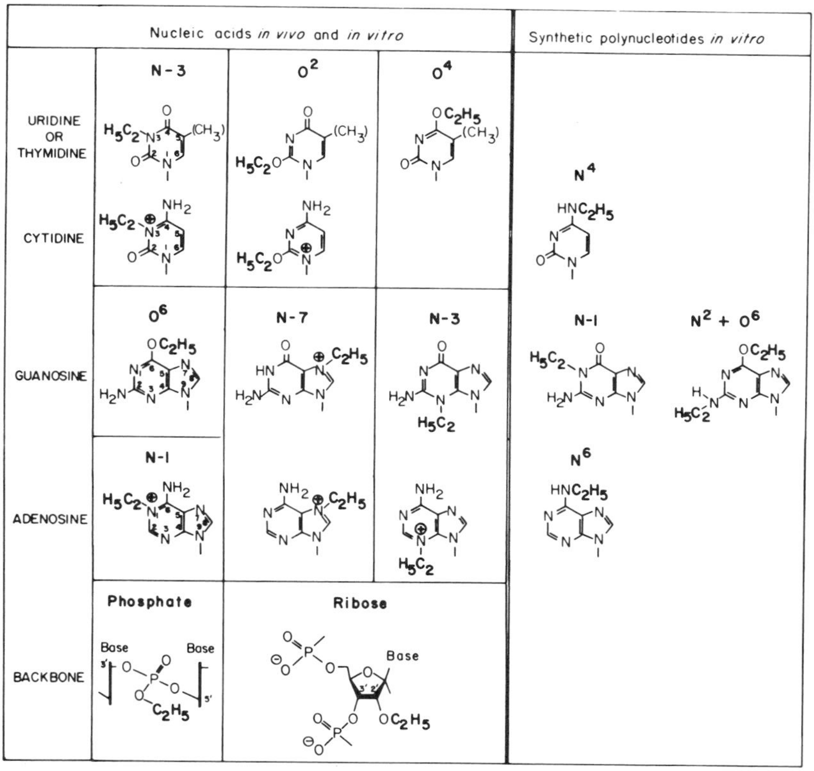

FIG. 1. Ethylated nucleosides (or bases) found as products of treatment of nucleic acids with ethylnitrosourea or *N*-ethyl-*N*′-nitro-*N*-nitrosoguanidine in neutral, aqueous solution. Four products found only after reaction of homopolynucleotides with the same alkylating agents are shown in the right side of the figure. Most of the sites that can be ethylated can also be methylated, but, in several instances, to a lesser extent (*34*).

been prepared using 1-methyl-6-thioguanosine, 7-methyl- and 3-methylxanthosines, 2,6-dimethyl-, 2-methyl-, N^6-hydroxy-, and 2-dimethylaminoadenosines, 1-methyl-, 7-methyl-, and 1,7-dimethylinosines, 5-ethyl- and 5-hydroxymethyluridines, ribosyl-6-methoxypurine and -6-methylpurine, N^4-acetyl-, N^4-hydroxy-, N^4-methoxy-, 5-ethyl-, and 5-hydroxycytidines.

3. Replacement by Complex Substituents

This list is restricted to substitutions studied in terms of base-pairing and excludes 2-(*N*-acetoxy)acetamidofluorene, benzo-[*a*]pyrene-7,8-dihydrodiol 9,10-epoxide, and aflatoxin modifica-

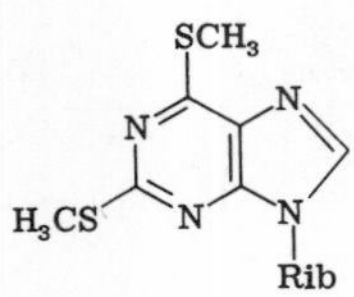

Ribosyl-2, 6-bis (methylthio)purine

30

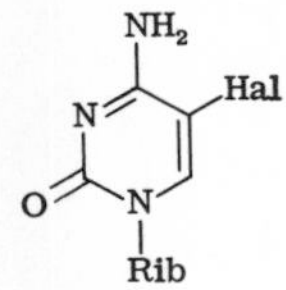

5-Halogeno-cytidine

31

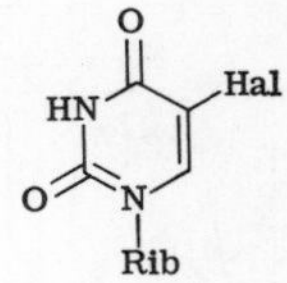

5-Halogeno-uridine

32

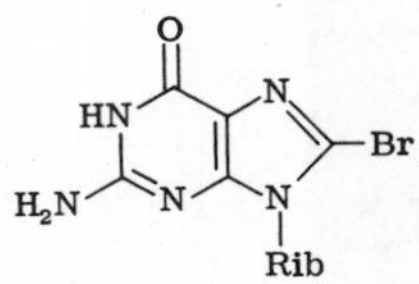

8-Bromoguanosine

33

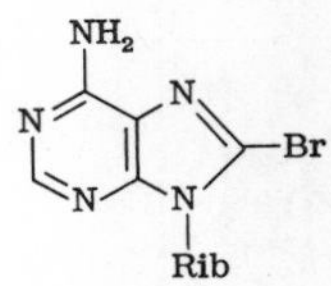

8-Bromoadenosine

34

tions, which are discussed by Grunberger and Weinstein in this volume. 1-Cyanoethylinosine, N^6-(Δ^2-isopentenyl)adenosine, and the methyl ester of 5-carboxymethyluridine (5-methoxycarbonylmethyluridine) are examples.

E. Addition Reactions

Addition products on the 5,6 double-bond of pyrimidines by water (UV irradiation), hydroxylamine, methoxyamine, bisulfite, semicarbazide, etc., form a separate class of derivatives. Since saturation of the pyrimidine 5,6 double-bond appears to prevent polymerization of the corresponding diphosphate by polynucleotide phosphorylase, these derivatives are only briefly discussed.

F. Cyclization

In 3,2′-*O*- and -*S*-cycloadenosine [35], 8,5′-*O*- and -*S*-cycloadenosine [36], and 6,2′-*O*-cyclouridine [37], the ribose is cyclized to the noncoding side of the base, while several other cyclic compounds such as 3,N^4-ethenocytidine and 1,N^6-ethanoadenosine have an additional ring on that side.

G. Ribose and Phosphate Modifications

The 2′ oxygen atoms of cytidine, uridine, guanosine, and adenosine [38] have been methylated and ethylated, while the 2′ hydroxyls of nucleosides have been replaced by azido or amino groups[3]

[3] See article by Sprinzl and Cramer in Vol. 22 of this series.

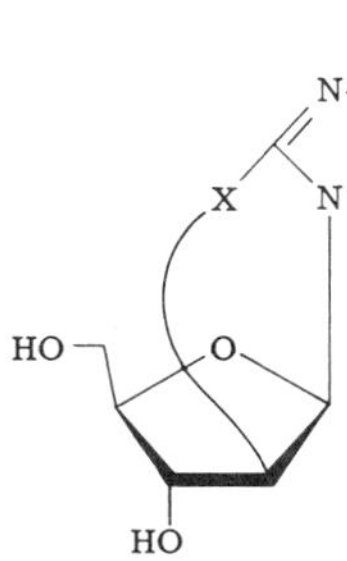

X = O or S

8, 2′-*O*(*S*)-
Cycloadenosine

35

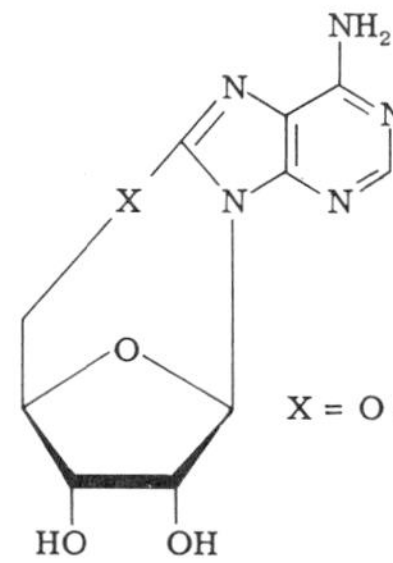

X = O or S

8, 5′-*O*(*S*)-
Cycloadenosine

36

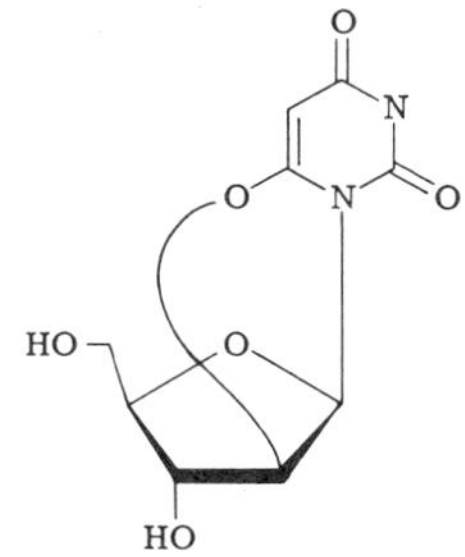

6, 2′-*O*-
Cyclouridine

37

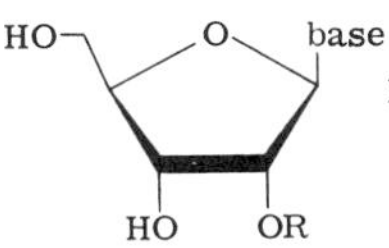

R = alkyl

2′-*O*-Alkyl-
nucleoside

38

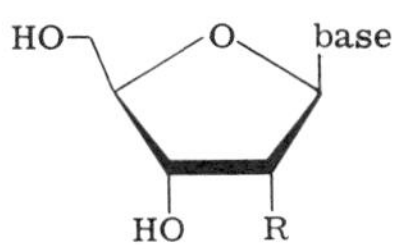

R = H, N_3, NH_2, halo

2′-Modified
nucleoside

39

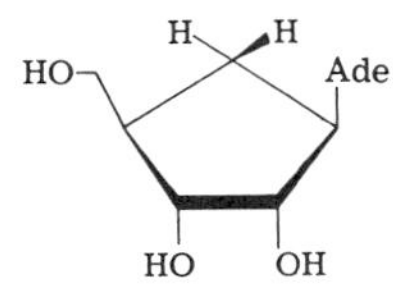

Aristeromycin

40

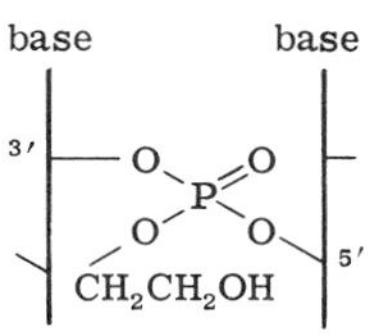

2-Hydroxy-
ethylphosphate

41

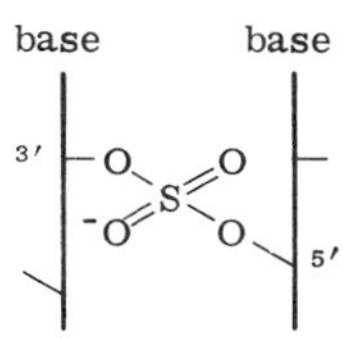

Sulfate

42

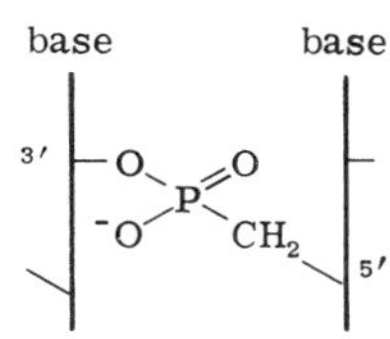

5′-Deoxy-5′-
phosphonate

43

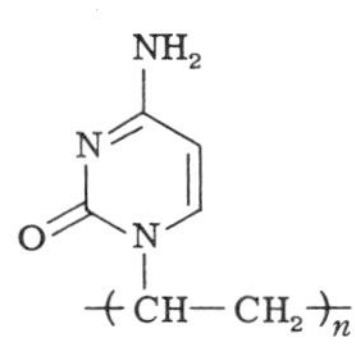

1-(Polyvinyl)cytosine

44

[39]. Aristeromycin has a carbon in place of the ring oxygen in the ribose moiety of adenosine [40]. The phosphate group can be modified in several ways. Alkylation can lead to substitution of a methyl or hydroxymethyl group for hydrogen [41]. Sulfur or carbon may be introduced by synthetic procedures, yielding sulfate [42] and phosphonate [43] derivatives. In 1-(polyvinyl)cytosine [44], both the ribose and the phosphate have been replaced.

H. Naturally Occurring Modified Nucleosides

Many of the derivatives in the preceding sections are found in various nucleic acids, the majority being in tRNAs [reviewed by Jukes (*35*)].[4] tRNAs also contain hypermodified bases in the "wobble" position of the anticodon, and these are discussed when pertinent. Modified nucleosides occurring in nucleic acids from numerous sources, in addition to tRNAs [as reviewed by Dunn and Hall (*36*)][4] are 1MeA, 2MeA, N^6MeA, N^6Me_2A, 2′MeA, 3MeC, 5MeC, 2′MeC, 1MeG, N^2MeG, 7MeG, N^2,N^2,7Me_3G, 2′MeG, 5MeU, 5,6H_2U, 2′MeU, and Ψ.

III. Complex Formation between Polynucleotides

A. Secondary Structure

Homopolymers of modified nucleotides generally exhibit differences in secondary structure compared to the parent homopolymer. These differences can be illustrated by poly (Ψ), which has much secondary structure, whereas poly(U) has almost no secondary structure; or by poly(s^2U) [27], which has a high stacking force, thus forming a stable self-helical molecule (*37*). On the other hand, poly(5-hydroxycytidylate [20] exhibits little secondary structure compared to poly(C).

The ability of modified polymers to form hydrogen bonds with complementary polymers is influenced by numerous factors, such as self-complementarity, stacking forces, and base orientation, as well as steric hindrance due to the substituent and the number of possible hydrogen bonds (see Section VI). Most of these factors are diminished, or even absent, when the modified base is incorporated in a polymer containing a high proportion of unmodified bases, as is the case in naturally occurring nucleic acids. Since this review is primarily directed toward the biological role of modified nucleosides, work on

[4] See the article by Singhal and Fallis on p. 227.

heteropolymers containing relatively small amounts of such nucleosides is stressed. However, the properties of homopolymers have been studied to a greater extent than those of heteropolymers, and data on homopolymers are valuable in assessing the physical–chemical changes produced by the incorporation of modified nucleosides. In some cases where a modification prevents a homopolymer from participating in complex formation or translation or transcription, the presence of 50% or more of an unmodified base in the polymer enables the modified base to function, although not necessarily in a predictable manner.

B. Comparison of Homo- and Heteropolymers

Table I lists the various modified polynucleotides that have been studied in terms of their ability to form complexes (*1, 2, 4–29, 37–68*). The data are in part clear and self-evident. However, in several instances, homopolymers of modified nucleotides form no complexes while heteropolymers containing the same nucleosides do form double-stranded structures. It is not clear in most of these cases whether the modified base in these heteropolymers is actually base-paired. Brimacombe and Reese (*27*) believe that m^4C in poly(C) does base-pair with poly(I) [albeit with a lower t_m than poly(C)·poly(I)], while m_2^4C in poly(C) is described as "looped out" in the complex with poly(I). Similarly, Ikehara and Hattori (*18, 21*) find that m^2G and m_2^2G form products they consider to be 1 : 1 complexes with poly(C) when they represent less than 50% of a poly(G). However, they do not postulate a double-stranded helix with loops but rather that m_2^2G is kept in the helical structure by intrastrand stacking forces.

TABLE I

FORMATION OF COMPLEXES OF POLYNUCLEOTIDES CONTAINING MODIFIED NUCLEOSIDES

Modified nucleoside	In copolymer with	Complexes with poly-						Other	References
		(A)	(G)	(U)	(C)	(I)	(X)		
Adenosines									
1-Methyladenosine	30% A[a]			−		−	+		(*11*)
	41% A[a]			−					(*38*)
	85% A[a]			+					(*38*)
	—						+		(*13*)
2-Aminoadenosine [**19**]	—			+					(*39*)
	—					+			(*40*)
	—			+				(s^4U)	(*41*)
2-Methyladenosine	—			+[b]					(*22*)
2,N^6-Dimethyladenosine	—			+[b]					(*42*)

(*Continued*)

TABLE I (*Continued*)

Modified nucleoside	In copolymer with	Complexes with poly- (A)	(G)	(U)	(C)	(I)	(X)	Other	References
2-Dimethylamino-adenosine	—			+[b]				(br^5U)[b]	(*43*)
N^6-Methyladenosine	30% A			−		−	+		(*11*)
	—						+		(*13*)
N^6-Methyldeoxy-adenosine	82% A							(T)	(*44*)
2-Hydroxyadenosine ("iso-guanosine") [**12**]	—	−	−	−	−	(+)	−		(*46*)
N^6-Hydroxymethyl-adenosine	—			−		−	+		(*8*)
N^6-Hydroxyethyl-adenosine	—						+		(*8*)
N^6-[Δ^2-Isopentenyl]-adenosine	—			−					(*45*)
N^6,N^6-Dimethyladenosine	—			−					(*26*)
"3-Iso-adenosine" [**11**]	—	−		+	−	+	+		(*13*); (*12*
8-Aza-9-deazaadenosine (formycin) [**2**]	—			+				(I)	(*47*)
8-Oxyadenosine [**22**]	—			+					(*48*)
2′-*O*-Methyladenosine [**38**]	—			+					(*49*)
2′-*O*-Ethyladenosine [**38**]	—			+				(Um)	(*49*)
2′-Azido-2′-deoxy-adenosine [**39**]	—			+					(*50*)
Purines (*as nucleosides*)									
2-Aminopurine [**14**]	—			+	−	−		(s^4U)	(*51*); (*41*
2-Amino-N^6-methoxy-purine	40–70% A			−					(*52*)
2,6-Bis(methylthio)-purine [**30**]	50% G				+				(*20*)
6-Methylpurine	—			−					(*53*)
6-Methoxypurine	—			−					(*53*)
6-Methylthiopurine	—			−					(*53*)
Purine (nebularin) [**17**]	—			−					(*53*)
7-Deazapurine (7-deazanebularin) [**10**]	—			−					(*54*)
	36–50% A			+					(*54*)
Inosines									
Inosine [**16**]	—	+			+			[c]	(*28*)
1-Methylinosine	70% I				−				(*14*)
7-Methylinosine	—	−			+			(br^5C)	(*11*)
1,7-Dimethylinosine	—	−			−				(*11*)
2-Methylthioinosine	—	+		−	−	−		(7-Deaza-A)	(*19*)

(*Continued*)

TABLE I (*Continued*)

Modified nucleoside	In copolymer with	Complexes with poly- (A)	(G)	(U)	(C)	(I)	(X)	Other	References
6-Thioinosine [**25**]	—				–				(*55*)
8-Aza-9-deazainosine (laurusin; deamino formycin; formycin B) [**1**]	—							Oligo (A)	(*56*)
Xanthosine [**15**]	—	+	–	+	–	+		[c]	(*13*)
2′-Deoxy-2′-fluoroinosine [**39**]	—				+				(*57*)
2′-*O*-Methylinosine [**38**]	—				+			(Cm)	(*49*)
Guanosines[e]									
1-Methylguanosine	—	–			–	–		Not (br^5C)	(*14*)
	60% G					+			(*14*)
N^2-Methylguanosine	—				–				(*14*)
	30% G				–				(*18*)
	45–75% G				+				(*18*)
N^2-Dimethylguanosine	—	–		–	–	–			(*16*)
	10–16% G				–				(*18*)
	50–85% G				+				(*18*)
	84% I				+				(*21*)
	65% A			+					(*21*)
O^6-Methylguanosine	—			–	–				(*58*); (*59*)
O^6-Ethylguanosine	—			–	–				(*58*); (*59*)
6-Thioguanosine [**24**]	—			–	(+)				(*29*)
	50% G[a]				+				(*29*)
7-Methylguanosine	—	–			+			(br^5C)	(*11*)
8-Aminoguanosine [**23**]	—	–			+				(*60*)
8-Bromoguanosine [**33**]	77% A			+					(*17*)
Uridines									
3-Methyluridine	—	–							(*1*)
	75% U	+							(*1*)
	95% U	+							(*1*)
3,5-Dimethyluridine	—	–							(*23*)
5-Halo [methyl,hydroxy]uridine	—	+							(*2*); (*10*)
5-Ethyluridine	—	+							(*6*)
5-Hydroxymethyluridine	67% U	+							(*23*)
5-Ribosyluracil (pseudouridine) [**3**]	—	+							(*9*)
2-Thiouridine [**27**]	—	–							(*37*)
4-Thiouridine [**28**]	—	+							(*24*)
	—							(2APuo)[d]	(*41*)
	—							(2,6-A_2Puo)[d]	(*41*)
	50% C		+						(*61*)

(*Continued*)

TABLE I (*Continued*)

Modified nucleoside	In co-polymer with	Complexes with poly- (A)	(G)	(U)	(C)	(I)	(X)	Other	References
2,4-Dithiouridine [29]	—	–							(*37*)
	—	–							(*25*)
2′-*O*-Methyluridine [38]	—	+						(Am)	(*49*)
2′-Amino-2′-deoxy-uridine [39]	—	–							(*62*)
2′-Azido-2′-deoxy-uridine [39]	—	+							(*63*)
2′-Fluoro-2′-deoxy-uridine [39]	—	+							(*57*)
Cytidines									
3-Methylcytidine	85% C[a]					+			(*27*)
N^4-Methylcytidine	—					–			(*27*)
	61% C					+			(*27*)
N^4-Acetylcytidine	80% C[a]					+			(*11*)
N^4-Hydroxycytidine	—	–		–					(*5*)
N^4,N^4-Dimethylcytidine	—					–			(*27*)
	87% C					+			(*27*)
5-Methylcytidine	—					+			(*4*)
5-Ethylcytidine	—					+			(*64*)
5-Hydroxycytidine [20]	—	–	–			–			(*65*)
5-Bromocytidine [31]	—					+			(*15*)
	—					+			(*10*)
5-Iodocytidine [31]	—					+			(*15*)
2-Thiocytidine [26]	—					+			(*66*)
	75% C					+			(*66*)
2′-*O*-Methylcytidine [38]	—					+		(Im)	(*49*)
	—							not (dI)	(*7*)
2′-*O*-Ethylcytidine [38]	—					+			(*67*)
2′-Amino-2′-deoxy-cytidine [39]	—					+			(*62*)
2′-Azido-2′-deoxy-cytidine [39]	—					+			(*62*)
2′-Deoxy-2′-fluoro-cytidine [39]	—					+			(*57*)
1-Vinylcytosine [44]	—					+			(*68*)
	90% U					+			(*68*)

[a] Chemically modified polymer.

[b] Evidence presented for Hoogsteen or reverse Hoogsteen bonding.

[c] Poly(I) and poly(X) complex with a variety of modified polynucleotides. Data are give by Michelson *et al.* (*28*).

[d] 2-APuo and 2,6-A_2Puo are abbreviations for the nucleosides of 2-aminopurine (6-deox guanosine [14]) and 2,6-diaminopurine, respectively.

[e] "Iso-guanosine" is listed under adenosines as 2-hydroxyadenosine.

The homopolymer poly(s^6G) [**24**] forms an unstable complex with poly(C), attributed to the fact that s^6G has a high tendency to complex with itself. Poly(G,s^6G), in a 1 : 1 ratio, forms a 1 : 1 complex with poly(C) that has a relatively high t_m of 85°C (*29*). This is explained as resulting from the influence of the sulfur on stacking rather than true base-pairing, as the complementary hydrogen bonding scheme proposed shows only two hydrogen bonds for the $s^6G \cdot C$ pair. It is postulated that the sulfur is not hydrogen-bonded to the 4-amino group of the cytidine. When 2,6-bis(methylthio)purine [**30**] is present in a copolymer with G, a 1 : 1 complex can be formed with poly(C) (*20*). In this case, the modified nucleotide is presumed to be held in the double-strand solely by stacking forces, and no hydrogen bonds appear possible.

Homopolymers of m^1G and of m^3U do not complex in any measurable way with other polynucleotides. However, in heteropolymers, complexes exhibiting a low t_m are described. The complex of poly(U,m^3U) (3 : 1) with poly(A) has a t_m of 34°C [poly(U)•poly(A) has a t_m of 58.5°C] (*1*), and that of poly(G,m^1G) (1.6 : 1) with poly(C) has a t_m comparable to oligo(G)·poly(C) (*14*). In both cases, the low t_m is taken as an indication that m^1G and m^3U are looped out of the double-strand. No hydrogen bonds are formed, and the stacking ability of the two modified nucleosides is believed to be too low to keep them in the helical form.

1-Methyladenosine in a homopolymer, or in a heteropolymer with adenosine, complexes with poly(X) (*11, 13*). No complex is formed with poly(U) except in a copolymer with 85% A (*38*). Thrierr and Leng (*38*) do not propose a specific hydrogen-bonded structure for the complex of poly(A,m^1A) (85 : 15) with poly(U), but instead discuss a helix with loops. It should be noted that the heteropolymers were obtained by chemical modification of poly(A), and it is likely that derivatives other than m^1A were also formed (*69*). In the case of poly(m^1A) itself, Michelson and Monny (*13*) did not report whether the homopolymer can complex with polynucleotides other than poly(X).

Poly(7-deazanebularin) [**10**] (DN) does not complex with poly(U) or poly(brU) but Ward and Reich (*54*) find DN to complex with poly(U) in copolymers in which the ratio of A to DN is 1 : 1 or 0.36 : 1. Again the t_m of the complex is lower than that of poly(A)·poly(U).

One other modified nucleoside studied in both homo- and heteropolymer forms differs from the examples given above. Poly(m^6A) should be able to form a normal hydrogen-bonded 1 : 1 complex with poly(U); however, it forms only a very weak complex with poly(U) (*26*), and it also complexes with poly(X) (*13*). On the other hand, when in a heteropolymer with 30% A, m^6A complexes only with poly(X), not with poly(U) (*11*). This unexpected behavior of polymers containing m^6A illustrates the complexity of base-pairing; this is discussed in detail in Section VI (p. 182).

C. Polynucleotides Modified on Sites Not Involved in Watson–Crick Base-Pairing

The remaining homopolymers in Table I can be grouped in terms of their ability to complex with polynucleotides. All but one of the many 5-substituted uridines and cytidines base-pair with poly(A) or poly(I), respectively (*2–4, 10, 15, 64*). Only poly(5-hydroxycytidine) [**20**] is incapable of complexing with poly(A), poly(G), or poly(I) (*65*). This is attributed to the fact that it has little secondary structure in neutral or acid solution. Polymers of 7-alkyl purines (m^7G and m^7I) behave similarly to the unmodified polymers and form 1 : 1 complexes with poly(C) (*11*).

Alkylation of the 2′-oxygen atom of ribose [**38**] does not prevent complex formation with the normal complementary polynucleotide. However, the t_m is lowered for poly(Cm) · poly(I) but increased for poly(Im) · poly(C) and poly(Um) · poly(A) (*49*). Ethylation of the 2′-O gives a greater stability to complexes (*49, 67*). Other replacements for the 2′-OH [**39**] have varying effects on complex formation. Poly(2′-fluoroinosine) · poly(I), poly(2′-fluorouridine) · poly(A), poly(2′-azidouridine) · poly(A), poly(2′-azidoadenosine) · poly(U) are all more stable than the corresponding unmodified double helices (*57, 63*).[5] Poly(2′-fluorocytidine) and poly(2′-azidodeoxycytidine) complex with poly(I) (*57, 62*), but the t_m is lower than that of poly(C)·poly(I). The azido derivative has a lower degree of stacking than poly(C), but no similar reason is expressed for the fluoro derivative. Replacement of the 2′-oxygen by an amino group [**39**] prevents complex formation, apparently due to steric hindrance (*62*) or to a change of the sugar to a more rigid conformation.

The three 8-substituted purine polynucleotides studied are poly(8-oxyadenosine) [**22**], which forms 1 : 1 and 2 : 1 complexes with poly(U) (*48*), poly(8-aminoguanosine) [**23**], which complexes only at alkaline pH with poly(C) (*60*), and poly(A,8-bromoguanosine) (77 : 23), which forms a complex with poly(U) in which it is probable that the 8-bromoguanosine [**33**] residues are looped out (*17*). The effect of the high secondary structure of poly(G) seems to be increased by the pres-

[5] 2′-Deoxy is omitted from each of these names, for clarity [Ed].

ence of an amino group in position 8. There is a strong tendency for the amino group in poly(8-aminoguanosine) to form a third interbase hydrogen bond, and only when this is prevented at higher pH can normal Watson–Crick bonding be observed (*60*). Since 8-BrGDP could not be polymerized except with ADP or GDP (*17*), there is no direct information on whether homopolymers of this derivative can in fact base-pair with poly(C).

The polymers discussed above are all modified on positions not involved in conventional Watson–Crick base-pairing. In general, such polymers form complexes with the usual complementary polynucleotide. In the few cases where this is not so, the various investigators have plausibly argued that the modification causes the polynucleotide to have too much secondary structure (8-aminoguanosine), or too little secondary structure (5-hydroxycytidine), or introduces steric factors (2′-amino nucleosides).

D. Polynucleotides of Analogs

Pseudouridine [**3**], "iso-adenosine" [**11**], formycin [**2**], and laurusin [**1**] all are nucleosides rearranged on the portion of the base not directly involved in Watson–Crick base-pairing. Homopolymers of two of these analogs form complexes with the expected complementary polynucleotide [(poly(Ψ)·poly(A); poly(iso-A)·poly(U); poly(iso-A)·poly(X)] (*9, 12, 13*). Although the distance between the N-glycosyl bonds is that of a pyrimidine–pyrimidine base-pair, the complex formation between iso-adenosine and uridine seems to be primarily due to the unchanged order of hydrogen bonds.

The polymer of formycin,[2] an adenosine analog, does not complex with poly(U), but this is attributed to the base being in the *syn* conformation, not to the lack of potential hydrogen bonds (*47*). Inasmuch as formycin triphosphate can substitute for ATP, an alternating copolymer of formycin and U could be prepared (*70*). This polymer had the same t_m as poly(A·U), indicating that formycin forms normal hydrogen bonds with U. An even more complicated polymer is polylaurusin (polydeaminoformycin), which complexes with an oligonucleotide of 8,2′-*S*-cycloadenosine [**35**] (*56, 71*).

Iso-guanosine [**12**], 2-hydroxyadenosine, has a very high preference for self-association, a fact noted already for other purine derivatives; only protonated poly(iso-guanosine) is able to complex weakly with poly(I) at pH 2.6–3.3 (*46*).

E. Polynucleotides Modified on Sites Involved in Formation of Watson–Crick Base-Pairs

Modification at the N-1, C-2, and N^6 of adenine, at the N-1, N^2, and O^6 of guanine, at the O^2, N-3, and O^4 of uracil, and at the O^2, N-3, and N^4 of cytosine are expected to disturb normal base-pairing (Fig. 2). In most cases, any substituent blocking a hydrogen acceptor site or substituting for a proton prevents Watson–Crick base-pairing, although Hoogsteen or reverse Hoogsteen base-pairing can sometimes be demonstrated. However, in Section III, B it is indicated that 1-methyl-, N^2-methyl-, and N^2-dimethylguanine, 3-methyluracil, and N^4-dimethylcytosine in heteropolymers do not form hydrogen bonds, although the unmodified ("carrier") base in the polymer does participate in Watson–Crick base-pairing. *In vivo,* modified bases exist in heteropolymers that should prevent the modified base from odd hydrogen bonding. Thus, in this section we do not discuss complex formation when it is clear from the investigator's data that Watson–Crick base-pairing does not occur. A footnote to Table I indicates these non-Watson–Crick complexes.

The modifications that can be made on the positions listed above that still allow Watson–Crick base-pairing include deamination of the N^2 of guanosine (xanthosine), deamination of the N^6 of adenosine (inosine), isomerization of adenosine to ribosyl-2-aminopurine, addition of an amino group to the 2-position of adenosine (2-aminoadenosine), and possibly substitution of the O^4 of uridine by sulfur (4-thiouridine).

Two modifications that have greater effects than predicted are substitution of one hydrogen of the 4-amino group of cytidine and of the 6-amino group of adenosine. Poly(N^4-hydroxycytidine) does not complex with poly(I), poly(U), or poly(A), thus not acting like C or U (5). Since poly(N^4-hydroxycytidine), when deaminated to poly(U), complexes normally with poly(A) (5), it is difficult to atttribute the lack of complex formation to any factor other than steric hindrance. In contrast, a heteropolymer of cytidine and N^4-acetylcytidine forms a 1 : 1 complex with poly(I) (*11*), with a t_m identical to that of poly(C)·poly(I). The conclusion is reached that N^4-acetylcytidine residues are hydrogen bonded in normal fashion. N^4-Methylcytidine in a heteropolymer with cytidine (39 : 61) is probably hydrogen-bonded to poly(I) (*27*), but the t_m of the complex (37°C) is considerably lower than that of the unmodified complex (52°C), which leave the issue unresolved.

As discussed in Section III, poly(m^6A) base-pairs with poly(X) (*11, 13*), but not unequivocally with poly(U) (*26*). Both poly(N^6-hydroxymethyladenosine) and poly(N^6-hydroxyethyladenosine) also

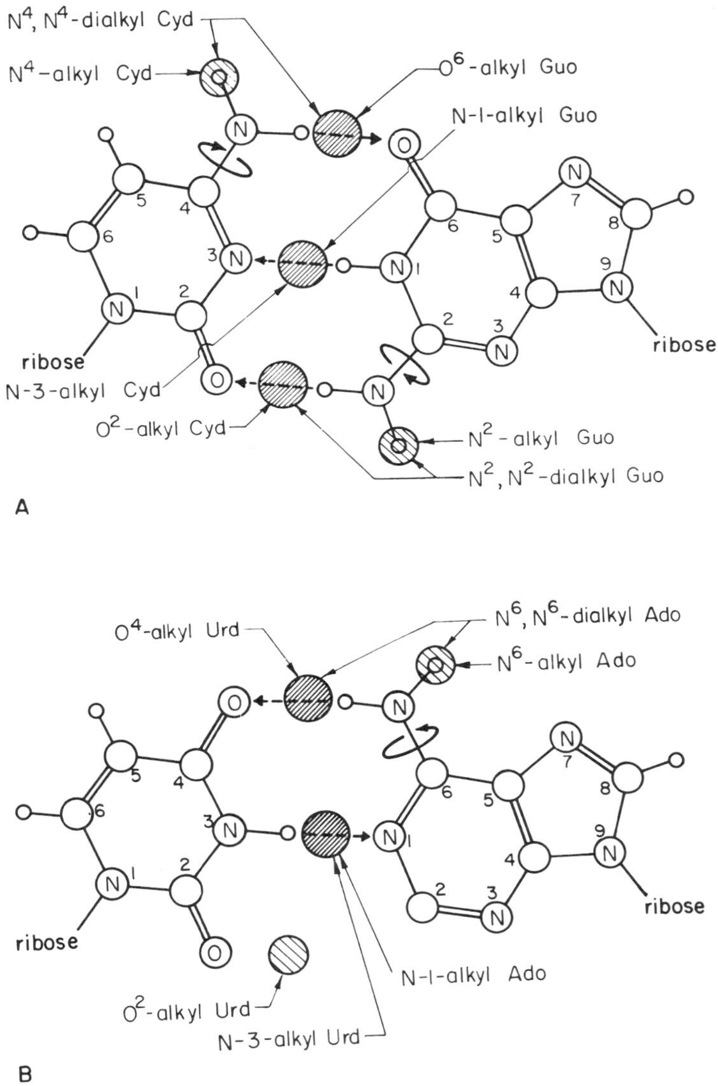

FIG. 2. Schematic representation of Watson–Crick base-pairs with various sites modified. (A) G·C base-pair. (B) A·U base-pair. The large, heavily shaded circles represent positions of alkyl group on sites involved in Watson–Crick hydrogen bonding. The lightly shaded circles represent positions of alkyl groups not generally involved in Watson–Crick hydrogen bonding. Rotational freedom of exocyclic amino groups, with or without alkyl groups, are indicated by circular arrows. The names of nucleosides resulting from alkylation of different sites are given in the figure.

form 1 : 1 complexes with poly(X), but not with poly(U) (*8*). Similarly, formaldehyde modification of poly(A) to form increasing amounts of the N^6-hydroxymethyl derivative leads to decreasing interaction with poly(U) (*72, 73*). These results are attributed to orientation of the N^6 substituent causing steric hindrance. However it appears that the m^6A residues in poly[d(A,m^6A)] (9 : 1) do in fact base-pair with poly(dT) (*44*).

From the conflicting data regarding base-pairing of N^4-modified cytidine and N^6-modified adenosine, it seems that no general conclusion can be drawn without information on the orientation of each substituent (see Fig. 2).

IV. Codon–Anticodon Interaction

This section is quite broad and is arbitrarily divided into three subsections, all related to the effect of modification on the ability of a nucleoside in a polynucleotide, codon, or anticodon to participate in protein synthesis. Table II presents data for both the messenger activities of polynucleotides and the codon-stimulated binding of tRNAs to ribosomes (*8, 9, 12, 31, 32, 37, 47, 53, 70, 74–102*).

A. Polynucleotides Containing Modified Bases

There are several modifications that do not affect the normal coding function of polymers. 5-Halo- and 5-methyluridines and 5-halo- and 5-methylcytidines all direct polyphenylalanine or polyproline synthesis, respectively (*84, 92, 93, 97, 98*). The only polynucleotide of this class tested for all other possible messenger activities was 5-fluorouridine, which gave negative results for amino acids having A, G, or C in the first two codon positions (*93*). Similarly poly(fl^5C) does not act as poly(U) (*97*), and poly(br^5C) does not act as poly(A) (*92*). This unchanged character of 5-halo pyrimidines was expected inasmuch as polymers of these derivatives all complex with the same complementary polynucleotide as do unmodified poly(U) and poly(C) (Table I). The only 5-substituted pyrimidines that are poor messengers are 5-hydroxyuridine and 5-hydroxycytidine, but poly(5-hydroxyuridine) directed some polyphenylalanine synthesis (*92*), and poly(5-hydroxycytidine) directed some polyproline synthesis (*98*).

Six derivatives of adenosine that function in polynucleotides as adenosine are 3-deazaadenosine, 7-deazaadenosine (tubercidin), formycin, nebularin, 7-deazanebularin, and 2-aminopurine. However, formycin in a homopolymer is inactive for lysine incorporation but is active in heteropolymers (*47, 70*). Both poly(3-deaza-A) and poly(7-

TABLE II
EFFECT OF MODIFIED NUCLEOSIDES ON RIBOSOME BINDING OF tRNA OR TRANSLATION

Modified nucleoside	Tested in[a]	Substitutes for A	G	U	C	References
Adenosines						
2-Methyladenosine	Homopolymer	−				*(74)*
N^6-Methyladenosine	Homopolymer	−		−		*(8)*
N^6-Hydroxyethyladenosine	Homopolymer	−				*(8)*
	Heteropolymer	+				*(8)*
N^6,N^6-Dimethyladenosine	Codon	−	−	−	−	*(75)*
"3-Iso-adenosine" [**11**]	Homopolymer	−	−	−	−	*(12)*
	Heteropolymer	−	−	−	−	*(12)*
1-Deazaadenosine [**4**]	Homopolymer	−				*(76)*; *(74)*
3-Deazaadenosine [**5**]	Homopolymer	+				*(76)*; *(74)*
7-Deazaadenosine [**6**]	Codon	+	−		−	*(75)*
	Homopolymer	+			−	*(74)*
	Homopolymer	±	−	−		*(76)*
8-Aza-9-deazaadenosine (formycin) [**2**]	Codon	+				*(74)*
	Homopolymer	+				*(74)*
	Homopolymer	−				*(70)*; *(47)*
	Heteropolymer	+	−	−	−	*(77)*
	Heteropolymer	+				*(70)*; *(47)*
8-Bromoadenosine [**34**]	Codon	−				*(74)*
	Homopolymer	−				*(74)*
8-Oxyadenosine [**22**]	Codon	(±)				*(74)*
	Homopolymer	±				*(74)*
Purines (as nucleosides)						
2-Aminopurine [**14**]	Homopolymer	+	+			*(78)*
6-Methylpurine	Homopolymer	−				*(53)*
6-Methoxypurine	Homopolymer	−				*(53)*
6-Methylthiopurine	Homopolymer	(±)				*(53)*
Purine [**17**] (nebularin)	Codon	+				*(53)*
	Homopolymer	+				*(53)*
	Homopolymer	+	±	−	−	*(76)*
7-Deazapurine [**10**] (7-deazanebularin)	Codon	+	±			*(79)*
	Homopolymer	+				*(79)*
	Heteropolymer	+	+			*(79)*
Inosine and xanthosine						
Inosine [**16**]	Doublet		−			*(80)*
	Codon	−	−	−		*(81)*
	Codon		±			*(80)*
	Codon			−	−	*(82)*
	Heteropolymer		+			*(83)*
Xanthosine [**15**]	Heteropolymer	−	−	−	−	*(8)*; *(84)*

(Continued)

TABLE II (*Continued*)

Modified nucleoside	Tested in[a]	Substitutes for A	G	U	C	References
Guanosines						
6-Thioguanosine [**24**]	Homopolymer	–	–	–	–	(*85*)
8-Azaguanosine [**7**]	Codon		+			(*86*)
	Heteropolymer	–	±			(*87*)
Uridines						
3-Methyluridine	Doublet	–				(*88*)
	Doublet			–	+	(*89*)
	Codon			–		(*90*)
	Codon			–		(*91*)
	Homopolymer			–		(*8*)
	Heteropolymer			–		(*84*)
	Heteropolymer	–		–		(*8*)
5-Methyluridine (ribothymidine)	Doublet			+		(*32*)
	Codon			+		(*90*)
	Homopolymer			+		(*92*)
*O*²-Ethyluridine	Doublet			±	±	(*89*)
*O*⁴-Methyluridine	Doublet			±	±	(*89*)
5-Fluorouridine [**32**]	Homopolymer	–		+	–	(*93*)
	Homopolymer			–		(*84*)
	Heteropolymer	–	–	+	–	(*93*)
	Heteropolymer			+		(*84*)
5-Bromouridine [**32**]	Doublet			+		(*32*)
	Codon	–		+	–	(*91*)
	Codon			+		(*90*)
	Homopolymer			+		(*92*)
5-Chlorouridine [**32**]	Homopolymer			+		(*92*)
5-Iodouridine [**32**]	Codon	–		+	–	(*91*)
	Homopolymer			+		(*92*)
5-Hydroxyuridine [**21**]	Homopolymer			±		(*92*)
Dihydrouridine	Codon			–		(*94*)
	Heteropolymer	–	–	–	–	(*95*)
5-Ribosyluracil (pseudouridine) [**3**]	Homopolymer			–		(*9*)
	Heteropolymer			+		(*9*); (92)
2-Thiouridine [**27**]	Codon			+	–	(*96*)
	Homopolymer			–		(*37*)
4-Thiouridine [**28**]	Codon			–		(*96*)
	Homopolymer			±		(*37*)
2,4-Dithiouridine [**29**]	Codon			–		(*96*)
	Homopolymer			–		(*37*)
4-Deoxyuridine [**18**]	Codon			±		(*31*)
	Homopolymer				–	(*53*)
3-Deaza-4-deoxyuridine	Codon			–		(*31*)
	Homopolymer				–	(*31*)
3-Deazauridine [**8**]	Codon			±		(*31*)
6-Azauridine [**9**]	Doublet			–		(*32*)

(*Continued*)

TABLE II (*Continued*)

Modified nucleoside	Tested in[a]	Substitutes for				References
		A	G	U	C	
ytidine						
-Methylcytidine	Doublet			+	−	(*89*)
^{2}-Ethylcytidine	Doublet			+	−	(*89*)
-Fluorocytidine [**31**]	Heteropolymer			−	+	(*97*)
-Bromocytidine [**31**]	Homopolymer	?			+	(*98*)
	Homopolymer	−			+	(*92*)
-Hydroxycytidine [**20**]	Heteropolymer				±	(*98*)
-Thiocytidine [**26**]	Codon				±	(*96*)
	Homopolymer				−	(*99*)
	Heteropolymer				+	(*100*)
ibose or phosphate						
'-*O*-Methyladenosine [**38**]	Homopolymer	−				(*101*)
	Heteropolymer	±	−		−	(*101*)
'-*O*-Methyluridine [**38**]	Homopolymer			−		(*101*)
'-*O*-Methylcytidine [**38**]	Homopolymer				−	(*101*)
	Heteropolymer				+	(*101*)
risteromycin [**40**]	Heteropolymer	+				(*74*)
'-Deoxyuridine [**39**]	Doublet			−		(*80*)
	Codon			+		(*80*)
'-Deoxy-5'-phosphonouridine [**43**]	Codon			−	−	(*102*)
ridine 5'-phosphonate	Doublet			+		(*80*)
ytidine 5'-phosphonate	Doublet				+	(*80*)
ethyl 5'-uridylate	Doublet			−		(*80*)
ethyl 5'-cytidylate	Doublet				+	(*80*)
ridine 5'-methylphosphonate	Doublet			−		(*80*)
Hydroxyethyl 5'-uridylate [**41**]	Doublet			+		(*80*)
ridine 5'-sulfate [**42**]	Doublet			+		(*80*)
ytidine 5'-sulfate [**42**]	Doublet				+	(*80*)

[a] Homo- and heteropolymer are polyribonucleotides used as messengers in *in vitro* protein osynthesis experiments or for studies on specificity of tRNA binding. "Codons" are tri-ıcleoside diphosphates (NpNpN), and "doublets" are dinucleoside diphosphates (pNpN). studies on doublets the dinucleoside diphosphate substitutes for the first two nucleosides a codon. The "wobble" base is deleted.

deaza-A) are modified on sites not involved in Watson–Crick hydrogen bonding, and their ability to stimulate the binding of $tRNA^{Lys}$ to ribosomes (*74, 76*) is not surprising. It is surprising that polynebularin, which lacks the ability to form more than one hydrogen bond, stimulates the binding of $tRNA^{Lys}$ as much as does poly(A) (*53*). Additionally, there was a lesser ability to stimulate $tRNA^{Arg}$ (*76*), indicating that nebularin can, to a limited extent, act like guanosine.

7-Deaza-nebularin in a homopolymer also acts like adenosine (*79*), but in heteropolymers with A (at ratios to A of 1 : 1 and 1 : 3), it is apparent that it also acts like G and stimulates $tRNA^{Arg}$ binding to ribosomes (*79*). This is one clear example of ambiguity resulting from modifications that leave only one possible normal hydrogen-bonding site.

Another example of ambiguity is found with polymers containing 2-aminopurine. The homopolymer stimulates lysyl-, glutamyl-, and arginyl-tRNA binding, indicating that 2-aminopurine acts like A or G (*78*). The same polynucleotide forms complexes with poly(U) but not with poly(I) (Table I). The preference for substituting for poly(A) in complex formation is also shared by 7-deazanebularin, which (in copolymers with A) complexes with poly(U) (Table I).

Nebularin, 7-deazanebularin and ribosyl-2-aminopurine, considered as adenosine analogs, all lack the 6-amino group; in polynucleotides, all can substitute for G, but probably less well than for A.

There are very few other modified nucleosides that, in polymers, can function as messengers. Inosine can substitute for guanosine (*83*), while xanthosine has no activity (*8*, *84*). Pseudouridine in a heteropolymer, but not in a homopolymer, can direct polyphenylalanine synthesis (*9*, *92*). Similarly N^6-hydroxyethyladenosine can substitute for adenosine in a heteropolymer, but not in a homopolymer (*8*). Presumably the secondary structure of the homopolymers is unfavorable. 8-Oxyadenosine and 8-azaguanosine have diminished abilities to substitute for adenosine or guanosine, respectively (*74*, *87*). Of the various thio derivatives, poly(s^4U) directs phenylalanine incorporation, although to lesser extent than does poly(U) (*37*). Poly(6-methylthiopurine) has a very low, but detectable, effect on $tRNA^{Lys}$ binding (*53*).

Homopolymers of 2′-*O*-methyladenosine, -uridine, and -cytidine (Am, Um, Cm) are all inactive as messengers (*101*). When copolymerized with U, Cm stimulates incorporation of proline as well as does the control, poly(U,C) (*101*). In heteropolymers with C, Am is less effective in stimulating amino-acid incorporation (codons with A and C), but is nevertheless active, in contrast to the homopolymer (*101*).

In each of the instances where a homopolymer of a modified nucleotide is inactive while the heteropolymer containing that nucleotide is active, it is suggested that secondary structure of the homopolymer is responsible. In many cases where the homopolymer is inactive as a messenger, no comparable results are given for a heteropolymer, and it is quite possible that erroneous conclusions may be drawn regarding base-pairing. Candidates for heteropolymer study are those containing 6-thioguanosine, 8-oxyadenosine, and 5-hydroxyuridine, all of which, in homopolymers, complex with the complementary polynucleotide, but are inactive or poorly active as messengers.

B. Doublets and Triplets Containing Modified Bases

Following the successful use of triplets as codons to elucidate the amino-acid code, triplets containing modified bases were prepared and tested for their ability to stimulate specific tRNA binding to ribosomes. Generally, the codons were compared to those with normal bases, and only occasionally did investigators also look for possible ambiguity or for a completely changed character.

Whenever it was observed that a modified nucleoside in the first or second position of a codon could substitute for its unmodified counterpart, the same was almost always also true in polynucleotides. Examples in Table II are 7-deazaadenosine, formycin, nebularin, 7-deazanebularin, 8-azaguanosine, 5-methyl- and 5-iodouridine. Conversely, negative results with codons generally paralleled results with polymers; such is the case for 8-bromoadenosine, dihydrouridine, 2,4-dithiouridine, 3-methyluridine, and 2-thiocytidine. Some disagreements were noted as it was not clear whether inosine in a codon functioned as guanosine (*80, 81*). 2-Thiouridine acted as uridine in a codon (*96*) but not in a homopolymer (*37*), whereas 4-thiouridine could not substitute for uridine in a codon (*96*) but did so in a polynucleotide (*37*).

Holý *et al.* observed that dinucleoside diphosphates (pNpN) can act as codons (NpNpN) whenever the third base in the anticodon is not specific (*80*). The doublet pGpU stimulated $tRNA^{Val}$ binding and pGpC stimulated $tRNA^{Ala}$ binding, with no ambiguity. In doublets as in codons, br^5U and m^5U (T) substituted for U, while m^3U did not (*32*). Inosine did not substitute for uridine (*80*), and 6-azauridine (not previously tested) was also unable to function as uridine (*32*).

Turning their attention to the role of the phosphate and sugar in doublets, Holý *et al.* found that substitution of dU for U in the pGpU doublet (but not a triplet) was completely inactivating (*80*) as was 5′-deoxy-5′-phosphonouridine in the middle position of GUU (*102*). Neither modification should change the possible hydrogen bonding, but they perhaps do change conformation. Many substituted 5′-phosphates in doublets were tested, with electronegatively charged substituents increasing binding (phosphate, phosphite, sulfur), while bulky and hydrophobic electronegatively charged substituents (methyl phosphate, methanephosphonate) have little effect, as does a

3′-phosphate (*80*). Adding a 5′-phosphate to many codons greatly increased binding, and under these circumstances inosine acted like guanosine (*80*). The stabilizing effect of phosphates in codon–anticodon interaction probably explains why inosine substitutes for guanosine in a heteropolymer but not in a normal codon, except in the "wobble" position. Although there are no data, it is likely that a 5′-phosphate on a codon containing 4-thiouridine would enable this base to function like uridine, as poly(4-thiouridine) has a detectible ability to substitute for uridine in protein synthesis (*37*).

It was fortunate that doublets can be used as codons, as Singer and co-workers were able to prepare only doublets, but not triplets, containing the newly elucidated O-alkyl pyrimidines. These labile modified nucleosides were poor substrates for all the enzymes used for condensation, but the methods established by Grunberger *et al.* and Uchida and Funayama-Machida (*103, 104*) were applicable and a series of doublets was prepared using O^2-ethyl- and 3-methylcytidines, and O^2-ethyl-, O^4-methyl-, and 3-methyluridines (*89*). All doublets with pG in the first position were tested with two different tRNAs (Fig. 3). 3-Methyluridine did not act as uridine, which agreed with previous data obtained with triplets. However, pG-m^3U stimulated $tRNA^{Ala}$ binding, showing that m^3U acts like C. In contrast, m^3C did not act like C, but instead acted like U. Similarly, O^2-ethylcytidine strongly stimulated $tRNA^{Val}$ binding, thus also acting like U. Both O^2- and O^4-alkyluridines showed ambiguity and could partially substitute for U or C. All the cytidines and uridines modified on the coding side of the pyrimidine were, by this criterion, mutagenic (Fig. 3) (*89*).

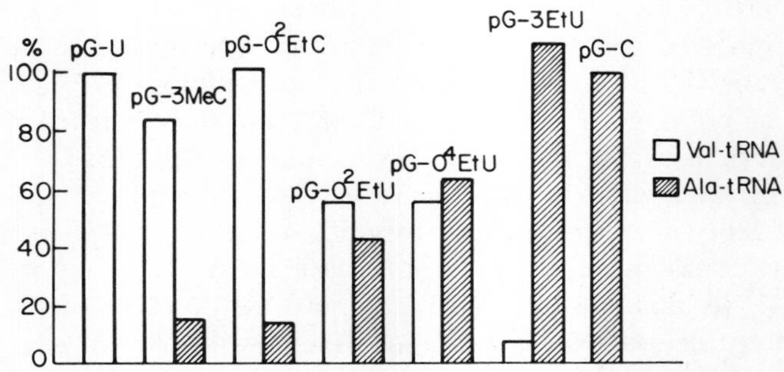

FIG. 3. Effect of alkyl pyrimidine-containing doublets in stimulating binding of Val-tRNA and Ala-tRNA to ribosomes. The amount of Val-tRNA (in picomoles) bound under the influence of pGpU (0.76 pmol) and of Ala-tRNA by pGpC (0.30 pmol) have each been given as 100%. The relative stimulation of binding, under the same conditions, with other doublets is shown by the height of the bars (*89*).

C. tRNA

At first glance, the interaction between tRNA and mRNA during the ribosomal process of translation looks quite simple: The three nucleotides of the codon search for three complementary nucleotides in the anticodon of the appropriate tRNA. But as more and more details about this process became known, it became evident that protein biosynthesis is considerably more complicated, and many points are presently unclear.

The sequences of about 120 tRNAs are now known (*105*),[4] and a discussion of the large number of modified bases found in tRNAs exceeds the limits of this review. However, there are some recent reports on the structure–function relationship of modified bases in tRNAs that are helpful in understanding their base-pairing abilities. In restricting this section to base-pairing we do not discuss problems such as specificity of aminoacylation, ribosomal binding sites, conformational changes, requirement of certain factors, or "capping" (*106*).

The first polynucleotide to yield to X-ray crystallography was $tRNA^{Phe}$ (*107–109*). Subsequently, details about the classical wobble pair G·U, which is present in the aminoacylation stem of $tRNA^{Phe}$, were elucidated, showing that, rather than no interaction or a tautomeric shift, there is a change in the sugar conformation such that the hydrogen at the N-1 of G interacts with the O^2 of U and the hydrogen at N-3 of U with the O^6 of G (*110*). This is the first demonstration of a "forced" base-pair—forced, as this interaction is likely to occur only because on both sides there are thermodynamically stable base-pairs.

Although there is a variety of modified bases in position 34 (the wobble position)[4] in the anticodon of tRNAs, it is very obvious that the two other positions, with only one exception, show no modifications. In tyrosine-specific tRNA, Ψ replaces U in two out of four reported sequences (*105*).[4] However, it is known from other studies that Ψ base-pairs like U (*9*); consequently, this is a very special case of modification.

Of all the codons for amino acids, only the methionine initiator tRNA has but one sequence of three nucleotides in the anticodon: all nine species sequenced have CAU in positions 34–36.[4] Obviously, to maintain specificity, no modification that may influence the base-pairing properties may be expected.

Tryptophan and methionine tRNAs are also limited to single sequences; only those modifications occur in the anticodon that do not change base-pairing, i.e., N^4-acetyl- and 2′-*O*-methylcytidine. Experiments discussed in earlier sections (see Tables I and II) indicate that these two substitute for cytidine and no other nucleoside in heteropolymers (*11, 49, 101*). N^4-Acetylcytidine may have additional functions and may influence the aminoacyl-tRNA synthetase to prevent charging with fMet (*111*). This is supported by an earlier finding that deamination of C (to U) in position 34 completely prevents aminoacylation (*112*). Cm in this position may have a similar function, although it is not present in all $tRNA^{Met}$ molecules (*113*). The two cysteine and twelve phenylalanine tRNAs sequenced have G or Gm in the wobble position, but none has A, although the code allows this (*105*).

From such comparisons of the anticodons of sequenced tRNAs, it seems that the old "two-letter code" of Roberts (*114*) should be reconsidered, with modifications. And, in fact, Mitra, Lagerkvist, and colleagues recently reported experimental evidence for this theory (*115, 116*). In addition, Weissenbach *et al.* (*117*) postulated pyrimidine–pyrimidine interaction, and Jank *et al.* (*118*) found the preferential occurrence of an I·G interaction, neither of which should be allowed under the conventional wobble rules.

Therefore, it may well be that the wobble hypothesis should now be viewed as stating that a hypermodified base in the wobble position simply functions to prevent the formation of a third base-pair. This is also suggested by Jank *et al.* (*118*). But, as Richards (*119*) summarizes, additional experimental data are needed to establish the "two-letter code" for certain amino acids.

Such additional support may already be provided by those X-ray data from Gassen's group (*120–122*). From their data, it can be argued that any modification of a uridine in the wobble position prevents proper base-pairing. These workers do not conclude this, but it is obvious from their data that a severe disorientation always results. We would call this "ambiguity," as we did the behavior of 3-methylpyrimidines and 1-methylpurines (*123*). (For details, see Section V.)

V. Transcription of Polynucleotides

Relatively few nucleic acid derivatives have been studied in terms of their ability to direct transcription. The principle of the method is to copolymerize, using polynucleotide phosphorylase, a normal nucleoside diphosphate ("carrier") with a relatively small amount of a modified nucleoside diphosphate. The resulting copolymers are transcribed with DNA-dependent RNA polymerase in the presence of Mn^{2+}. The use of Mn^{2+} rather than Mg^{2+} allows transcription of a ribopolynucleotide (*124*). We have performed such transcription experiments using all four NTPs with a different single NTP labeled in each of multiple tests. This we term "competitive conditions." We also used two NTPs, and this we term "noncompetitive." Ludlum and collaborators generally used two NTPs, or "noncompetitive" conditions, but occasionally a third triphosphate was included; these experiments Ludlum terms "competitive." Our experience is that the use of only two NTPs can give results that are not confirmed when all four normal NTPs are present. We have, in Table III, distinguished between the various experimental approaches and used brackets for results that are obtained only when a single labeled NTP is used in addition to the NTP complementary to the carrier (*98*, *100*, *123*, *125–131*). Table III includes those experiments from Ludlum's laboratory in which three

TABLE III
TEMPLATE ACTIVITY OF MODIFIED NUCLEOSIDES INTRODUCED INTO POLYNUCLEOTIDES[a]

Synthetic polynucleotide			In transcription, substitutes for[b]				
Modified nucleoside	%	In copolymer with	A	G	U	C	References
Adenosines							
1-Methyladenosine	8%	C	+	±	+	[+]	(*123*)
	9%	U	±			+	(*123*)
N^6-Methyladenosine	8%	C	+	[±]	[+]	[−]	(*123*)
	11%	U	+			−	(*123*)
2′-*O*-Methyladenosine [**38**]	100%	—	[−]				(*125*)
Inosine and Xanthosine							
Xanthosine [**15**]	13%	C	+	−	−		(*123*)
	~15%	U				−	(*123*)
2′-*O*-Methylinosine [**38**]	100%	—		[−]			(*125*)
Guanosines							
7-Methylguanosine	3%, 5%	C	−				(*126*)
	2%, 4%	U		+			(*126*)
O^6-Methylguanosine	5%	C	+	−	+		(*127*)
O^6-Methyl-2′-deoxyguanosine	0.4%	C	+	−			(*128*)
Uridines							
3-Methyluridine	7%	C	+	±	+		(*123*)
	4%	U				±	(*123*)
5-Fluorouridine	33%	A		−	+		(*123*)
O^2-Ethyluridine	4–8%	U		+		+	(*129*)
	8%	C	−	+	−		(*129*)
O^4-Methyluridine	4–13%	U		+		+	(*129*)
	7–13%	C	−	+	−		(*129*)
2′-*O*-Methyluridine [**38**]	100%	—			[+]		(*125*)

(*Continued*)

TABLE III (*Continued*)

Synthetic polynucleotide			In transcription, substitutes for[b]				
Modified nucleoside	%	In copolymer with	A	G	U	C	Reference
Cytidines							
3-Methylcytidine	6%, 10%	C	+				(*130*)
	4%	C	+	+	+		(*131*)
	13%	C	+	+	+	[+]	(*123*)
	17%	U				±	(*123*)
3-Ethylcytidine	4%, 6%	C	+				(*130*)
N^4-Hydroxycytidine	4%	C	−	−	+		(*131*)
N^4-Methoxycytidine	6%	C	−	−	+		(*131*)
5-Fluorocytidine [**31**]	20–67%	C			−	+	(*131*)
5-Bromocytidine [**32**]	100%	—	−	−	−	+	(*98*)
5-Hydroxycytidine [**21**]	15%	C	−	−	−	+	(*98*)
2-Thiocytidine [**26**]	100%	—				−	(*100*)
	10%	C	−	−	−	+	(*100*)
2′-*O*-Methylcytidine [**38**]	100%	—				[±]	(*125*)

[a] Experiments using chemically modified polynucleotides are not included, as it is difficul to eliminate the possibility of multiple reactions occurring.

[b] All experiments were performed with DNA-dependent RNA polymerase in the presenc of Mn^{2+}. Brackets are used to designate data from experiments in which only two nucleosid triphosphates were used, one complementary to the carrier and one other. This type c noncompetitive experiment, in contrast to the use of all four triphosphates, can give result that are not reproduced under the more biological or competitive situation. Experiment from Ludlum's laboratory (*130*) were generally performed with three nucleoside diphosphate

triphosphates are used. Only homopolymers of 2′O-methylated nucleosides were studied, and only the expected complementary NTP was used.

In some cases, [α-^{32}P]NTPs were used and nearest-neighbor analyses established that the labeled nucleotides were internally incorporated in the complementary polynucleotide. This type of experiment was of great importance in establishing that the 3-alkyl pyrimidines and 1-alkyl purines directed any nucleotide into a complementary copolymer (Table IV) (*123*). However, none of the experiments can be used to distinguish between normal base-pairing and other mechanisms leading to incorporation (Section VI).

As in other types of experiments, 5-fluorocytidine [**31**], 5-bromocytidine [**31**], and 5-hydroxycytidine [**20**] directed only incorporation of G, thus acting like C (*97, 98*). N^4-Hydroxycytidine and

TABLE IV
INCORPORATION OF NUCLEOTIDES IN TRANSCRIPTION OF POLY(C) CONTAINING m^3C, m^3U, m^1A, AND m^6A

Template	Modified nucleotide (%)	Nearest-neighbor analysis % of total ^{32}P incorporation[a]			
		Ap	Cp	Up	Summation[b]
Poly(C,m^3C)	11	1.5	2.7	2.3	6.5
Poly(C,m^3U)	12	1.6	0.2	3.8	5.6
Poly(C,m^1A)	12	1.3	0.6	3.8	5.7
Poly(C,m^6A)[c]	8	—	—	4.5	4.5

[a] Transcription was performed using [α-^{32}P]GTP and unlabeled ATP, UTP, and CTP (*123*). Since about 95% of the transcript is found as Gp, data for misincorporation of Gp are not directly obtainable when the "carrier" is poly(C), but it could be calculated that m^3C directed Gp into the transcript. Similar experiments using poly(U) as "carrier" and [α-^{32}P]ATP showed that m^1A directed incorporation of Gp, while m^6A did not.

[b] Total incorporation of nucleotides that are not complementary to the "carrier" does not include possible Gp incorporation directed by the modified nucleotide. See footnote *a*.

[c] No misincorporation is found in transcription unless only GTP and one other NTP are present ("noncompetitive" conditions).

N^4-methoxycytidine behaved as U or C, as a result of tautomerism. There is a considerable amount of other evidence that supports the tautomeric shift we observe (*132–134*). 7-Methylguanosine behaved as G, not as A (*126*), while 2-thiocytidine [**26**] paired like C, not like any other nucleoside (*100*).

Some results of transcription experiments do not wholly agree with other work on the base-pairing properties of modified nucleosides, but can be rationalized. For instance, xanthosine in a polynucleotide can complex with poly(A), poly(U), or poly(I) (Table I), while in protein synthesis poly(X) is inactive (Table II). However, poly(X) can be transcribed, incorporating only UMP. Poly(m^6A) complexes with poly-(X) but not with poly(U). In contrast, in transcription with all four NTPs present, m^6A acts only like A, as would be expected, since m^6A should retain the base-pairing ability of A (Table IV). If, instead of four NTPs, a copolymer of C and m^6A is transcribed using GTP and one other triphosphate, C or A can also be incorporated into the complementary strand (Table III). This ambiguity is not likely to occur *in vivo*, but the fact that apparent mispairing can be forced indicates that the presence of the alkyl group does have an effect.

Modifications on the hydrogen-bonding sites of nucleosides in templates would be expected to prevent or inhibit transcription, as normal hydrogen-bonding cannot occur. It was therefore unexpected that in polynucleotides m^1A, m^3C, and m^3U directed the incorporation of any nucleotide into a complementary polynucleotide (*123*) (Table IV). This type of ambiguity was also found for O^2-alkyl-U, O^4-alkyl-U, and O^6-alkyl-G, all of which can substitute for at least two different nucleosides (Table III). In polymers, O^2-alkyl-U and O^4-alkyl-U direct G and C incorporation while O^6-alkyl-G directs A and U incorporation. As illustrated in Fig. 2, none of these alkylated nucleosides is capable of hydrogen-bonding with its normal partner, and none is capable of forming more than one hydrogen-bond with more than one other nucleoside. Yet nucleotides are almost randomly directed into a complementary polymer. This is not an occasional event but occurs with almost the same frequency as expected from the amount of modified base in the template (Table IV). We conclude that the RNA polymerase does not recognize the modified base, but instead tries all NTPs for fit in order to continue transcription. In this process, almost any triphosphate can be incorporated. Any preference for two or three triphosphates is likely to be a result of stacking forces contributed by the neighboring nucleotides (*123*).

Support for the hypothesis that transcriptional errors are not specific and are polymerase-directed comes from the codon–anticodon ribosome binding data previously discussed (Table II; Fig. 3). Here we find that m^3U substitutes for C but not U, while in transcription m^3U substitutes for U, as well as for other nucleosides. Similary O^2- and O^4-alkyl-Us substitute for C or U in codons but apparently not for U in transcription.

Thus, substitutions on the N-1 of adenosine, the O^6 of guanosine, the N-3, O^2, and O^4 of uridine, and the N-3 and O^2 of cytidine all give rise to mutagenic compounds.

VI. Parameters of Base-Pairing

The central postulate of the very specific formation of Watson–Crick base-pairs is generally accepted. One would expect that spontaneous and chemically induced mutagenesis could also be explained by the parameters for base-pair formation. We concern ourselves here only with observations on those polymers containing modified nucleosides that use natural nucleoside triphosphates for transcription and tRNA for translation, respectively.

The parameters of base-pairing appear to be (*a*) the ability to form a certain number of hydrogen bonds, (*b*) tautomerism, (*c*) sugar conformation, (*d*) base orientation, (*e*) steric hindrance, (*f*) appropriate enzyme activity, and (*g*) participation of neighboring groups. Each of these parameters we discuss separately.

A. Number of Hydrogen Bonds

The observation that (G·C)-rich DNA and double-stranded RNA have a considerably higher t_m than [A·U(T)]-rich nucleic acids is considered to be due to an additional hydrogen bond in the G·C pair. However, this additional bond does not increase the specificity, nor does it lead to a substantially increased lifetime of the complex between mRNA and tRNA. Other factors also influence complex stability (*28*), and it is difficult to draw conclusions for the *in vivo* situation from only one determinable parameter, namely t_m.

The influence of the number of hydrogen bonds on the specificity of base-pairing has been investigated, using a variety of modified polynucleotides. We summarize the data collected in Tables I–III as follows.

1. The addition of an amino group to adenosine to form 2-aminoadenosine [**19**] increases the thermodynamic stability (*39–41, 135*), while the specificity to act as adenosine is unchanged. This is also the case for naturally occurring 2-amino-2′-deoxyadenosine which replaces 2′-deoxyadenosine in the DNA of S-2L cyanophage (*135*). As expected, the t_m of this phage DNA is substantially higher than that of other DNAs. The actual function of "2-aminoadenosine" is not understood.
2. Studies of a series of deaza derivatives of adenosine (*76*) indicate that the nitrogen at position 1 is necessary for the nucleoside to be recognized as adenosine whereas replacement of the N-3 or the N-7 nitrogen by carbon has no effect, presumably because these changes do not influence the ability to form the characteristic adenosine hydrogen bonds.
3. Loss of the amino groups of adenosine (*53, 76*) and guanosine has a less dramatic effect. Thus nebularin [**17**] in polymers acts like adenosine, although only one hydrogen bond per pair can be formed. Inosine [**16**] acts only as guanosine in all *in vitro* translation or transcription experiments, but the homopolymer, poly(I), forms a complex with poly(A). This is not surprising since the geometry of an I·A pair is that of an A·U pair, except that the distance between the N-glycosyl bonds is increased. However, the formation of a I·A pair is not expected to occur during transcription or replication since polymerases

construct nucleic acids with a constant distance between N-glycosyl bonds (*136*).

4. The absence of the N-3 of uridine has been tested in only one experiment. 3-Deazauridine in codons had a very limited ability to replace uridine, although it was recognized that only one hydrogen bond could be formed with A (*31*).

5. Absence of the 4-amino group of cytidine or of the 4-oxygen of uridine yields ribosylpyrimidone [**18**], which does not act as cytidine but as a weak uridine (*31*). Ribosylpyridone (3-deaza-4-deoxyuridine) shows no binding activity or specificity. Here obviously hydrogen bonding to a single keto group is not strong enough for base-pairs to be formed (*31*).

6. Probably the most interesting derivative in terms of changed base-pairing is the product of removal of the 6-oxygen of guanosine, ribosyl-2-aminopurine [**14**], which is also the product of the removal of the 6-amino group of 2-aminoadenosine [**19**]. Ribosyl-2-aminopurine [**14**] can form two hydrogen bonds and it could theoretically behave like G or A. In complex formation, it acts like A (*41, 51*), but in messenger activity it can substitute for both A and G (*78*). This is the first reported example of ambiguity.

From this fairly complete series of deletions of functional groups, one may draw the conclusion that one hydrogen bond may be enough to form a specific base-pair. However, this hydrogen bond must be the specific one between the N-3 of the pyrimidine and the N-1 of the purine. When a hydrogen is present at the N-1 of a purine, it will act like guanine. If N-1 is a hydrogen acceptor, it acts like adenine. Similarly, a hydrogen at N-3 of a pyrimidine leads to uracil activity, while acceptor activity at N-3 yields cytosine activity.

This basic rule is followed with all naturally occurring nucleic acids, including the thio nucleosides. Only when this acceptor–donor orientation is disturbed by additional steric factors do we observe exceptions.

B. Tautomerism

Tautomerism describes a change in the position of specific hydrogens, which influences the hydrogen acceptor and donor ability. This is made clear by the studies and calculations of Topal and Fresco (*137*), who suggested that the keto–enol tautomerism of unmodified bases is an important factor in spontaneous mutations. Richards (*138*) pointed out that the 5-halo pyrimidines [**31, 32**] may exhibit an enhanced keto–enol tautomerism, which could explain their ability to enhance the rate of spontaneous mutations (*98*). This appears to be a theoretical conclusion, since in most systems the 5-substituted pyrimidines behave like their unmodified parent nucleosides (Tables I–III) (*139, 140*). However, no *in vitro* testing system is sensitive enough to measure the increase in the tautomer from 5×10^{-5} to 5×10^{-4}, which is the order of magnitude given by Topal and Fresco (*137*) for tautomeric equilibria. Such small amounts of "wrong" tautomers can be detected only *in vivo*. Classical examples of a chemically induced tautomeric shift resulting in similar levels of both forms are N^4-hydroxy- and N^4-methoxycytidines, which in transcription act like U as well as C (see Section IV).

C. Steric Factors

1. Sugar Conformation

The most impressive demonstration of the flexibility of the nucleic acid structure is the natural occurrence of the G · U "wobble" base-pair in $tRNA^{Phe}$ [see Section III (*135*]. This base-pair is possible only if the ribose is allowed to change from the preferred conformation. The X-ray structure of $tRNA^{Phe}$ gave definitive proof for this conformation (*110*). High-resolution NMR studies[6] (*141*) also showed a wobble base-pair with a forced sugar conformation in the double helix, poly[d(G · T)]. However, it was made clear that this helix has very little stability and is unlikely to occur in a Watson–Crick helix unless it is thermodynamically forced into the helical structure.

The changed sugar conformation in the "C-nucleosides" pseudouridine [**3**] and formycin [**2**] does not prevent base-pairing. Ψ acts like U and formycin acts like A (*9, 74, 77*). Aristeromycin [**40**], which resembles adenosine but with ribosyl replaced by a cyclopentyl radical, also acts like A (*74*). In general, the substitution of the 2′-hydroxyl group by hydrogen, as in DNA, has no effect on the base-pairing abilities. Little effect is observed for all 2′-O-alkyl or other 2′-substituted nucleosides (*39, 49, 50, 57, 63, 67*). There is, however, one exception: A 2′-amino group unexpectedly prevents the formation of a Watson-Crick double helix (*62*). Presumably additional hydrogen bonds are formed that change the structure of the polymer by influencing the sugar conformation.

2. Base Orientation

Results obtained using chemically modified polynucleotides are often questionable, as most chemical reactions are not absolutely

[6] See article by Kearns in Vol. 18 of this series.

specific. In the same way, results obtained from synthetic polymers may be questioned when the nucleoside is polymerized in the *syn*-conformation. This property probably accounts for the fact that poly(8-bromoguanosine) [**33**] has not been synthesized; however, the monomer can be incorporated into a heteropolymer with A as "carrier." Similar properties have been reported for formycin [**2**], 8-bromoadenosine [**34**], and 8-hydroxyadenosine [**22**]. Studies with polycyclic hydrocarbons indicate that the *syn–anti* conformation plays a critical role in the mechanism of carcinogenesis.[7]

3. Added Alkyl Groups

Alkylation of nucleic acids in cells has been widely used in studies on chemical carcinogenesis. Simple alkylating agents such as dimethyl sulfate alkylate primarily the nitrogens, while N-nitroso alkyl compounds modify oxygens more than nitrogens (*142*), and many products that interfere with Watson–Crick base-pairing are thus formed (*34*).

(i) The products of monoalkylation of the amino groups of C, A, and G retain the base-pairing specificity of the parent nucleosides. As illustrated in Fig. 2, the monoalkylated amino group has rotational freedom. Thus, in experiments on polymer complex formation (Table I), and in transcription under competitive conditions when all four nucleosides are present (Table III), N^4-alkyl-C, N^6-alkyl-A, and N^2-alkyl-G behave like the parent nucleosides. However, when transcription experiments are performed with N^6-methyl-A containing polymers in the presence of only two triphosphates, a completely different result is obtained. Now N^6-methyladenosine also directs the incorporation of adenosine and, to a smaller extent, that of cytidine (Table III (*123*). This behavior of N^6-methyl-A may be explained by the existence of two isomers [called *cis* and *trans* isomers by Engel and von Hippel (44)]. The *trans* isomer is obviously the preferred conformation, and under competitive conditions a normal A · U(T) base-pair is formed. If no U is present, the *cis*-isomer may determine which base is accepted in the complementary strand. Since the methyl group in the *cis* isomer interferes sterically, a nonspecific incorporation is observed (Table III).

In the case of substitution of both hydrogens of the exocyclic amino groups, the data on transcription available from the literature are incomplete. It is not clear whether the presence of dialkyl adducts in heteropolymers prevents base-pairing (*18*). The exocyclic amino groups still have their rotational freedom, which would allow them,

[7] See the article by Grunberger and Weinstein on p. 105.

theoretically, to avoid steric hindrance in base-pairing, leaving the donor or acceptor activity at the N-1 of A or G and N-3 of C unchanged.

(ii) Alkylation of the base oxygens has two different effects: addition of a sterically interfering alkyl group, which should be capable of rotational freedom; and the disappearance of the N-3 hydrogen of uridine or the N-1 hydrogen of guanosine and the formation of a cationic pyrimidine ring system in cytidine (*34*). Thus all four known O-alkyl bases have a changed hydrogen donor or acceptor activity. Therefore, it is not surprising that each O-alkyl nucleoside loses the base-pairing specificity of the appropriate parent nucleoside.

Although three of the O-alkyl nucleosides appear to be able to base-pair only with a single nucleoside, experimental evidence indicates that O^2- and O^4-alkyluridines substitute for both guanosine and cytidine, while O^6-alkylguanosine substitutes for both uridine and adenosine (Table III). It is not possible to explain the ability of O-alkyl nucleosides to substitute for more than one nucleoside by any single base-pairing scheme. The influence of O-alkylation must therefore be twofold: removal of hydrogen leading to the loss of the base-pairing character of the parent nucleoside; and the presence of a sterically bulky alkyl group preventing a new specificity. It should be noted that, in codon–anticodon experiments (*89*), the O-alkyl uridines again behave ambiguously, but in this case they substitute for U and C. The difference between transcription and codon binding data may be attributed to the different influence of polymerases and ribosomes.

O^2-Ethylcytidine behaves only like U, not like C, in ribosome binding (*89*). This result, like those of other O-alkyl nucleosides, cannot be explained by any simple base-pairing scheme. More work is needed in order to understand this behavior of O^2-alkylcytidine, but the lability of the N-glycosyl bond of this cationic nucleoside (*143*) has not allowed the preparation of polynucleotides containing O^2-alkylcytidine.

(iii) The lack of base-pairing specificity found with O-alkyl nucleosides is even more obvious if the ring nitrogens on the base-pairing side are alkylated. 3-Methyluridine, 3-methylcytidine, and 1-methyladenosine lose almost all specificity in transcription experiments, along with all hydrogen acceptor–donor abilities (see also Fig. 2). Studies on the influence of removal of hydrogen-bonding capability clearly illustrate the importance of the hydrogen bond between the pyrimidine N-3 and the purine N-1. In addition to removing donor and acceptor functions, the alkyl group prevents the correct positioning of a nucleoside triphosphate in transcription, and of tRNA in translation.

Thus it is not unexpected that all of these N-alkylated nucleosides show at least partial ambiguity and are devoid of all specificity for base-pairing (*123*). Unfortunately, 1-methylguanosine cannot be classified in the same way, as, even in heteropolymers, it strongly inhibits transcription (*123*). Whether this inhibition is complete and thus constitutes a lethal modification, or whether it only slows the process of transcription to below the level of detection of *in vitro* experiments, remains to be clarified.

4. Sulfur Substitution

The substitution of sulfur for oxygen does not generally change base-pairing properties. If other factors such as secondary structure do not intervene, all monothio derivatives of U [**27, 28**], C [**26**], and G [**24**] do not differ in base-pairing ability from the unmodified nucleosides (*24, 29, 41, 96, 100*). The only known dithio derivative, 2,4-dithiouridine [**29**], does not substitute for uridine (*25, 37*), but it seems not to have been tested in heteropolymers.

Any influence on base-pairing of the sulfur in naturally occurring thionucleotides in tRNA may therefore also be ruled out. Although many hypermodified 2-thiouridines occur in the wobble position of anticodons, they may only help to prevent the formation of normal base-pairs. The mechanism of this is not yet clear, but it may be due to the enhanced stacking force of all thionucleosides (Section IV, C).

The importance and influence of the enhanced stacking force of 5-methyl-2-thio-uridine is unequivocally demonstrated in $tRNA^{fMet}$ of a thermophilic bacterium (*144, 145*).[4] With one single replacement of oxygen by sulfur in the TψC-loop, the thermal stability of the required tertiary structure is provided and the bacterium is able to grow at temperatures up to 80°C; its $tRNA^{fMet}$ has a t_m of 86.5°C, 9°C higher than that of the corresponding $tRNA^{fMet}$ of *E. coli* (*144*).

The role of 4-thiouridine [**28**] in position 8 of *E. coli* tRNAs[4] is, however, still not known (*146*).

D. Influence of Neighboring Bases and Enzymes

As already pointed out (in Section III), the formation of base-pairs is often dependent on the secondary structure (stacking) of the polynucleotide. This is particularly the case for the sulfur-containing nucleosides, which have a high degree of stacking (*100*). When negative results were obtained with their homopolymers in transcription experiments, heteropolymers containing them could often be transcribed successfully (see Tables II and IV).

Interactions between nucleotides that appear insignificant may become pertinent under special circumstances, such as the recognition of oligonucleotide sequences by proteins or other nucleic acids. For example, the occurrence of the "forced" G · U wobble base-pair was discussed above (Section IV, C). Many other tertiary interactions in $tRNA^{Phe}$ (*108, 109*) were the first proof for the natural occurrence of non-Watson–Crick base-pairs. But it should be pointed out that all of these interactions exist only as a part of the complete tertiary structure, and such may also be true for other RNAs such as ribosomal RNAs and possibly DNA (*147*). The importance of tertiary interactions can also be illustrated by the influence of a single methyl group at the 2′-hydroxyl of guanosine in the D-loop of mammalian $tRNA^{Ser}$. The presence or absence or this methyl group appears to be responsible for organ specificity (*148*).

Neighboring bases also have a strong influence on the base-pairing properties of an individual base in experiments on codon–anticodon interactions. The ambiguity in the behavior of the wobble base can be attributed to a new specificity introduced by the neighboring bases. Neighboring bases can be either in the anticodon itself (*149*) or the hypermodified base next to the anticodon (*150*). Another possible explanation for the wobble ambiguity is that modification in this position eliminates all specificity (Section III, C). In codon–anticodon interaction, the fact that the minimal size for codon activity is a 5′ dinucleoside diphosphate indicates the importance of the sugar-phosphate backbone in positioning and binding (*80, 88, 89, 151*).

The influence of DNA and/or RNA polymerases in transcription is important in base recognition (*152*). The error rate of these enzymes has been studied extensively, primarily by Loeb and colleagues (*153, 154*). One reason for the error rate *in vivo* is attributed to the natural tautomerism of the bases (*137*). Environmental changes can also contribute to transcriptional errors. For example, an increased error rate is seen in the presence of Mn^{2+} or Co^{2+} (*153*), which presumably substitute for Mg^{2+} in the Mg-triphosphate complex.

The polymerases are likely to have a control function during the process of replication or transcription. They demand a specific distance between the N-glycosyl bonds, thus allowing only pyrimidine–purine interactions (*136*). Recent experiments from our laboratory show that, under certain circumstances, the enzyme is capable of placing a purine opposite a purine, presumably without hydrogen bonding. This is found when templates containing 1-methyladenine are transcribed (*123*).

Another experimental approach that offers convincing evidence that polymerases can transcribe templates lacking bases capable of hydrogen bonding is that of Shearman and Loeb (*155, 156*). Poly[d(A · T)] and poly[d(G · C)] that were partially depurinated were prepared. The resulting apurinic sites can be considered as "holes" in the template. One might have speculated that this would lead to a total loss of template activity, yet the DNA-polymerase did not release the template at these sites, but instead incorporated any triphosphate into the complementary strand, although with varying efficiency. It is speculated that stacking forces direct the triphosphate incorporation (*156*).

VII. Concluding Remarks

Modified nucleosides, including analogs that do not have large substituents [such as 2-(*N*-acetoxy)acetamidofluorene[7]] on the non-Watson–Crick sites, generally can base-pair with the same specificity as the comparable unmodified nucleosides. When normal base-pairing is prevented by such modification, it is ascribed to steric effects or to disturbed stacking interactions. Such inhibitory effects are greatly decreased if the modified nucleoside is a minor component in a heteropolymer, as is likely *in vivo*. Thus mispairing by modification on non-Watson–Crick sites has not been observed and would not be expected.

Another group of derivatives can also form hydrogen bonds with complementary nucleosides, but in these cases, the modification is such that Watson–Crick base-pairing with either of two nucleosides can occur. Thus, in translation, ribosyl-2-aminopurine, nebularin, and 7-deazanebularin can all act as A or G.

Changed base-pairing resulting from modification is found for inosine, xanthosine, and N^4-hydroxy- or N^4-methoxycytidine. These derivatives exemplify classical mutation resulting from chemical modification.

The last class of modified nucleosides includes derivatives that are modified on positions normally involved in Watson–Crick base-pairing, and they were long regarded as lethal modifications. That is, no base-pairing was assumed to take place even in a heteropolymer. Recent research has shown that modifications at the N-1 of purines, the N-3 of pyrimidines, and the oxygens of G, U, and C do allow transcription, but with little or no specificity for the base introduced into a complementary polynucleotide. In this respect, nucleosides incapable of normal base-pairing resemble depurinated sites in DNA, which also lead to random incorporation of any nucleotide. This total ambiguity differs from that of 2-aminopurine, the ambiguity of which arises from the fact that two specific hydrogen bonds can be formed with U or C. The role of an occasional modified nucleoside in the translation or transcription of a polynucleotide need not involve formation of any hydrogen bonds, but can occur through other forces, which include the participation of the polymerase, or the ribosome, or stacking forces contributed by adjacent bases.

Acknowledgments

The authors are grateful to Dr. H. Fraenkel-Conrat for his patience, interest, and helpful suggestions. This work was supported by Grant CA 12316 from the National Cancer Institute. Dr. Manfred Kröger is a fellow of the Deutsche Forschungsgemeinschaft.

References

1. W. Szer and D. Shugar, *Acta Biochim. Pol.* **8**, 235 (1961).
2. W. Szer and D. Shugar, *Acta Biochim Pol.* **10**, 219 (1963).
3. W. Szer, *BBRC* **20**, 182 (1965).
4. W. Szer and D. Shugar, *JMB* **17**, 174 (1966).
5. C. Janion and D. Shugar, *Acta Biochim. Pol.* **15**, 107 (1968).
6. M. Swierkowski and D. Shugar, *JMB* **47**, 57 (1970).
7. B. Zmudzka, M. Tichy, and D. Shugar, *Acta Biochim. Pol.* **19**, 149 (1972).
8. A. M. Michelson and M. Grunberg-Manago, *BBA* **91**, 92 (1964).
9. F. Pochon, A. M. Michelson, M. Grunberg-Manago, W. E. Cohn, and L. Dondon, *BBA* **80**, 441 (1964).
10. J. Massoulié, A. M. Michelson, and F. Pochon, *BBA* **114**, 16 (1966).
11. A. M. Michelson and F. Pochon, *BBA* **114**, 469 (1966).
12. A. M. Michelson, C. Monny, R. A. Laursen, and N. J. Leonard, *BBA* **119**, 258 (1966).
13. A. M. Michelson and C. Monny, *BBA* **129**, 460 (1966).
14. F. Pochon and A. M. Michelson, *BBA* **145**, 321 (1967).
15. A. M. Michelson and C. Monny, *BBA* **149**, 88 (1967).
16. F. Pochon and A. M. Michelson, *BBA* **182**, 17 (1969).
17. M. Ikehara, I. Tazawa, and T. Fukui, *Bchem* **8**, 736 (1969).
18. M. Ikehara and M. Hattori, *Bchem* **10**, 3585 (1971).
19. M. Ikehara and M. Hattori, *BBA* **269**, 27 (1972).
20. M. Ikehara and M. Hattori, *BBA* **281**, 11 (1972).
21. M. Ikehara and M. Hattori, *BBA* **287**, 9 (1972).
22. M. Ikehara, M. Hattori, and T. Fukui, *EJB* **31**, 329 (1972).
23. K. H. Scheit, *BBA* **134**, 17 (1967).
24. J. Simuth, K. H. Scheit, and E. M. Gottschalk, *BBA* **204**, 371 (1970).
25. P. Faerber and K. H. Scheit, *FEBS Lett.* **11**, 11 (1970).
26. B. E. Griffin, W. J. Haslan, and C. B. Reese, *JMB* **10**, 353 (1964).
27. R. L. C. Brimacombe and C. B. Reese, *JMB* **18**, 529 (1966).
28. A. M. Michelson, J. Massoulié, and W. Guschlbauer, This Series **6**, 83 (1967).
29. V. Amarnath and A. D. Broom, *Bchem* **15**, 4386 (1976).
30. A. J. Lomant and J. R. Fresco, This Series **15**, 185 (1975).
31. H. G. Gassen, H. Schetters, and H. Matthaei, *BBA* **272**, 560 (1972).

32. D. Grunberger, A. Holý, and F. Šorm, *BBA* **157**, 439 (1968).
33. D. Grunberger, A. Holý, and F. Šorm, *BBA* **161**, 147 (1968).
34. B. Singer, *J. Natl. Cancer Inst.* **61**, 1329 (1979).
35. T. H. Jukes, *Adv. Enzymol.* **47**, 375 (1978).
36. D. B. Dunn and R. H. Hall, *in* "Handbook of Biochemistry and Molecular Biology, Nucleic Acids" (G. D. Fasman, ed.), 3rd ed., Vol. 1, p. 65. CRC Press, Cleveland, Ohio, 1975.
37. W. Bähr, P. Faerber, and K. H. Scheit, *EJB* **33**, 535 (1973).
38. C. Thrierr and M. Leng, *EJB* **19**, 135 (1971).
39. F. B. Howard, J. Frazier, and H. T. Miles, *Bchem* **15**, 3783 (1976).
40. F. B. Howard, M. Hattori, J. Frazier, and H. T. Miles, *Bchem* **16**, 4637 (1977).
41. C. Janion and K. H. Scheit, *BBA* **432**, 192 (1976).
42. M. Hattori, M. Ikehara, and H. T. Miles, *Bchem* **13**, 2754 (1974).
43. F. Ishikawa, J. Frazier, F. B. Howard, and H. T. Miles, *JMB* **70**, 475 (1972).
44. J. D. Engel and P. H. von Hippel, *JBC* **253**, 927 (1978).
45. R. Thedford and D. B. Straus, *Bchem* **13**, 535 (1974).
46. T. Golas, M. Fikus, Z. Kazimierczuk, and D. Shugar, *EJB* **65**, 183 (1976).
47. D. C. Ward and E. Reich, *PNAS* **61**, 1494 (1968).
48. J. O. Folayan and D. W. Hutchinson, *BBA* **474**, 329 (1977).
49. F. Rottman, K. Friderici, P. Comstock, and M. K. Khan, *Bchem* **13**, 2762 (1974).
50. M. Ikehara, T. Fukui, and N. Kakiuchi, *NARes* **3**, 2089 (1976).
51. C. Janion and D. Shugar, *Acta Biochim. Pol.* **20**, 271 (1973).
52. C. Janion, *Acta Biochim. Pol.* **23**, 57 (1976).
53. J. Kohlschein, L. Hagenberg, and H. G. Gassen, *BBA* **374**, 407 (1974).
54. D. C. Ward and E. Reich, *JBC* **247**, 705 (1972).
55. A. D. Broom, M. E. Uchic, and J. T. Uchic, *BBA* **425**, 278 (1976).
56. M. Ikehara and T. Tezuka, *NARes* **1**, 907 (1974).
57. M. Ikehara, N. Kakiuchi, and T. Fukui, *NARes* **5**, 3315 (1978).
58. J. R. Mehta and D. B. Ludlum, *Bchem* **15**, 4329 (1976).
59. D. B. Ludlum, J. R. Mehta, and R. F. Steiner, *BBA* **475**, 197 (1977).
60. M. Hattori, J. Frazier, and H. T. Miles, *Bchem* **14**, 5033 (1975).
61. F. Pochon, C. Balny, K. H. Scheit, and A. M. Michelson, *BBA* **228**, 49 (1971).
62. J. Hobbs, H. Sternbach, M. Sprinzl, and F. Eckstein, *Bchem* **12**, 5138 (1973).
63. P. F. Torrence, A. M. Bobst, J. A. Waters, and B. Witkop, *Bchem* **12**, 3962 (1973).
64. T. Kulikowski and D. Shugar, *BBA* **374**, 164 (1974).
65. M. A. W. Eaton and D. W. Hutchinson, *BBA* **319**, 281 (1973).
66. P. Faerber, K. H. Scheit, and H. Sommer, *EJB* **27**, 109 (1972).
67. M. Kielanowska and D. Shugar, *NARes* **3**, 817 (1976).
68. P. M. Pitha and A. M. Michelson, *BBA* **204**, 381 (1970).
69. B. Singer and H. Fraenkel-Conrat, *Bchem* **8**, 3260 (1969).
70. D. C. Ward, A. Cerami, E. Reich, G. Acs, and L. Altwerger, *JBC* **244**, 3243 (1969).
71. S. Uesugi, T. Nagura, E. Ohtsuka, and M. Ikehara, *Chem. Pharm. Bull.* **24**, 1884 (1976).
72. C. L. Stevens, *Biopolymers* **13**, 1517 (1974).
73. R. F. Steiner and R. F. Beers, Jr., *BBA* **33**, 470 (1959).
74. E. Ohtsuka, T. Nagura, K. Shimokawa, S. Nishikawa, and M. Ikehara, *BBA* **383**, 236 (1975).
75. M. Ikehara and E. Ohtsuka, *BBRC* **21**, 257 (1965).
76. L. Hagenberg, H. G. Gassen, and H. Matthaei, *BBRC* **50**, 1104 (1973).
77. M. Ikehara, K. Murao, F. Harada, and S. Nishimura, *BBA* **174**, 696 (1969).

78. A. Wacher, E. Lodeman, K. Gauri, and P. Chandra, *JMB* **18**, 382 (1966).
79. D. Grunberger, D. C. Ward, and E. Reich, *JBC* **247**, 720 (1972).
80. A. Holý, D. Grunberger, and F. Šorm, *BBA* **217**, 332 (1970).
81. D. Grunberger, A. Holý, and F. Šorm, *BBA* **134**, 484 (1967).
82. T. Sekiya, M. Yoshida, and T. Ukita, *BBA* **149**, 610 (1967).
83. C. Basilio, A. J. Wahba, P. Lengyel, J. F. Speyer, and S. Ochoa, *PNAS* **48**, 613 (1962).
84. A. J. Wahba, R. S. Gardner, C. Basilio, R. S. Miller, J. F. Speyer, and P. Lengyel, *PNAS* **49**, 116 (1963).
85. C. K. Carrico, L. S. Cunningham, and A. C. Sartorelli, *BBRC* **78**, 1204 (1977).
86. D. Grunberger, A. Holý, and F. Šorm, *BBA* **161**, 147 (1968).
87. D. Grunberger, C. O'Neal, and M. Nirenberg, *BBA* **119**, 581 (1966).
88. D. Grunberger, A. Holý, and F. Šorm, *BBA* **157**, 439 (1968).
89. B. Singer, R. G. Pergolizzi, and D. Grunberger, *NARes* **6**, 1709 (1979).
90. J. Smrt, W. Kemper, T. Caskey, and M. Nirenberg, *JBC* **245**, 2753 (1970).
91. D. Grunberger, A. Holý, J. Smrt, and F. Šorm, *Collect. Czech. Chem. Commun.* **33**, 3858 (1968).
92. M. Grunberg-Manago and A. M. Michelson, *BBA* **80**, 431 (1964).
93. M. Grunberg-Manago and A. M. Michelson, *BBA* **87**, 593 (1964).
94. J. Smrt, J. Skoda, V. Lisý, and F. Šorm, *BBA* **129**, 210 (1966).
95. F. Rottman and P. Cerutti, *PNAS* **55**, 960 (1965).
96. R. Vormbrock, R. Morawietz, and H. G. Gassen, *BBA* **340**, 348 (1974).
97. M. Gleason and H. Fraenkel-Conrat, *PNAS* **73**, 1528 (1976).
98. G. E. Means and H. Fraenkel-Conrat, *BBA* **247**, 441 (1971).
99. H. R. Rackwitz and K. H. Scheit, *EJB* **72**, 191 (1977).
100. M. Kröger and B. Singer, *Bchem* **18**, 91 (1979).
101. B. E. Dunlap, K. H. Friderici, and F. Rottman, *Bchem* **10**, 2581 (1971).
102. A. Holý, D. Grunberger, J. Smrt, and F. Šorm, *BBA* **138**, 207 (1967).
103. D. Grunberger, A. Holý, and F. Šorm, *Collect. Czech. Chem. Commun.* **33**, 286 (1968).
104. T. Uchida and C. Funayama-Machida, *J. Biochem.* (*Tokyo*) **81**, 1237 (1977).
105. D. H. Gauss, F. Grüter, and M. Sprinzl, *NARes* **6**, rl (1979).
106. S. Altman, ed., "Transfer RNA," Cell Monogr. Ser., Vol. 2. MIT Press, Cambridge, Massachusetts, 1978.
107. S. H. Kim, G. J. Quigley, F. L. Suddath, A. McPherson, D. Sneden, J. J. Kim, J. Weinzierl, and A. Rich, *Science* **179**, 285 (1973).
108. J. D. Robertus, J. E. Ladner, J. T. Finch, D. Rhodes, R. S. Brown, B. F. C. Clark, and A. Klug, *Nature* **250**, 546 (1974).
109. S. H. Kim, *Adv. Enzymol.* **46**, 279 (1978).
110. H. Mizuno and M. Sundaralingam, *NARes* **5**, 4451 (1978).
111. L. Stein and L. H. Shulman, *JBC* **252**, 6403 (1977).
112. L. H. Shulman and J. P. Goddard, *JBC* **248**, 1341 (1973).
113. P. W. Piper, *EJB* **51**, 283 (1975).
114. B. R. Roberts, *PNAS* **48**, 897 and 1245 (1962).
115. S. K. Mitra, F. Lustig, B. Åkesson, U. Lagerkvist, and L. Strid, *JBC* **252**, 471 (1977).
116. U. Lagerkvist, *PNAS* **75**, 1759 (1978).
117. J. Weissenbach, G. Dirheimer, R. Falcoff, J. Sanceau, and E. Falcoff, *FEBS Lett.* **82**, 71 (1977).
118. P. Jank, N. Shindo-Okada, S. Nishimura, and H. J. Gross, *NARes* **4**, 1999 (1977).
119. E. G. Richards, *Nature* **273**, 488 (1978).

120. W. Hillen, E. Egert, H. J. Lindner, and H. G. Gassen, *FEBS Lett.* **94**, 361 (1978).
121. W. Hillen, E. Egert, H. J. Lindner, and H. G. Gassen, *Bchem* **17**, 5314 (1978).
122. E. Egert, H. J. Lindner, W. Hillen, and H. G. Gassen, *Acta Crystallogr., Sect. B* **34**, 2204 (1978).
123. M. Kröger and B. Singer, *Bchem* **18**, 3492 (1979).
124. S. K. Niyogi, *JMB* **64**, 609 (1972).
125. G. F. Gerard, F. Rottman, and J. A. Boezi, *BBRC* **46**, 1095 (1972).
126. D. B. Ludlum, *JBC* **245**, 477 (1970).
127. L. L. Gerchman and D. B. Ludlum, *BBA* **308**, 310 (1973).
128. J. R. Mehta and D. B. Ludlum, *BBA* **521**, 770 (1978).
129. B. Singer, H. Fraenkel-Conrat, and J. Kuśmierek, *PNAS* **75**, 1722 (1978).
130. D. B. Ludlum, *BBA* **213**, 142 (1970).
131. H. Fraenkel-Conrat and B. Singer, *in* Biological Effects of Polynucleotides (R. F. Beers, Jr. and W. Braun, eds.), p. 13. Springer-Verlag, Berlin and New York, 1971.
132. D. M. Brown, M. J. E. Hewlins, and P. Shell, *JCS (C)* p. 1925 (1968).
133. E. I. Budowsky, E. D. Sverdlov, and T. N. Spasokukotskaya, *FEBS Lett.* **17**, 336 (1971).
134. D. L. Sabo, E. Domingo, E. F. Bandle, R. A. Flavell, and C. Weissmann, *JMB* **112**, 235 (1977).
135. M. D. Kirnos, I. Y. Khudyakov, N. I. Alexandrushkina, and B. F. Vanyushin, *Nature* **270**, 369 (1977).
136. J. Ninio, This Series **13**, 301 (1973).
137. M. D. Topal and J. R. Fresco, *Nature* **263**, 285 and 289 (1976).
138. E. G. Richards, *Nature* **263**, 369 (1976).
139. E. S. Ramberg, M. Ishaq, S. Rulf, B. Moeller, and J. Horowitz, *Bchem* **17**, 3978 (1978).
140. J. Ofengand, J. Bierbaum, J. Horowitz, C.-N. Ou, and M. Ishaq, *JMB* **88**, 313 (1974).
141. T. A. Early, J. Olmsted, III, D. R. Kearns, and A. G. Lezius, *NARes* **5**, 1955 (1978).
142. B. Singer, *TIBS* (*Trends Biochem. Sci.*) **2**, 180 (1977).
143. B. Singer, M. Kröger, and M. Carrano, *Bchem* **17**, 1246 (1978).
144. K. Watanabe, M. Shinma, T. Oshima, and S. Nishimura, *BBRC* **72**, 1137 (1976).
145. K. Watanabe, T. Oshima, M. Saneyoshi, and S. Nishimura, *FEBS Lett.* **43**, 59 (1974).
146. J. A. McCloskey and S. Nishimura, *Acc. Chem. Res.* **10**, 403 (1977).
147. D. Johnson and A. R. Morgan, *PNAS* **75**, 1637 (1978).
148. H. Rogg, P. Müller, G. Keith, and M. Staehelin, *PNAS* **74**, 4243 (1977).
149. H. J. Grosjean, S. de Henau, and D. M. Crothers, *PNAS* **75**, 610 (1978).
150. L. S. Khan, P. O. P. Ts'o, F. von der Haar, M. Sprinzl, and F. Cramer, *Bchem* **14**, 3278 (1975).
151. M. Nirenberg and P. Leder, *Science* **145**, 1399 (1964).
152. M. J. Chamberlin, *in* "RNA Polymerase" (H. Losick and M. J. Chamberlin, eds.), p. 17. Cold Spring Harbor Lab., Cold Spring Harbor, New York, 1976.
153. M. A. Sirover, D. K. Dube, and L. L. Loeb, *JBC* **254**, 107 (1979).
154. S. S. Agarwal, D. K. Dube, and L. L. Loeb, *JBC* **254**, 101 (1979).
155. C. W. Shearman and L. L. Loeb, *Nature* **270**, 537 (1977).
156. C. W. Shearman and L. L. Loeb, *JMB* **128**, 197 (1979).

The Accuracy of Translation

MICHAEL YARUS

Department of Molecular, Cellular, and Developmental Biology
University of Colorado
Boulder, Colorado

Introduction

The variety of living things is fashioned from a group of rather similar molecules. That is, the ubiquitous nucleotides have much of their structure and many of their chemical properties in common (*1*). The same is true of some amino acids (e.g., leucine, valine, and isoleucine), though on the whole these show greater chemical variety than nucleotides. In fact, the list of chemically similar amino acids is larger than normally considered, since many biosynthetic intermediates are themselves α-amino acids (e.g., homocysteine and homoserine).

These similarities are the result of the evolution of the modern repertoire of "biomonomers" from common evolutionary precursors or from each other. It necessarily follows that no process can make a precise distinction among amino acids or nucleotides, since they offer only limited opportunity for distinctive interactions. In addition, even macromolecules fashioned from these subunits may not interact precisely, though the accumulation of small dissimilarities into a definitive distinction was undoubtedly one of the forces that selected large polymeric molecules as the agents of biosynthesis. Protein biosynthesis, in particular, is necessarily imprecise to some extent. The questions which one can usefully ask therefore, concern the nature of the errors, and how often and where they occur.

Progress in Nucleic Acid Research
and Molecular Biology, Vol. 23

ISBN 0-12-540023-3

The following sections develop the consequences of the chemical similarity of biomonomers in a more systematic and formal way, in order to aid discussion of the experimental evidence. The reader who has no need of a summary of accuracy from a chemical kinetic viewpoint could turn here to Section III, treating Sections I and II as an appendix to be used if needed.

I. The Velocities of Alternative Reactions

A wide variety of selections are carried out by schemes closely related to that diagrammed in Fig. 1. We first need equations for the rates at which different substrates react in this scheme. A familiar development based on the steady-state assumption for ES leads to:

$$\mathrm{ES} = \frac{\mathrm{E}k_1}{k_2 + k_3}\,\mathrm{S} \qquad \text{and to}$$

$$v = \frac{dp}{dt} = k_3\mathrm{ES} = \frac{\mathrm{E}k_3}{K}\,\mathrm{S} = \frac{VS}{K+S} \qquad \text{and} \qquad K = \frac{k_2 + k_3}{k_1} \tag{1}$$

the next-to-last of which is the familiar Michaelis–Menten equation for the initial velocity v, with maximal velocity V, and one substrate.

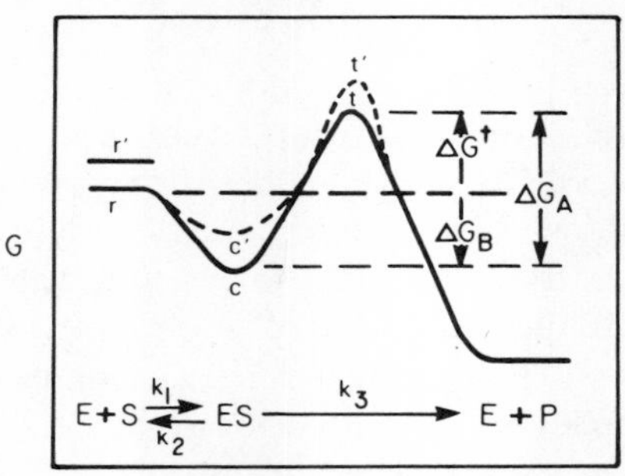

FIG. 1. The course of a simplified selection process. The free energy (G) of the reacting components, arbitrarily named E and S, is plotted versus a quantity describing the course of the reaction between two individual molecules (on the abscissa). Below this graph, the pathway is diagrammed in chemical notation and the elementary rate constants, k, are defined. Each species or group of reacting species (E, S, ES, P) is named below the part of the graph that indicates their (collective) G. The lower-case letters on the graph, r, c, t for reactants, complex, and transition state, respectively, are used in the text to refer to the path along the graph and also to denote the point on that path closest to the letter. For the path rct (the solid lines), $\Delta G^{\dagger}$ is the free energy of activation of E and S in the overall reaction leading to P, ΔG_B (<0) is a free energy of binding between E and S, and ΔG_A (<0) is the free energy required to activate the ES complex in order to reform the chemical bonds leading to P. Analogous definitions apply to the alternative paths.

If there are two possible substrates with distinct products, each is a competitive inhibitor from the point of view of the other and

$$v_j = \frac{V_j S_j}{K_j(1 + S_i/K_i) + S_j} \tag{2}$$

in which the two substrates are distinguished by the subscripts i and j. This is also a familiar equation, often used in the study of enzymic competitive inhibition. Note that when two potential substrates with two distinguishable products are simultaneously acted upon by the same enzyme in the same reaction mixture, we have, using Eq. 2:

$$\frac{v_j}{v_i} = \frac{(V_j/K_j)S_j}{(V_i/K_i)S_i} \tag{3}$$

By use of the steady-state assumption, which leads directly to forms like $v = E(k/K)S$ in Eq. 1 for every substrate in a simultaneous reaction mixture, it can be seen that Eq. 3 is general for any two simultaneous substrates, no matter how many others there are. The rather simple Eq. 3, then, expresses the accuracy of the scheme in Fig. 1 under the condition in which an enzyme is actually choosing between substrates i and j. If i is the correct substrate, Eq. 3 yields the relative rate of production of the erroneous product, or "error rate."

An alternative way to speak of accuracy (Eq. 4) is based instead on Eq. 1.

$$\frac{v_j}{v_i} = \frac{V_j S_j/(K_j + S_j)}{V_i S_i/(K_i + S_i)} \tag{4}$$

This compares reactions conducted separately, with no interaction (competition) between them.

I now wish to argue, following Fersht (2), that Eq. 3 is the superior form for the error rate. The two types of definitions are both in wide use, though they give different results. Confusion between them has led to serious mistakes in published work. First, Eq. 3 is better because it is analytically simpler. Second, and even more fortunate, it is also more realistic since substrate selections do indeed occur *in vivo* in the simultaneous presence of both substrates. Third, Eq. 4 behaves spuriously in several respects. Consider what happens as $S_i = S_j$ is raised. This will increase the relative error rate, according to Eq. 4, since one will ultimately saturate the enzyme with S_j and therefore suppress any distinction based on the fact that $K_j > K_i$. But this is quite wrong; as Eq. 3 shows, this change in concentrations in an actual reaction mixture will have no effect at all. Fourth, the ratio V/K or, equivalently, k_3/K, which appears in Eq. 3, is a chemically significant quantity, related to the net free energy of activation for the overall reaction (see below). Fifth, V/K is easier to obtain experimentally; it can easily be measured, for example, for slow noncognate aminoacyl-tRNA synthesis under conditions in which the separated constants required for Eq. 4 are impossible to obtain (*3*).

The relative error during competition is always less than that calculated for separate reactions.

$$\frac{(V_j/K_j)S_j}{(V_i/K_i)S_i} < \frac{(V_j/K_j + S_j)S_j}{(V_i/K_i + S_i)S_i} \tag{5}$$

as long as $K_j > K_i$, so that j is being discriminated against with regard to saturation of E. This effect becomes more and more significant as S_j and S_i get larger. In molecular terms, the real reaction is more accurate than the separated ones (*4*) because the favored substrate tends to exclude this disfavored one from E by competing more successfully for E. This can be a large effect: under the biochemically reasonable conditions ($S_i = S_j = 100K_i = K_j$), the calculated error rate for Eq. 3 is about $^1/_{50}$th that for Eq. 4.

We now use Eq. 3 to construct a general discussion of the behavior of systems that achieve their specificity in an informative way.

Suppose that $k_{3,i} = k_{3,j}$, so that there is no discrimination in velocity, and $k_3 \ll k_2$, so that the production of product does not disturb binding, which therefore comes to equilibrium. Comparing velocities at $S_i = S_j$

$$\frac{v_j}{v_i} = \frac{k_{1,j}/k_{2,j}}{k_{1,i}/k_{2,i}} \tag{6}$$

We have assumed that this system discriminates only at an equilibrium binding step, and indeed, the error rate v_j/v_i in Eq. 6 is simply the ratio of the true equilibrium dissociation constants for S_j and S_i. In Fig. 1, j follows path rc′t′ and i follows path rct; this clearly produces the difference in $\Delta G^\dagger$ which explains the selection against j. If we now allow product to be formed faster and faster (k_3 becomes significant, then large), the error rate (Eq. 3) remains constant if the preequilibrium step makes its distinction on the basis of "on" rates ($k_{1,j} < k_{1,i}$; $k_{2,j} \cong k_{2,i}$), but errors will ultimately increase rapidly if the complex formation step made its distinction on the basis of the dissociation (or half-life or "stickiness") of ES($k_{2,i} < k_{2,j}$; $k_{1,j} \cong k_{1,i}$). Thus a general rule: When the discrimination resides in the first step in schemes like that in Fig. 1, the error rate is never improved when the overall reaction is faster.

The relative error rate, Eq. 6, reflects the full difference in free energy of binding of i and j.

$$K^* = k_2/k_1 = e^{\Delta G_B/RT} \tag{7}$$

in which K^* is a true dissociation constant ($\Delta G_B < 0$). Thus

$$\frac{v_j}{v_i} = \frac{K_i{}^*}{K_j{}^*} = e^{(\Delta G_{B,i} - \Delta G_{B,j})/RT} \tag{8}$$

Equation 8 has been derived for a special set of circumstances, but it can be recast as another very general rule: an ordinary scheme cannot provide an error rate less than that calculated in Eq. 8 if the exp($\Delta G/RT$) term is calculated for the point at which the plots of G vs reaction coordinate for the two reactants are most widely separated (5). Equation 8 is this calculation for the assumptions leading to Eq. 6, that is calculated at cc′ (or tt′) in Fig. 1. To put this point another way, Eq. 8 shows that we can conceivably have difficulty making a distinction between two amino acids (for instance) which are so similar that the predicted error rate will be unacceptable. Pauling (6) originally suggested that the exponential in Eq. 8 should be only 1/20 or so for the one methylene group that can be used to select isoleucine instead of valine. We return to this problem below.

Now consider the more complicated type of selection in which i and j may be distinguished both in the binding and chemical transformation steps. Consideration of Fig. 1 shows that $\Delta G\dagger$, which determines the overall rate of the reaction of free E and S, is

$$\Delta G\dagger = \Delta G_A + \Delta G_B \tag{9}$$

Since binding and activation need not be independent events (they can involve the same groups of atoms), their interaction must be carefully considered. The second-order rate constant, which is determined by $\Delta G\dagger$, is that for the rate of reaction of free E and S to give P. We have already determined its form in Eq. 1. By the transition-state theory of reaction rates,

$$k_3/K \propto e^{-\Delta G\dagger/RT} = e^{-\Delta G_A/RT}\, e^{-\Delta G_B/RT} \tag{10}$$

and the error rate for two substrates (Eq. 3) can be considered simply in terms of the differences in their binding and activation. This justifies the claim (above) that V/K or k_3/K is a basic constant for the reaction.

An illustration, which is also relevant to the discussion of ribosomes (Section V) is provided by the case of "strain" or "induced fit." This refers to a type of mechanism in which one reactant is changed by binding the other; for example, accommodation to the shape of a substrate may swing a catalytic or strongly binding group into place on an enzyme. This mechanism couples the binding step to the catalytic one; the introduction of strain sacrifices some of the free energy of binding to pull the enzyme (or the substrate) toward the transition state for both. We wish to determine whether this can help distinguish sterically similar substrates.

Figure 2 diagrams the situation; the same enzyme exists in two forms, with an essential group locked in place (Fig. 2, lower diagrams) or free to move in response to a sterically "correct" substrate called i (Fig. 2, upper diagrams). By hypothesis, we make the free energies of i

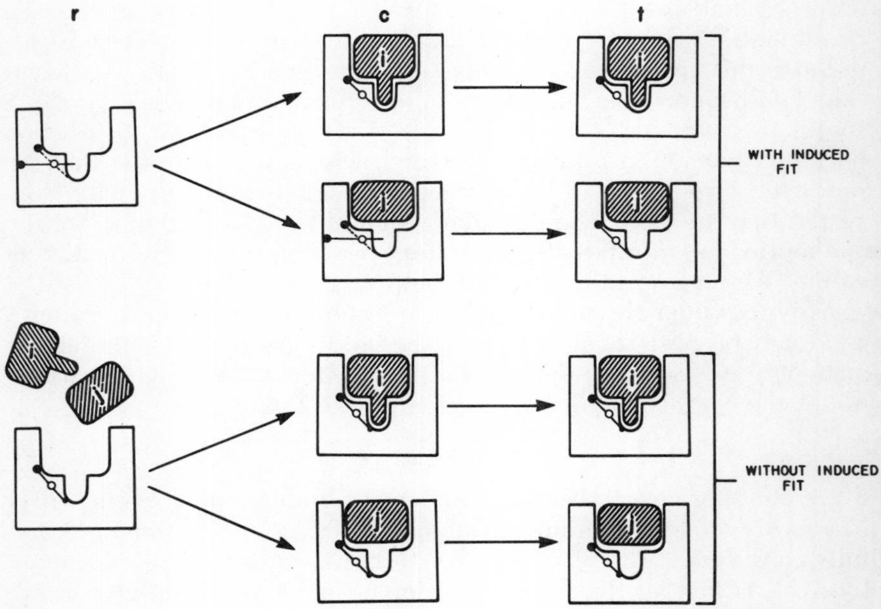

FIG. 2. A hypothetical induced-fit mechanism. The letters i and j refer to two substrates, the letters r, c, and t at top of the columns to reaction pathways defined in Fig. 1. The filled circle is an essential group. In the lower two rows the essential group is locked in a functional orientation; in the upper rows it is free to move, but prefers to be away from the active site. See Section I for other details. The dashed positions in the upper rows represent the freedom of movement of the critical groups. In passing from c to t above and in Fig. 1, the complex is activated for the reaction leading to P. This can involve many changes in E and S, but, for simplicity, only those involving the mobile group are shown.

and j in solution, of their binding, and of their activation the same, since we do not wish to consider differences in the substrates, but only the effect of the induced conformational change in E. In particular, the "tongue" projecting from i, the desired substrate, actuates the fit, but does not otherwise participate in binding.

Let us consider a simple question first. If the mobile group on E participates only in binding, does induced fit help distinguish i and j? The answer is no, as is obvious from inspection of Fig. 2. Consider the complex column (c). When the group has spontaneously rotated into place, in the j complex, the free energies of the i and j complex are the same. The intrinsic ΔG_B's are the same by hypothesis, the free energy to move the group must be the same, and the energy of interaction with the substrate is the same since the mobile group approaches identical positions of i and j. The path to the bound state will not matter, as long as the final free energies are the same, and they are: the positions of all atoms that matter in the two cases are identical. The bottom two rows of column (c) remind us that i and j were not distinguished by hypothesis by the nonmotile enzyme, since the only essential change has been locked in. However, the enzyme in the lower two rows of Fig. 2 does start at a higher G than in the top, because locking the group in position costs free energy. The curves in Fig. 1 will start at r′ rather than r, but that does not change the result.

Suppose the mobile group is catalytically active, instead of possessing an affinity for the substrate. Does this change the result? The G of the i and j complexes (Fig. 2, diagrams marked "without induced fit") at t in Fig. 1 is again the same by hypothesis. Once again (Fig. 2, diagrams marked "with induced fit") it can be seen that the t complexes are the same in the induced-fit case; i has traversed the path rc′t in Fig. 1, and j has reached the transition state via rct. What is gained in binding for j is lost because spontaneous activation of the enzyme (movement of the group) was not coupled to binding ($\Delta G_{A,i} < \Delta G_{A,j}$). There is, at best, no net change at t, which fixes the overall rate of reaction of i and j at equality. Therefore, i and j are not distinguished. For a more general discussion of this topic, see Fersht (2, 7).

Induced fit is an extreme example of the rule that the error rate cannot be greater than that calculated from the maximum free-energy difference between paths. Though there is a finite difference at cc′ in Fig. 1, it produces no improvement in the error rate at all, despite the apparent plausibility of the mechanism suggested. Indeed, it will worsen the error rate unless the transduction of binding free energy into progress toward the transition state is completely efficient.

II. Multiplying the Precision

Solutions to the problem of improving the error rate when the intrinsic differences between substrates are small were first stated in general form by Hopfield (*3a*) and, with a different emphasis, by Ninio (*8*, *9*).

Ninio focuses on systems in which complexes of correct and incorrect product molecules have different stabilities. If the utilization of these complexes, once synthesized, can be delayed for a time, the error rate will decline because the relative amount of (incorrect/correct) complex or product declines as they decay at different exponential rates. This can be quite costly because of destruction of the desired product (when the two are similar), and no system has yet been found that provides such a delay. We therefore turn to Hopfield's ideas.

Stated simply, if we could apply the same selection several times serially, we could multiply even a small precision to whatever level was required. However, it is surprisingly hard to do this. The induced-fit scheme above, which is an attempt to use serial binding and catalytic steps, fails owing to coupling between the two. Kurland (*10*) has emphasized that reverse reactions will defeat equilibrium schemes, because the mistaken product and the correct one are usually equally stable thermodynamically. Therefore the serial use of equilibria, however selective, cannot help (*5*). Hopfield saw that a practical scheme could be constructed by breaking discriminations into sections: in each section we are limited by the above difficulties, but by inserting an essentially irreversible step between sections, or by conducting one discrimination far from equilibrium, they can operate independently and the precision can be multiplied. A practical way to break equilibrium constraints and achieve irreversibility is to hydrolyze a triphosphate. Thus Scheme I (*5*) can discriminate in the formation of ES and then ES*, a high-energy intermediate, which can selectively break down to S for the disfavored substrate, and to P for the favored one.

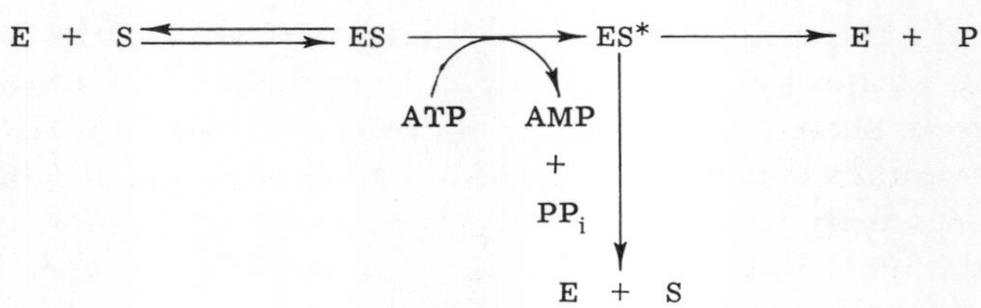

(11)

SCHEME I

The substrates, already distinguished in the binding step (the formation of ES), and therefore predominantly consisting of the correct complex, are fed to the activation step. ES* may break down tens or hundreds of times as frequently for the mistaken product as it does for product. The ES* of the correct substrate may have a similar bias for product formation. Note, however that even the selected substrate complex must break down to some extent. The resulting combination of the binding step to the left of the driven one and the kinetic discrimination to the right can have an error that approaches the square of that suggested by the general form of Eq. 8 (5). This was called "kinetic proofreading" (5).

The original discussion (5) used error fractions related to separate reactions (Eq. 4 above) and expressions for relative error that took no account of conservation of mass. It also suggested an explicit formulation of the model for aminoacyl-tRNA synthetases that is probably incorrect in detail (*11*), and the original simplifying assumptions about rates (no discrimination in the rate of production of product) has been found to be numerically inconsistent with later data on synthetase behavior (*12*). However, none of this should obscure the correctness of the essential idea.

The distinctive feature of a potential proofreading scheme is its consumption of the high-energy intermediate, and therefore of triphosphate, in the reactions involving the substrate to be rejected. This aspect of the scheme is usually tested experimentally (*13*). In order to reject the incorrect substrate, many moles of triphosphate must be consumed for every mole of mistaken product produced. However, this is not the same as finding that the error rate declines as triphosphate supply increases. Triphosphates can have other effects on accuracy (*10*). For example, product may be produced slowly in order to preserve the accuracy of a prebinding step (discussed in Section I). But the reverse of product production must also be slow with respect to production to avoid equilibration (Section II). These can be contradictory requirements, which can be resolved in favor of accuracy by coupling product production to the consumption of a triphosphate.

Similarly, the greater instability of a mistakenly activated intermediate does not prove proofreading, though it is consistent with it. It must also be shown that the rate of breakdown of the mistake will compete kinetically with the transformation to and consumption of the mistaken product. The combination of these two results is required to show that kinetic proofreading can occur and be effective.

Accuracy, in this scheme, literally and directly costs energy. We will see below that aminoacyl-tRNA synthetases can proofread (*13;*

Section IV). The energy cost, however, need not be spent primarily in hydrolysis of mistakes, which will (usually) be infrequent because of the preliminary discrimination of the first step. Instead, the mistaken hydrolysis of the desired substrate is often where the price is paid (*14*), and is presumably therefore the part of the mechanism selected for energy economy. These costs, as calculated (*14*) from existing data on aminoacyl-tRNA synthetase systems, range from 0.01 to 0.50 additional ATP per correct AA-tRNA inserted into protein. The systems for which measurements exist apparently do not operate at the lowest possible error level, which would require infinite ATP expenditure. Instead, they are poised near the point at which reduction in error per ATP expended from more editing declines (*14*), suggesting that they may be selected for this quality.

III. The Overall Error Rate of Translation

The available measurements relevant to this topic concern the incorporation of a single amino acid at an inappropriate site in an otherwise accurately constructed protein. This is not the only conceivable type of error, nor is it even known to be the most frequent one. Other errors are discussed in the last sections of this essay; this section treats only errors of substitution. They are presumed to occur by synthesis of a misacylated tRNA or by miscoding, that is, ribosomal transfer of the growing peptide chain to a tRNA mismatched to the message.

What was historically the first, and is still the most explicit, measurement was performed by Loftfield (*15*) and Loftfield and Vanderjagt (*16*). They measured incorporation of radioactive valine into two tryptic-chymotryptic peptides of rabbit hemoglobin, which should contain isoleucine, but not valine. The hemoglobin was isolated from suspensions of reticulocytes, and the protein digest was resolved by repetitive electrophoresis. The level of valine incorporated corresponded to an average of about 10^{-4} valines per amino-acid position, if the 9 possible sites of substitution in the two peptides are all used [see the discussion in Yarus (*17*)].

Edelmann and Gallant (*18*) measured cysteine incorporation into a gel electrophoretic band of *E. coli* flagellin, a protein that normally contains no cysteine. The extracellular location of the bacterial flagellum and its high sedimentation coefficient allow easy purification of the protein. Nevertheless, they observed 6×10^{-4} mol of cysteine per flagellin molecule. Since flagellin has about 450 amino acids, this number implies an error rate smaller than that for valine in rabbit hemoglobin; potentially much smaller. Edelmann and Gallant (*18*) argue that arginine codons are being mistranslated as cysteine, but it is not established that the predominant basal error measured is ribosomal, nor whether it is the eighteen arginine codons of flagellin in which the error occurs, or some greater or lesser number. Therefore the error rates implied range from 6×10^{-4} per amino-acid position (one error-prone site) to 1.3×10^{-6} per amino-acid position (all codons equally error-prone). Perhaps the most likely number is between 10^{-5} and 10^{-4} [compare the discussion in Yarus (*17*)].

We need more measurements of the absolute error rate for substitution error for several reasons. (*a*) We have, in the above experiments, only a small sample of possible substitutions. Since the two existing measurements available may differ substantially, they themselves suggest that we may not have a representative sample. (*b*) The two data are also from two different organisms with radically different modes of existence, and two proteins with different functions. Neither is an enzyme, or is produced in small numbers. (*c*) Ribosomal errors have already shown a distinct context dependence (see Section V). Even if the relevant context for a codon is as small as the neighboring codons, there are thousands of contexts, and no codon has yet been examined in more than about 1% of its contexts. (*d*) Both measurements are for a single amino acid at a greater or smaller number of positions. But what is biologically relevant is the total error rate for all possible amino acids per position. Therefore, the nature of the question we wish to answer calls for a larger sample of errors, so that the error rates for all possible errors may be estimated. (*e*) Finally, both measure the substitution frequency in a stable, completed protein. Since aberrant, error-containing proteins can be selectively degraded [reviewed by Goldberg and Dice (*182*)], the error rate in nascent proteins could be much higher than these numbers. However, cellular maintenance of appreciable levels of homogeneous proteins containing $>10^3$ amino acids, e.g., β-galactosidase, implies total substitution error rates (all amino acids) of less than a few $\times 10^{-3}$ per amino-acid position. Where they higher, an improbably large fraction of the protein made would have to be degraded.

IV. The Error Rate of Aminoacyl-tRNA Synthesis

Because I reviewed this topic recently (*17*), the conclusions are summarized briefly in the following text and in Fig. 3. My intention is to present a plausible and current picture, not necessarily a final one. Much remains to be proved, and I hope that the reader will consult the

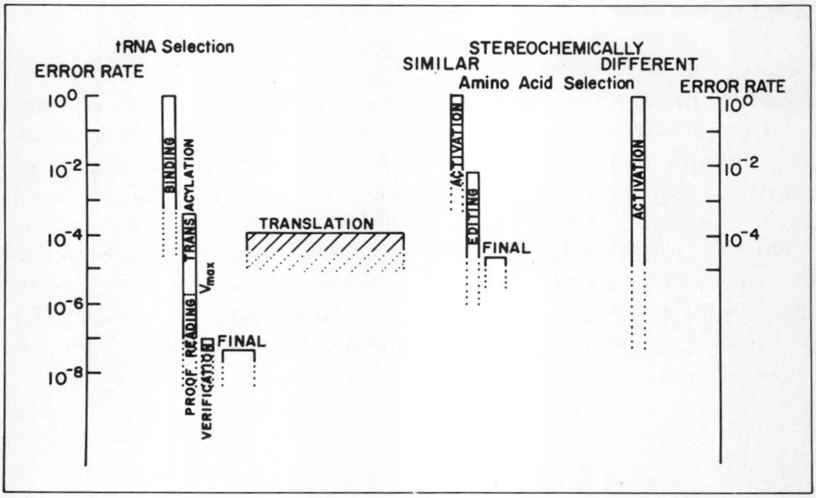

FIG. 3. The sources and magnitude of the relative error rate of aminoacyl-tRNA synthesis. These are averaged values from characterized systems, which therefore refer to no particular reactions. There is an unavoidable bias toward systems where mistakes can be detected and measurements can be made. Error rates are therefore probably maximized; the open-ended representation of each of the sources of specificity is meant to indicate this and other uncertainties. The specificity referred to as "binding" is that resulting from K_m; this is probably allowable because the rate-limiting steps in misacylation are probably late, so that the tRNA binding step equilibrates. V_{max} specificity is provisionally divided into actual transacylation and subsequent proofreading. "Verification" is hydrolysis of a misacylated tRNA by a second aminoacyl-tRNA synthetase cognate to the tRNA (Section IV). An error rate of 1 corresponds to synthesis of correct and misacylated tRNA at the same rate. Sources of specificity are listed (top downward) in order of occurrence during aminoacyl-tRNA synthesis. The band marked "translation" (center) represents the apparent error rate for a single amino-acid substitution, during production of stable proteins, by the translation apparatus as a whole.

Amino acid selections (right side) are divided into two groups. Rejection of those that are very different in size and/or chemical character can be done with adequate specificity by not binding or activating them. The subsequent fate of the few molecules of those that may be activated is unknown. Noncognate isosteric and chemically similar amino acids can be activated, and subsequently destroyed by editing or proofreading processes. "Activation" means formation of aminoacyladenylate. "Editing" refers to selective destruction of the misactivated amino acid (see text).

other text for details and additional references, and will note well the cautionary statements below.

First, on the left of Fig. 3 is the accuracy of the tRNA selection process. Equations 12 and 13 represent reactions that can accomplish the rejection of noncognate tRNAs by straightforward specificity in complexation and reaction.

$$E_i(AMP\text{-}AA_i) + tRNA_k \rightleftarrows E_i(AMP\text{-}AA_i)(tRNA_k) \qquad (12)$$

Equation 12 represents the binding step of Fig. 3, to form the quaternary complex which transacylates; noncognate tRNAs are rejected because they do not bind well.

$$E_i(AMP\text{-}AA_i)(tRNA_k) \rightleftarrows E_i(AA_i\text{-}tRNA_k) + AMP \qquad (13)$$

Equation 13 represents the transacylation step (Fig. 3), which is usually slow for a noncognate tRNA, probably because the acceptor end has a slightly aberrant position. The misacylated tRNA may now be proofread or "edited."

$$E_i(AA_i\text{-}tRNA_k) + H_2O \rightarrow E_i + AA_i + tRNA_k \qquad (14)$$

Equation 14 represents a guess about the mechanism of a proofreading step: the misacylated tRNA is selectively hydrolyzed before it dissociates from the enzyme.

This might occur selectively for a noncognate tRNA because its slightly aberrant position can force its amino acid toward the hydrolysis site of the synthetase (*7*). There is some evidence for excess ATP consumption and thus proofreading (e.g., Eq. 14) during reactions 12 and 13 (*13*), but no evidence as to the pathway.

That fraction of the AA_i-$tRNA_k$ dissociating from E_T before hydrolysis (by Eq. 14) is not necessarily committed to protein synthesis; it may be hydrolyzed as in Eq. 15.

$$E_k(AA_i\text{-}tRNA_k) + H_2O \rightarrow E_k + AA_i + tRNA_k \qquad (15)$$

This has been called "verification" (*4*). This proposal has been criticized because aminoacyl-tRNA is utilized too rapidly in protein synthesis to be recaptured by a second synthetase (*19*). However, this assumes that protein synthesis proceeds at the rate characteristic of fast logarithmic growth. But such growth and protein synthesis cannot have occurred during by far the greater part of the evolution of, e.g., *E. coli*, since such high growth rates cannot be maintained for evolutionarily significant periods. Accordingly, Eq. 15 may have an expanded role under conditions of slow growth; the size of the "verification" bar in Fig. 3, however, corresponds to fast growth. Binding of aminoacyl-tRNA to elongation factor Tu(GTP) would also protect AA_i-$tRNA_k$ (Eq. 15) from hydrolysis (*12*). Thus E_k (Eq. 15) must compete with T factor for misacylated tRNA. We do not know how much aminoacyl-tRNA can be free of T factor *in vivo* when protein synthesis is slow: this is another reason for preferring smaller values for error reduction due to verification.

In any case, the overall error rate for tRNA mis-selection (marked "final" in Fig. 3 on the left) is very securely known to be low compared to the overall error rate for translation (Fig. 3, center). Therefore, this process should limit the precision of translation only under unusual conditions.

Amino-acid selection usually proceeds by activation of the amino acid through aminoacyl-adenylate formation.

$$E_i + AA_j + ATP \rightleftarrows E_i(AA_j\text{-}AMP) + PP \tag{16}$$

Equation 16 includes opportunities for selective binding and catalysis of a noncognate amino acid that are quite sufficient to reject it, unless the noncognate closely resembles the cognate amino acid. In this case (central right side of Fig. 3), the specificity of the first step, which is mostly due to selective binding, is augmented to the level required for translation as a whole by later editing or proofreading. For example, valine is activated by isoleucyl-tRNA synthetase (see Introduction) of *E. coli* (*20*), but can be removed from the tRNA after formation of Val-$tRNA^{Ile}$ (Eq. 18; *11, 21*).

$$E_i(AA_j\text{-}AMP) + tRNA_i \rightleftarrows E_i(AA_j\text{-}tRNA_i) \tag{17}$$

$$E_i(AA_j\text{-}tRNA_i) + H_2O \rightarrow E_i + AA_j + tRNA_i \tag{18}$$

In principle, transacylation (Eq. 17) could also contribute to a low error rate if it were slower for AA_j than AA_i; but the available evidence suggests that there is little, if any, specificity at this step (e.g., *22*). This needs further investigation, but is the reason why transacylation does not appear as a contributor to amino acid selection in Fig. 3.

Editing by hydrolysis of a newly formed misacylated cognate tRNA (Eq. 18) is the only proven mechanism for lowering the error rate when synthetases activate isosteric noncognate amino acids. This is therefore what is intended by "editing" on the right of Fig. 3, but other mechanisms are conceivable. For example, the cognate tRNA might provoke direct breakdown of the adenylate (*11*).

$$E_i(AA_j\text{-}AMP) + tRNA_i \rightarrow E_i + AA_j + AMP + tRNA_i \tag{19}$$

The possible existence of other editing or proofreading mechanisms needs more investigation.

The overall relative error rate (marked "final" on the right in Fig. 3) at which noncognate, but stereochemically similar, amino acids are stably transferred to cognate tRNA can approach the overall error rate of protein biosynthesis (Fig. 3). Thus, in contrast to tRNA selection, error in amino acid selection may well limit the precision of translation of some codons.

V. The Error Rate of Ribosomal tRNA Selection

The matching of aminoacyl-tRNAs with codons on the ribosome cannot be less accurate than translation as a whole. This has been difficult to understand, because the free energy of addition of one standard base-pair to an RNA helix is -1.2 to -4.8 kcal/mol, depending on the nearest neighbors involved (*23*). This corresponds to a difference in binding constant of 7.4- to 3000-fold. In the lower part of this range, then, the difference in messenger binding a tRNA whose anticodon matches two adjacent bases and one that matches three can be only about one order of magnitude. Furthermore, there are always several conceivable mismatches. The result is that, if codons and anticodons pair as do small oligonucleotides at equilibrium, the total error rate in these worst cases would be greater than 0.1. This surely must be prevented.

As one possible solution, the tRNA anticodon might be constructed so as to pair more accurately than oligonucleotides. Trinucleotides do bind more strongly to anticodons in tRNA than expected (*24, 25*), but not strongly enough to allow measurements of two nucleotide complementarities. However, the discovery of very strong complexes between complementary tRNA anticodons (*26*) has allowed many measurements on partially complementary pairs (*27*). The results show, however, that not only are anticodon loops in tRNAs no more accurate than oligonucleotides, but certain mismatches (e.g., terminal 2-thiouridine opposite uridine) actually approach the stability of the ordinary base-pairs. Thus the various contributors to the unusual stability of tRNA anticodon pairs (*28*) have not enhanced the potential discrimination among competitors; quite the contrary.

Therefore, we are left with the ribosome, whose detailed construction might make possible more accurate base-pairing. For example, it might strengthen hydrogen bonds by excluding H_2O, or might provide a pairing site sterically incompatible with certain mispairs. This deserves further consideration but so far it appears that the washed, codon-instructed ribosome, selecting at equilibrium, is usually no more accurate than are ordinary oligonucleotides (*29*). It brings codons and tRNAs together in a strongly bound complex by adding a large nonspecific contribution to the binding free energy, but does not increase the differences in binding constant between similar anticodons.

At the next level of complexity, we may ask whether the addition of factors and energy increases the accuracy of the ribosomes. Since ribosomes are capable of the elementary steps of oligopeptide synthesis without factors and GTP (*30*), though at low rates, the comparison ($\pm$ elongation factors T and G) can be made directly. The miscoding

[Leu substitution for Phe with poly(U)] is difficult to detect in the factor-free system because total synthesis is only 1% of that with factors. However, it is possible to conclude that the error does not dramatically decrease with addition of factors and GTP; it must remain the same or go up (*30*). The error rate attained for a Leu base mismatch with factors and GTP in these experiments is about 0.1, not better than unaided oligonucleotides.

When all other components of a crude *E. coli* extract are also present, wide experience (e.g., *31, 32*) continues to suggest error rates [Leu for Phe with poly(U)] of 1 to 5%, though some errors, probably also requiring single mismatches (serine, isoleucine) are now at satisfactory levels around 10^{-4} (*31*).

Thus the puzzle: there is an inadequate free-energy difference between the bindings of various partial codon · anticodon pairs and the completely complementary one. If we assume that the steps after aminoacyl-tRNA binding that lead to peptide bond formation are similar for all aminoacyl-tRNAs, this plausible and apparently innocuous assumption forces the conclusion that the ultimate error rate is limited by the error-prone binding step (see Sections I and II). The addition of various equilibrium discriminations on the ribosome, including the use of induced fit, will not solve this problem, though such schemes have been suggested (*33*). The discussion of induced fit in Section II of this review (Figs. 1 and 2) can be recast with the "substrate" taken to be the ribosomal codon · anticodon complex, and the "enzyme" taken to be the remainder of the tRNA. This corresponds to the model of Kurland *et al.* (*33*). The conclusion is unaltered: the original discrimination intrinsic to the codon · anticodon complex is not amplified by the introduction of induced fit.

Furthermore, tRNA and ribosomes, as well as the simpler complete *in vitro* systems do not seem to amplify the specificity at all. The most complex *in vitro* (simple polymer) translation systems can be somewhat more accurate. This suggests that the rapidly cycling ribosome in its complete *in vivo* milieu, is, up to now, the sole possessor of the accuracy of translation. Clearly, an *in vitro* system that is highly accurate would markedly advance this line of study.

The reader should realize that ribosomal inaccuracy is sometimes attributed to actual mispairs between codon and anticodon (*33a*). This is an alternative point of view, but if it is applied to both *in vivo* and *in vitro* data, it requires that the *in vitro* ribosomal environment vastly increase the frequency of such rare mispairs. For the moment, this seems improbable, and so is not pursued further here.

Nevertheless, it is quite clear that the ribosomal milieu has been fashioned by opposing evolutionary forces to have its present error rate, since the accuracy of the ribosome can easily be adjusted in both directions by mutation. Mutants of 30 S ribosomal protein S12 (*34*), recognizable by resistance to streptomycin [*str A;* see the summary in Gorini (*35*)], make the ribosome less error-prone *in vitro* (*36*) and *in vivo* (*37, 38*). Mutants of 30 S ribosomal protein S4, called *ram*, make the ribosome more error prone *in vitro* (*39*) and *in vivo* (*38*). Because these phenomena have been extensively discussed [for a review, see Gorini (*40*)], I do not pursue them further here, save for emphasis of a few points peculiarly relevant to our current discussion. First, these mutations, particularly the error *reduction* observed for *str A*, show that ribosomal proteins limit the accuracy of translation *in vivo* (and also *in vitro*). Thus, along with the activation of isosteric amino acids, we must consider ribosomal tRNA selection as one of the potential limits to accuracy in translation. Second, the similar effects of *str A* and *ram* mutants *in vivo* and *in vitro*, along with analogous results with streptomycin [summary in Gorini (*35*)], connect the two types of measurements and validate the study of accuracy *in vitro* under what must admittedly be anomalous conditions in some respects.

The nature of the ribosome contribution to error remains undefined. Error is also increased by raising the pH (*41*), raising the magnesium and the monovalent ion concentration[1] (*36*), lowering the temperature (*31, 42*), or adding organic solvent (*43*) to an *in vitro* translation system. These effects are presumably ribosomal, but await a unified, or even a fragmentary, explanation. It may be significant that all could be thought of as effects on the effective charge of a neutral or mildly basic anionic group.

[1] There is a common feeling, expressed both in print and even more often informally, that ribosomes are "accurate" at low [Mg^{2+}] and "inaccurate" at higher [Mg^{2+}]. But, in fact, there should not be qualitative or very large quantitative differences in the accuracy of translation under the two conditions. The difference between high and low [Mg^{2+}] is small, e.g., only a factor of two, as at 20 m*M* and 10 mM. The [Mg^{2+}] presumably acts by binding to some component(s) of the translational apparatus, perhaps the ribosome. The extra free energy made available for changing the critical component by increasing the [Mg^{2+}] twofold is only 0.4 kcal per mole of Mg^{2+} bound in the change. Or, even more simply, the effect of doubling [Mg^{2+}] can be no more than to double the equilibrium constant for the production of the low-accuracy form of the ribosome per critical Mg^{2+} ion bound. Thus it is improbable that an extreme change is produced by this manipulation, even if the change is produced by cooperative binding of several Mg^{2+} ions.

The differences that certainly do occur on raising [Mg^{2+}] from 10 to 20 mM are instead the result of increasing the abundance of an inaccurate component that must already exist spontaneously with substantial probability at "low" [Mg^{2+}]. Therefore, the effect is probably moderate and enhances an existing tendency to error.

The types of coding errors detected during peptide synthesis *in vivo* and *in vitro* are summarized in Table I (*18, 32, 38, 43a–49*). The *in vitro* data tabulated include the values of those variables known to affect error rate significantly.

The data *in vitro* [see also Woese (*50*)] lend themselves to some generalizations.

1. All common mistakes are interpretable as errors at a single position.
2. Mistakes can be detected in all three positions. There is no compelling evidence for a most or least accurate position.
3. Pyrimidines in the codon's 5′ and central positions can often be "read" as the other pyrimidine or A, more rarely as G.
4. Purines are read more accurately. Notice that this implies a radical asymmetry in the ribosome: a message U can be mistaken for C, implying it can be in opposition to a G in an anticodon. But G in the message is not easily mistaken as U; thus codons and anticodons are subjected to different rules.
5. The mistakes are strikingly context-dependent. For example, poly(U) and poly(C) are uniquely error-prone messages; in a more complex context (the alternating copolymer messages), no pyrimidine misreading can be detected until provoked by streptomycin. The errors that then appear are a very limited subset of those that occur with pyrimidine nucleotide homopolymers. For example, no mistakes were detected when a central pyrimidine was flanked with two purines. The last three sets of entries (*in vitro*) can also be viewed as examples of a tendency for tRNAs to require only the 5′ and center bases of the codon for effective reading. Lagerkvist (*51*) has suggested that this may be a general tendency.

But in the original experiments of this kind (*46*), the single, purified Val-tRNAs used were the only Val-tRNAs present. Therefore there was no competition from any potentially better matched codon · anticodon combination. The observed ability of a single Val-tRNA to translate GUN (N is any nucleotide) may not be surprising under these circumstances. When a cognate aminoacyl-tRNA is absent, the translation apparatus apparently easily accepts other aminoacyl-tRNA species, both *in vitro* in extracts containing a temperature-sensitive synthetase (*52, 53*) and *in vivo* during amino acid starvation of rel$^-$ cells (*48, 49*). *In vitro* synthesis proceeds at 16 to 50% of the usual normal rate in the apparent absence of a required aminoacyl-tRNA (*52, 53*) and the apparent error rate *in vivo* is >1% at arginine codons (*54*) and >10 to 20% (!) at histidine codons (*48, 49*) during amino acid starvation.

This points to the availability of the aminoacyl-tRNA cognate to a codon as an important determinant of specificity, and therefore presumably to the importance of competition for the A site of the ribosome. To put this another way, the ribosome itself, under these conditions, does not have the intrinsic selectivity required to reject a poorly matched aminoacyl-tRNA.

The presence of Val-tRNAs that match three of three positions does in fact depress use of those that match at two positions out of three (55). Examination of the translation of the leucine codons, UUA/G, and CUN by single tRNA species, under conditions in which all Leu-tRNAs were present, (*47*) also reveals a much more restricted range of action than suggested by two-of-three translation (see Table I). Most incorporation occurs at expected pairings, but unexpected oppositions [e.g., C · C and probably C · A and G · A at the third codon position (anticodon base listed first)] are still found to allow efficient incorporation.

The most extraordinary property of the total data of Table I is the obvious tendency to favor certain mistakes, yet without obvious correspondence to simple mechanisms. However, the tendency of U to be mistaken as A (to pair as U) is reminiscent of the unexpected stability of U-containing pyrimidine–pyrimidine mispairs in anticodon–anticodon interactions (27). The formulation and testing of an error mechanism from these data remains an important unmet challenge.

The *in vivo* data of Table I argue strongly for misreading at the second and third positions of termination codons by nonsense suppressor tRNAs. These are frequencies at or above the level of error of ordinary translation, and so (*a*) they confirm the significance of mistakes in tRNA selection *in vivo;* and (*b*) they also confirm the primary role of mistakes at a single position (as also *in vitro*).

The last entries all refer to enhanced error rates that prevail when a normal aminoacyl-tRNA is not available. The nature of the misreading has been deduced, assuming that only one position can be in error and that misreading follows the pattern of *in vitro* errors shown in the topmost part of the table. These substitutions are accordingly less certain, but taken together with the *in vivo* data on nonsense codons, they show that mistakes at all positions are possible, as they are also *in vitro*. Misreading in the first position is lowest in the existing experiments, but more *in vivo* data, especially on defined substitution at defined loci, are clearly needed. This is especially so because of the apparently strong context and composition effects on error that are already evident.

Again, then, how is the low error rate evident *in vivo* attained by a

TABLE I

CODING THAT VIOLATES STANDARD BASE-PAIRING OR "WOBBLE" RULES

Codon read	Read as	Error rate × 10^2	Mg^{2+}	pH	Monovalent ion	Temp.	Comment	Reference
			A. *In Vitro*					
UUU	CUU or UUA/G	5	12.5 m*M*	7.8	50 m*M* NH_4^+	35°	Poly(U)message	*43a*
UUU	AUU	8						
UUU	UCU	4						
UUU	UAU	26						
CCC	CAC	1	12.5 m*M*	7.8	50 m*M* NH_4^+	35°	Poly(C)message	*43a*
CCC	CUC	1						
CCC	GCC	1						
CCC	ACC	17						
AAA, III	—	—	12.5 m*M*	7.8	50 m*M* NH_4^+	35°	No misreading at similar sensitivity	*43a*
UCU or CUC	CCU or CCC	12 or 21[a]	13 m*M*, 0.5 m*M* Mn^{2+}	8.0	45 m*M* K^+	37°	Alternating copolymer message; no misreading without Sm (4 μg/ml)	32
UCU or CUC	UUU or UUC	30 or 40						
CUC	CAC	4						
UGU	CGU	56						
UGU	AGU or UCU	7						
GUG, CAC, ACA, AGA, GAG	—	—	13 m*M*, 0.5 m*M* Mn^{2+}	8.0	45 m*M* K^+	37°	Not misread even in presence of Sm	32

UGA	UGG	?	21 m*M*	8.0	80 m*M* NH_4^+, 50 m*M* K^+	37°	Poly(U,G,A) message	*44*
UGU	UGG	?	11 m*M*	7.8	100 m*M* NH_4^+	37°	Poly(U,G) message	*45*
GUG	GUU/C	?	11 m*M*	7.3	50 m*M* K^+	37°	MS2 coat protein message	*46*
GUA	GUU/C							
GUU	GUA							
GUC	GUA							
CUC	CUG	?	5–7 m*M*	7.0	100 m*M* NH_4^+	37°	RNA phage coat and synthetase message	*47*
CUA	CUG							
CUU	CUG							
CUA	CUU/C							
				B. *In Vivo*				
UAA	UAG	0.015	—	—	—	—	Based on β-gal activity: misreading due to single suppressor tRNA	*38*
UGA	UAA	1.8						
UGA	—	—	—	—	—	—	UGA not read as UAG	*38*
CGU/C	UGU/C[b]	0.04	—	—	—	—	Starvation of rel⁻	*18*
CAU/C	CAA/G[b]	1.3–13	—	—	—	—	During amino acid starvation in rel⁻	*48, 49*
CAU/C	CAA/G[b]	23						
AAU/C	AAA/G[b]	?	—	—	—	—	Asparagine starvation in CHO cells	*49*

[a] Calculated for both possible substitutions.

[b] Nature of substitution guessed by requiring single mistake to produce observed changes of aberrant protein.

translation apparatus whose critical elements, codon–anticodon interactions and the ribosome, manifestly appear error-prone? It is undoubtedly true that many *in vitro* experiments overemphasize some errors. High [Mg^{2+}] is sometimes required, simple homopolymer messengers are sometimes used, low concentrations of cognate aminoacyl-tRNAs are supplied, and so on: all elevate the error rate. Were these conditions corrected, or not required, the error rate would converge with that *in vivo*. In the discussion above, I have also emphasized the mistakes that occur, without stressing that many possible mistranslations are never observed. But the problem of accounting for an error rate *uniformly* as low as we presume to exist *in vivo* would remain. I wish to suggest that his is achieved by improving the precision of coding by several editing or proofreading steps, very much in the way that the precision of aminoacyl-tRNA synthesis is enhanced.

The first, and most straightforward, of these has already been mentioned: the degradation of completed proteins containing errors (*182*, *56*).

The second is the rejection, by the ribosome, of nascent chains that contain an error. This process has been suggested by Menninger (*57*) as a result of his studies of a mutant of peptidyl-tRNA hydrolase, an apparently cytoplasmic enzyme that releases amino acids and tRNA from peptidyl-tRNAs. A temperature-sensitive mutant of this enzyme is a conditional lethal; the destruction of such apparently aberrant molecules is therefore a required function (*58*). After the mutant is shifted to high temperature, some species of tRNA are rapidly accumulated in this blocked form (*59*). This seems a sufficient reason for the lethal phenotype (*60*), and also suggests that release is a common event. Menninger has calculated (*59*) assuming that the frequency of release is not altered in the mutant at high temperature, that release occurs following ca. 10^{-3} of peptidyl transfers. He suggests that the released chains may selectively be those whose terminal tRNA did not match the message.

Such a frequency of premature release implies that only a fraction of the ribosomes that begin a long chain like β-galactosidase should reach the carboxyl terminus. There is evidence that this is so. Manley (*61*) has detected, both *in vivo* and *in vitro*, nine discrete amino-terminal fragments with the antigenic character of β-galactosidase, apparently representing prematurely ejected chains. These fragments alone account for 31% of the ribosomes that initiate synthesis. Some of these could be due to messenger fragmentation, but if any substantial fraction are consequences of errors of substitution, the error rate *in vivo* could be higher than that found in stable completed chains. The rate of rejection of peptidyl-tRNAs from the ribosome is also responsive to streptomycin and the *str A* mutation, as it might be if it is an important error rejection pathway (*62*). The locus and mechanism of peptidyl-tRNA dissociation are still unclear, but studies of the ribosomal P site suggest that it has the specificity to be the locus of a second check of the codon–anticodon match (*29*). That is, the P site is capable of detecting codon–anticodon resemblance at all three codon positions, and aminoacyl-tRNA binding constants at the codon-instructed P site vary accordingly over about three orders of magnitude. This could multiply the specificity of coding, since occupation of the P site is separated from that of A site by the irreversible step of translocation.

Third, there may be editing at the A site itself (*63*). The aminoacyl-tRNA is inserted into the ribosome as a ternary complex with elongation factor Tu (GTP) (*64*). In fact, it requires 30 to 60 GTPs to bind a molecule of a mismatched tRNA [Leu-$tRNA_2^{Leu}$, which reads CUU/C with poly(U)], to the A site, but only 1 or 2 GTP to bind the cognate tRNA (Phe-$tRNA^{Phe}$; R. C. Thompson, personal communication). This suggests that there is an activated state before committed AA-tRNA binding from which mismatches can be selectively rejected. When proofreading is combined with a 10- to 20-fold rejection of the mismatched tRNA on the basis of binding (*29;* R. C. Thompson, personal communication), one has a total selectivity of several hundredfold for the correct tRNA. Some mispairs [e.g., Leu-$tRNA_1^{Leu}$ with poly(U)] do not evoke excess GTP hydrolysis [cf. Thompson and Stone (*63*)] and must therefore be excluded by other mechanisms. It has not yet been shown that aminoacyl-tRNA rejection from the A site is fast enough to compete with peptide bond formation, but streptomycin and elevated [Mg^{2+}] both depress the excess consumption of GTP, as would be consistent with the notion that proofreading at the A site determines error rates (R. C. Thompson, personal communication).

The occurrence of all forms of editing and proofreading employing activated intermediates is limited in principle by measurements of the number of triphosphates required to convey an AA-tRNA into peptide linkage. Thus, consumption of more than two GTP (one for AA-tRNA binding and one for translocation) per residue could indicate energy expenditure for fidelity. *In vitro* measurements of this type have most recently been performed on isolated *E. coli* polysomes by Cabrer *et al.* (*65*). The average (±SEM) of eight experiments was 2.1 (±0.1) GTP per peptide bond (*65*). This suggests that this complete *in vitro* system (pH 7.8, 110 to 160 mM NH_4Cl, 11 mM $Mg(OAc)_2$) expended less than 10% of its energy in proofreading. However, there was always a background of GTP consumption at least equal to that due to peptide synthesis ("uncoupled hydrolysis"), which was subtracted as a blank. This might have concealed a somewhat larger editing contribution. Therefore we can only be sure that polysomes translating a variety of *E. coli* messages *in vitro* expend less than two extra GTPs per AA-tRNA in all other processes, including editing. Since AA-tRNAs (and their EF-T complexes), which are not cognate to the codon in the A site, necessarily outnumber the cognates by roughly twenty to one, the consequent sorting during AA-tRNA selection must usually not engage a proofreading mechanism.

The division of tRNA-codon errors into those constrained by the physical chemistry of tRNA–messenger interaction, the discovery of others that may in fact be constrained by the milieu inside the ribosome, and the characterization of the class just discussed, which must be edited after occurrence, should be an emergent project of great interest in the future study of translational precision.

VI. Errors in Initiation

Up to this point, we have considered only substitution errors—those in which an amino acid is incorporated at an inappropriate site. However, one cannot be sure that this is the major class of error of translation. For example, consider the initiation of protein synthesis: were a noninitiation region of a message to be used either in phase or out of phase, the result would be an abbreviated, perhaps unstable, and certainly elusive peptide, which would be invisible to most methods of analysis. No experiment seems ever to have been directed at the rate of this process, but it must happen.

The frequency of initiation, even at normal sites in f2 RNA, is negatively controlled by RNA structure (*66*). There must be some internal methionine codons, in contexts otherwise suitable for initiation, that do not function in the native message structure. In fact, partial denaturation by HCHO treatment of f2 RNA enhances an apparently aberrant initiation that is detectable (1% of total fMet dipeptides) even in the untreated RNA (*66*). HCHO treatment also induced several amino-terminal dipeptides not otherwise seen. If R17 phage RNA is fragmented, then for fragments smaller than a certain size (ca. 100 nucleotides), specificity of initiation is lost, and many aberrant sequences are recovered from ribosomal initiation complexes as RNase protected fragments (*67*). Therefore, normal conformational mobility will also occasionally expose these sequences or, perhaps more often, unfolding behind or ahead of a passing ribosome will be slow enough to permit capture of one of these sequences in an illegitimate initiation event.

Something very like this must happen *in vivo* during synthesis of "restart" fragments of, e.g., the lac repressor, on messages containing amber codons (*68*). In this case, even leucine and valine codons can serve as initiators. While these internal reinitiations require the upstream amber mutation and perhaps the ribosomal pause that it generates, the ribosome's movement is stochastic, not smooth and uniform (*69*), and natural pauses should occur.

Clearly, if such events total as little as 1% of normal initiation, they would be major contributions to the total misincorporation of amino acids. Study of this process might also illuminate the requirements for normal initiation and translational control.

VII. Other Errors of Elongation

The 50 S peptidyltransferase not only makes the peptide bond, it also functions in chain termination, apparently by transferring the completed peptide to water, when so directed by a termination codon and a release (R) factor (*70, 71*). The R-factor · ribosome complex is quite discriminating with regard to the ring nitrogen of the uridine in termination codons; a methyl group at this position blocks termination (*72*).[2] However, the purified protein R factors alone, when tested in solution, show very poor ability to discriminate termination codons from other normal nucleotide sequences (*73*). It is therefore quite possible that structural information can be misread as termination signals, and an amino-terminal protein fragment released as a result. This is one way to generate the fragments of β-galactosidase detected by Manley (*61*). The synthesis of large proteins is also selectively depressed during starvation for some amino acids in rel strains (*48, 54*). This is sometimes attributed to "premature termination," and may indeed be R factor acting instead of the missing aminoacyl-tRNA, but no evidence yet connects this observation with the normal apparatus of termination. More data on the frequency of accidental termination under normal *in vitro* or *in vivo* conditions are essential.

We may also wonder how often the ribosome spontaneously enters an erroneous translation frame, and scrambles the distal amino-acid sequence produced in that transit of the message. This process might be detected as leakiness in frame-shift mutants, and, in fact, Atkins *et al.* (*74*) have shown that 16/16 different lac$^-$ frame-shift mutations in

[2] See the article by Singer and Kröger on p. 151.

β-galactosidase have detectable activity. This activity is due to ribosomal error *in vivo*, since it responds in the expected way (up) to the *ram* mutations and streptomycin; and is reduced by *str A*. The available data do not allow any judgment on the natural frequency of such errors, since the selection of the mutants limited β-galactosidase to <0.1% of wild type, but the leakiness detected in the mutants does range downward from this (*74*).[3] Single frame shifts in both erroneous frames leak, indicating that the ribosome can step out of frame in both directions.

However, it is not entirely certain that the leak is due to a frame shift during normal translation in the vicinity of the mutated sequence. The error could occur at a termination codon generated by the frame shift, which would be 20 codons away, on average. This is less likely, however, since it does not seem probable that such a mechanism would generate active enzyme in 16 out of 16 cases. The phenomenon of phase error close to or during termination may nevertheless have been observed in other experiments, and this leads to the next (and final) topic.

VIII. Errors and Aberrations in Termination

Adventitious terminators, that is, nonsense mutants, can certainly be ignored or misread, since all nonsense mutants permit a background of activity under the control of the *str A* and *ram* genes (*38*). In some cases this is clearly due to ambiguous reading of the terminator by nonsense suppressor (*38*) or normal tRNAs (*44*), instead of an R factor. In fact, the R-factor competes with aminoacyl-tRNA for the termination signal, since antibodies to R-factors increase nonsense suppression (*75*). Therefore, propagation through an amber mutant by translation of the terminator codon should be responsive to tRNA and R-factor levels, and to other changes in the milieu that affect their relative activities.

[3] It should be noted that Barnett *et al.* (*78*) provide an upper limit on ribosomal frame shift and reinitiation. Suppressors (pluses or minuses) of frame shifts, when isolated and tested alone, were rIIB$^-$. About 1% of normal rIIB function is required to form a plaque on K12(λ) (B. Singer and L. Gold, personal communication). While the original mutation cannot be isolated if it is "leaky," almost no constraints are put on the efficiency of frame-shifting near a mutation selected as a frame-shift suppressor. Thus spontaneous frame-shifting in the neighborhood just ahead of a great variety of frame-shift mutations in the rII region of T4 must happen in <1% of ribosomal transits. The same conclusion applies to the frequency of subsequent termination and reinitiation. This suggests that the frame-shift leakiness detected in *lac* at ≤0.1% is within an order of magnitude of the maximum level ordinarily attainable (*74*).

However, natural terminators can also leak. The UGA-containing terminator of the Qβ coat protein gene is ineffective *in vivo* and *in vitro* and several percent of a longer product are produced (*76*, *77*). Fully 7% of the *in vitro* product of the synthetase genes of f2, MS2, or R17 phages has a molecular weight of 66,000 rather than the normal 63,000. Atkins (*78*) has shown that addition of one normal tRNA will increase this product, and another, corresponding to codons near the normal UAG terminator, will reduce it again. A similar effect and pair of tRNAs has been identified (*78*) for the similar readthrough of the terminator of the MS2 phage coat protein. It is argued that the modulation of the effect by normal tRNAs is most easily explained if the phase shift is due to an aberrant (e.g., two-base rather than three) translocation near the termination codon under the direction of particular tRNAs.

There are apparently similar cases in the classical work of Barnett *et al.* (*79*) on phase shifts *in vivo* in the rIIB cistron of T4 bacteriophage.[3] Combinations of two "plus" mutations nevertheless have detectable function when they bridge certain barriers (out-of-phase nonsense codons). Not all barriers cause this apparent plus shift. The barrier is essential, because mutation of the intercalated nonsense codon returns the double-plus phase shift to a nonfunctional state. Barnett *et al.* (*79*) suppose that either there may be a phase shift during infrequent anomalous propagation through the barrier, that is, during mistranslation of the terminator codon, or, alternatively, there may be termination and subsequent reinitiation in a corrected phase.

The relative frequency of frame-shifting and spontaneous termination/reinitiation seems ripe for clarification, and it appears that at least one frequent and novel translational error would thereby be delimited.

The study of termination by suppressor tRNAs has also provided the arena for discussion of another basic issue, that of the effect of factors outside the codon and anticodon on the efficiency of a given AA-tRNA in translation. Anything that selectively affects translational effectiveness of an AA-tRNA must also affect the error rate at its codons.

It is clear *a priori* that changes in a tRNA distant from the anticodon could alter its efficiency as an adaptor, and perhaps turn a marginal translation into an effective one. In fact, a mutation in the D arm of $tRNA^{Trp}$ confers the ability to read UGA (*80*). That is, the mutated UGG "reader," which still retains its original anticodon and ability to read UGG, can now also accept a C · A pair in the "wobble" position (*44*). There is evidence (*45*) that the same UGA suppressor tRNA can also translate UGU. Therefore, a mutation in the D arm has changed the spectrum of allowable pairings at the third codon position. This novel coding spectrum cannot be detected in solution when the UGA suppressor's anticodon loop pairs with oligonucleotides (*81*) or another tRNA (*82*). Therefore it is natural to suppose that the mutated tRNA has new interactions with the translation apparatus: with the ribosome, with factors, or with other tRNAs. In fact, tRNAs possess large free energies of interaction with ribosomal A and P sites independent of codon–anticodon resemblance (*29*). It is therefore plausible that these affinities could be varied to expand or contract the contribution required (for successful translation) from the codon–anticodon region. A second change in the same region of the D arm (cross-linking between thiouridine at position 8 and cytidine at position 13) also affects this tRNA: it selectively decreases the reading of UGA (*83*). The result is that the mutated and cross-linked $tRNA^{Trp}$ now behaves much as does the unmutated, uncross-linked wild-type $tRNA^{Trp}$. It seems clear that the disposition and/or flexibility of D arm can affect the disposition, flexibility, and activity of the anticodon region in the ribosome.

In addition, the efficiency of the same termination suppressor tRNA varies from site to site in messages (*84*), even when the wild-type amino acid is restored by suppression at all positions. More subtly, since ochre (UAA) suppressors also wobble to translate amber (UAG) codons, by using both ochre and amber mutations at various sites in messages it can be shown (*85–87*) that the ratio of ochre/amber translation varies for the same tRNA from site to site. Therefore, the efficiency of translation and meaning of these tRNAs is "context dependent"; it is determined by local message sequence outside the codon, since the codon is the same in all positions. A fundamental ambiguity is still unresolved, however: Is the context effect mostly exerted on the tRNAs or on the release factor that competes with them?

In any case, there seem to be contexts that intrinsically favor termination, or, alternatively, suppression by an aminoacyl-tRNA. A site at which a suppressor tRNA acts efficiently is also likely to be sensitive to streptomycin suppression (*88*). Some correlation can be found between the nature of the context effect and the nucleotide that is 3′-adjacent to the terminator codon (*87*). But since biochemical studies of suppression suggest that a terminator trinucleotide alone is sufficient for release factor action, as it is also for tRNA binding (*89*), we must assume that the contextual information is not essential, but rather modulates, peptide release.

In summary, the consideration of the ambiguous and varying translation of termination signals reminds us again (cf. Sections III and V) that there are probably a uncharacterized influences of ribosomal or messenger structure or neighboring tRNAs on the error rates at a specific codon. Error rates therefore, may have strictest meaining only in the contexts in which they are measured.

ACKNOWLEDGMENTS

I would like to acknowledge useful suggestions by Linda Breeden, David Draper, Larry Gold, Britta Singer, and Larry Soll and the support of USPHS RG GM25627.

REFERENCES

1. M. Yarus, *ARB* **38**, 841, (1969).
2. A. R. Fersht, *Proc. R. Soc. London, Ser. B* **187**, 397 (1974).
3. M. Yarus, *Bchem* **11**, 2352 (1972).
4. M. Yarus, *PNAS* **69**, 1915 (1972).
5. J. J. Hopfield, *PNAS* **71**, 4135 (1974).
6. L. Pauling, *Festschr. Prof. Dr. Arthur Stoll Siebzigsten Geburtstag* p. 597 (1958).
7. A. R. Fersht, "Enzyme Structure and Mechanism." Freeman, San Francisco, California, 1977.
8. J. Ninio, *Biochimie* **57**, 587 (1975).
9. J. Ninio, *Biochimie* **59**, 759 (1977).
10. C. G. Kurland, *Biophys. J.* **22**, 373 (1978).
11. A. R. Fersht and M. M. Kaethner, *Bchem* **16**, 1025 (1977).
12. S. M. Mulvey and A. R. Fersht, *Bchem* **16**, 4731 (1977).
13. T. Yamane and J. J. Hopfield, *PNAS* **74**, 2246 (1977).
14. M. Savageau and R. Freter, *Bchem* **18**, 3486 (1979).
15. R. B. Loftfield, *BJ* **89**, 82 (1963).
16. R. B. Loftfield and D. Vanderjagt, *BJ* **128**, 1353 (1972).
17. M. Yarus, in "Transfer RNA; Structure Properties, and Recognition" (J. Abelson, D. Soll, and P. Schimmel, eds.), p. 501. Cold Spring Harbor Lab., Cold Spring Harbor, New York, 1978.
18. P. Edelman and J. Gallant, *Cell* **10**, 131 (1977).

18a. A. Goldberg and J. Dice, *ARB* **43**, 835 (1974).

19. J. Bonnet and J. P. Ebel, *FEBS Lett.* **39**, 259 (1974).
20. A. N. Baldwin and P. Berg, *JBC* **241**, 839 (1966).
21. E. W. Eldred and P. R. Schimmel, *JBC* **247** 2961 (1972).
22. A. R. Fersht and M. M. Kaethner, *Bchem* **15**, 3342 (1976).
23. P. N. Borer, B. Dengler, I. Tinoco, and O. Uhlenbeck, *JMB* **86**, 843 (1974), and references therein.
24. G. Högenauer, *EJB* **12**, 527 (1970).
25. O. Uhlenbeck, I. Baller, and P. Doty, *Nature* **225**, 508 (1970).
26. J. Eisinger, *BBRC* **43**, 854 (1971).
27. H. J. Grosjean, S. de Henau, and D. M. Crothers, *PNAS* **75**, 610 (1978), and references therein.
28. H. J. Grosjean, D. G. Soll, and D. M. Crothers, *JMB* **103**, 499 (1976).
29. M. Peters and M. Yarus, *JMB* **133** (in press).
30. C. P. Gavrilova, O. E. Kostiashkina, O. E. Koteliansky, N. M. Rutkevitch, and A. S. Spirin, *JMB* **101**, 537 (1976).
31. W. Szer and W. Ochoa, *JMB* **8**, 823 (1964).
32. J. Davies, D. S. Jones, and H. G. Khorana, *JMB* **18**, 48 (1966).

33. C. G. Kurland, R. Rigler, M. Ehrenberg, and C. Blomberg, *PNAS* **72**, 4248 (1975)
33a. M. D. Topal and J. R. Fresco, *Nature* **263**, 289 (1976).
34. M. Ozaki, S. Mizushima, and M. Nomura, *Nature* **222**, 333 (1969).
35. L. Gorini, *Nature* **234**, 261 (1971).
36. J. Davies, W. Gilbert, and R. Gorini, *PNAS* **51**, 883 (1964).
37. P. Strigini and L. Gorini, *JMB* **47**, 517 (1970).
38. P. Strigini and E. Brickman, *JMB* **75**, 659 (1973).
39. R. A. Zimmerman, R. J. Garvin, and L. Gorini, *PNAS* **68**, 2263 (1971).
40. L. Gorini, *in* "The Ribosome" (M. Nomura, A. Tissières, and P. Lengyel, eds.), p. 791. Cold Spring Harbor Lab., Cold Spring Harbor, New York, 1974.
41. M. Grunberg-Manago and J. Dondon, *BBRC* **18**, 517 (1965).
42. J. L. Manley and R. G. Gesteland, *JMB* **125** 433 (1978).
43. A. G. So and E. W. Davie, *Bchem* **3**, 1165 (1964).
43a. J. Davies, L. Gorini, and B. D. Davis, *Mol. Pharmacol.* **1**, 93 (1965).
44. D. Hirsh and L. Gold, *JMB* **58**, 459 (1971).
45. R. H. Buckingham and C. C. Kurland, *PNAS* **74**, 5496 (1977).
46. S. L. Mitra, F. Lustig, B. Akesson, V. Lagerkvist, and L. Strid, *JBC* **252**, 471 (1977).
47. E. Goldman, W. M. Holmes, and G. W. Hatfield, *JMB* **129**, 567 (1979).
48. P. H. O'Farrell, *Cell* **14**, 545 (1978).
49. S. Parker, J. W. Pollard, J. D. Friesen, and C. P. Stanner, *PNAS* **75**, 1091 (1978).
50. C. Woese, "The Genetic Code," p. 136. Harper, New York, 1967.
51. U. Lagerkvist, *PNAS* **75**, 1759 (1978).
52. P. C. Tai, B. J. Wallace, and B. D. Davis, *PNAS* **75**, 275 (1978).
53. W. M. Holmes, G. W. Hatfield, and E. Goldman, *JBC* **253**, 3482 (1978).
54. B. Hall and J. Gallant, *Nature NB* **237**, 131 (1972).
55. F. Lustig, B. Akesson, T. Axberg, P. Elior, S. V. Mitra, and U. Lagerkvist, personal communication from Ulf Lagerkvist (1978).
56. A. L. Goldberg, *PNAS* **69**, 422 (1972).
57. J. R. Menninger, *Mech. Ageing Dev.* **6**, 131 (1977).
58. A. G. Atherly and J. R. Menninger, *Nature NB* **240**, 245 (1972).
59. J. R. Menninger, *JBC* **253**, 471 (1978).
60. J. R. Menninger, C. Walker, P. F. Tan, and A. G. Atherly, *Mol. Gen. Genet.* **121**, 307 (1973).
61. J. L. Manley, *JMB* **125**, 407 (1978).
62. A. R. Caplan and J. R. Menninger, *FP* **35**, 1352 (1976).
63. R. C. Thompson and P. J. Stone, *PNAS* **74**, 198 (1977).
64. Y. Ono, A. Skoultchi, J. Waterson, and P. Lengyel, *Nature* **223**, 697 (1969).
65. C. Cabrer, M. J. San-Millan, D. Vasquez, and J. Modolell, *JBC* **251**, 1718 (1976).
66. H. F. Lodish, *JMB* **50**, 689 (1970).
67. J. A. Steitz, *in* "RNA Phages" (N. Zinder ed.), p. 319. Cold Spring Harbor Lab., Cold Spring Harbor, New York, 1975.
68. J. G. Files, K. Weber, and J. H. Miller, *PNAS* **71**, 667 (1974).
69. V. Talkad, E. Schneider, and D. Kennell, *JMB* **104**, 299 (1976).
70. E. Scolnick, G. Milman, M. Rosman, and T. Caskey, *Nature* **225**, 152 (1970).
71. T. Caskey, A. C. Beaudet, E. M. Scolnick, and M. Rosman, *PNAS* **68**, 3163 (1971).
72. J. Smrt, W. Kemper, T. Caskey, and M. Nirenberg, *JBC* **245**, 2753 (1970).
73. M. R. Capecchi and H. A. Klein, *CSHSQB* **34**, 469 (1969).
74. J. F. Atkins, D. Elseviers, and L. Gorini, *PNAS* **169**, 1192 (1972).
75. M. R. Capecchi and H. A. Klein, *Nature* **226**, 1029 (1970).
76. K. Horiuchi, R. Webster, and S. Matsuhashi, *Virology* **45**, 429 (1971).

77. A. Wiener and U. Weber, *Nature NB* **234**, 206 (1971).
78. J. F. Atkins, *in* "Transfer RNA; Structure, Properties, and Recognition" (J. Abelson, P. Schimmel, and D. Soll, eds.). Cold Spring Harbor Lab., Cold Spring Harbor, New York, 1978.
79. L. Barnett, S. Brenner, F. H. C. Crick, R. G. Shulman, and R. J. Watts-Tobin, *Proc. R. Soc. London, Ser. B* **252**, 487 (1967).
80. D. Hirsh, *JMB* **58**, 439 (1971).
81. G. Högenauer, *FEBS Lett.* **39**, 310 (1974).
82. R. H. Buckingham, *Nucleic Acids Res.* **3**, 965 (1976).
83. J. Vacher and R. H. Buckingham *JMB* **129**, 287 (1979).
84. W. Salser, *Mol. Gen. Genet.* **105**, 125 (1969).
85. H. Yahata, G. Ocada, and A. Tsugita, *Mol. Gen. Genet.* **106**, 208 (1970).
86. S. Feinstein and S. Altman, *JMB* **112**, 453 (1977).
87. S. Feinstein and S. Altman, *Genetics* **88**, 201 (1978).
88. M. Fluck, W. Salsee, and R. Epstein, *Mol. Gen. Genet.* **151**, 137 (1977).
89. C. Caskey, E. Scolnick, T. Caryk, and M. Nirenberg, *Science* **162**, 135 (1968).

Structure, Function, and Evolution of Transfer RNAs (with Appendix Giving Complete Sequences of 178 tRNAs)

RAM P. SINGHAL AND
PAMELA A. M. FALLIS

Department of Chemistry
Wichita State University
Wichita, Kansas

[1] The authors plan to update the tRNA sequence list (Appendix A) periodically. Researchers are requested to send new tRNA sequences and changes in the known structures to the first author (RPS).

Progress in Nucleic Acid Research
and Molecular Biology, Vol. 23

ISBN 0-12-540023-3

I. Introduction

Transfer RNA is a relatively small molecule, uniquely designed to function as an intermediary between the information-carrying mRNA and the most functional macromolecule of the cell, the protein. The integrity of newly synthesized protein depends upon the abilities of the tRNAs to form esters specifically with their cognate amino acids, and to recognize the specific codons of the mRNA. This high degree of specificity derives from the sequence and three-dimensional structure of tRNA. Paradoxically, tRNAs must be sufficiently unique to be recognized individually by only cognate aminoacyl-tRNA synthetases and yet sufficiently alike to go through a common ribosomal machinery during protein synthesis.

In this study, a statistical analysis of 129 tRNAs from eukaryotic, prokaryotic, and viral sources is carried out in an attempt to answer the following questions. (*a*) What are the invariant (conserved) nucleoleotides in tRNAs? With this information, can a primordial structure reflecting the common features of each tRNA be derived? (*b*) Is the tertiary structure of yeast $tRNA^{Phe}$ representative of the tertiary bonds in other tRNAs? (*c*) What is the nature of the modified nucleotides in tRNAs? What, if any, are the differences between prokaryotic and eukaryotic modifications? (*d*) How does the T-ψ-C sequence, a feature of most tRNAs, vary in certain tRNAs? Finally, (*e*) can we predict one or more specific interaction sites between tRNA and the cognate synthetase after determining the variant (nonconserved) and the invariant residues between the isoacceptor sequences?

A. General Structure of tRNAs

The structure of tRNA, with emphasis on different aspects, has been reviewed by a number of investigators (for recent reviews, see references *1–9*). The following discussion is an attempt to summarize and update information on tRNA structure.

A generalized structure of tRNAs consists of four base-paired segments: the amino-acid-acceptor (AA) stem, the dihydrouridine (D) stem, the anticodon (AC) stem and the TΨCG (Ψ) stem (Fig. 1). The stems constitute two independent helices. The bases are paired, with few exceptions, by Watson–Crick base-pairing. The number of base-pairs is specific for each stem: the AA stem has seven, the D stem three or four, the AC and Ψ stems five base-pairs each. Extending from the helical stems are non-hydrogen-bonded loop regions: the D loop, the AC loop, and the Ψ loop. The Ψ and AC loops each contain seven nucleotides. The D loop contains two variable regions, alpha and beta,

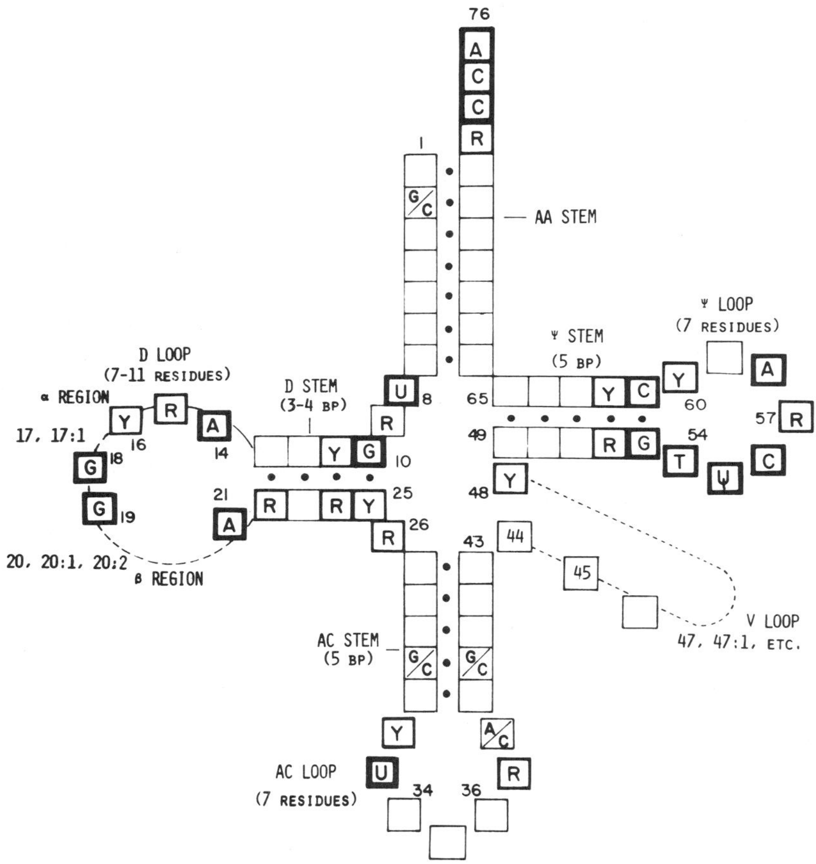

FIG. 1. A general structure of tRNAs; 16 invariant (conserved) and 20 semiconserved nucleotides (R, purine; Y, pyrimidine; G/C, G, or C; A/C, A, or C). Residues in boxes with dark borders are the invariant nucleotides present in both eukaryotic and prokaryotic tRNAs.

each containing one to three nucleotides in addition to 5 constant purine nucleotides found in all structures. The most distinct point of variability among tRNAs is the "variable" loop (also called the "extra" or the "small" loop, here called the V loop). The V loop may be short, containing 3, 4, or sometimes 5 nucleotides, or it may be long, containing as many as 21 nucleotides and creating a fifth double-helical stem of three to seven base-pairs and a loop of three to five residues. Based on the particular variability in the D-stem and V-loop regions, tRNAs may be assembled in four classes. Those tRNAs in class D_4V_5 contain four nucleotides in the D stem and 5 nucleotides in the variable loop; those in class D_3V_5, class $D_{3-4}V_4$, and class D_3V_n generally contain 3 nucleotides in the D stem and 5, 4, and an inconsistent number of nucleotides in the V loop, respectively. The basic two-dimensional "cloverleaf" structure is thus characteristic of all tRNAs, yet they may vary in length from 75 to 93 nucleotides.

B. Assignment of Residue Numbers in the tRNA Sequences

In this work, the tRNA sequences are numbered such that each tRNA can be compared directly with yeast $tRNA^{Phe}$, whose three-dimensional structure is known.[2] In order to begin with residue number one and end with residue number 76 in each tRNA sequence, three gaps in the sequential numbering are introduced at residues 17, 20, and 47, respectively. Thus, the residues in the α and β regions of the D loop are numbered 17, 17 : 1, and 17 : 2, and 20, 20 : 1, 20 : 2, and 20 : 3, respectively. Similarly, "extra" residues of the V loop are numbered 47, 47 : 1, 47 : 2, 47 : 3, etc. Exceptions to this scheme are the histidine tRNAs from *E. coli* and *S. typhimurium*, which contain one G before position 1 that is paired with the fourth position from the 3′ end. This G residue is assigned number ϕ. Since $tRNA^{Gly}$ of *S. cerevisiae* and $tRNA^{Val}$ of *T. utilis* contain only three nucleotides in the V loop, position 46 is omitted in these two tRNAs. Purine and pyrimidine nucleotides inside the heavily bordered boxes of the cloverleaf structure (Fig. 1) are found in more than 90% of the tRNAs whose structures have been determined (see Appendix A).

II. Invariant Bases of tRNAs

A. tRNAs in General

Invariant purine and pyrimidine nucleotides occurring with >60% and >90% probability are examined in 129 tRNAs, and the results are shown in Table I.[3] Twenty-three invariant residues have so far been

[2] See articles in earlier volumes of this series by Clark (Vol. 20), Kearns (Vol. 18), Kim (Vol. 17), and Cramer (Vol. 11) concerning this three-dimensional structure.

[3] When the study began, 129 tRNA sequences had been elucidated (63 prokaryotic, 57 eukaryotic, and 9 phage tRNAs). Glycine tRNAs involved in cell-wall synthesis, and the mutant tRNAs were not considered appropriate for comparison, and are not included in the statistical studies. The 129 tRNAs studied were (using the code numbers in Appendix A): 1A-A5, 1R, 2R, R4-R6, 1N, 1D, 2D, C2, 1Q, 2Q, 4Q, 1E-E4, 1G, 2G, 7G-G10, 1I, I2, 1L, 2L, 4L-L6, 1K-K3, IiM-iM5, iM7, iM9, iM10, 1M-M4, 1F-F6, F9, F11, F12, F14, 1P, 1S-3S, S5, S6, S9, S10, 1T, 2T, T4, 1W, 2W, W5, WG, 1Y, 2Y, Y4, 1V-V11.

TABLE I
NATURE OF INVARIANT RESIDUES IN 129 tRNAs[a]

Probability >90%			Probability >60%	
No. of invariants[d]	Nature[c]	Position[b]	Nature[c]	No. of invariants[d]
		1	G	94
113	G/C	2		
		4	R	76
		5	R	81
129	U,^{4}S	8		
114	R	9		
123	G,m^2G	10		
120	Y	11*	C	93
		12	Y	84
		13	C	77
128	A/m^1A	14		
128	R	15*	G	102
118	Y	16*	hU/U	86
129	G/Gm	18		
129	G	19		
		20	hU	80
129	A	21		
116	R	22*	G	86
		23	R	89
115	R	24*	G	91
118	Y	25		
120	R	26		
		27	Y	95
		29	R	88
123	G/C	30*	G	91
		31	R	83
129	Y	32*	C	91
122	U	33		
118	R	37*	A	88
113	A/C	38*	A	88
128	G/C	40*	C	91
		41	Y	89
		42	R	100
		43	R	102
		44	R	91
		46	G/m^7G	86
127	Y	48		
		49	R	97
122	R	52*	G	100
129	G	53		
113	T	54		
120	ψ	55		
129	C	56		
129	R	57		

(Continued)

TABLE I (*Continued*)

Probability >90%			Probability >60%	
No. of invariants[d]	Nature[c]	Position[b]	Nature[c]	No. of invariants[d]
129	A/m¹A	58		
		59	R	86
117	Y	60*	U	91
129	C	61*		
122	Y	62*	C	100
		63	Y	77
		65	Y	94
		67	G/C	84
		68	Y	80
		69	G/C	78
		70	G/C	89
		71	G/C	100
		72	C	95
114	R	73*	A	90
129	C	74		
129	C	75		
129	A	76		

[a] These include 63 prokaryote, 57 eukaryote, and 9 phage tRNAs, but no cell wall-synthesizing glycine tRNA.

[b] Slash sign, /, indicates presence of one or the other residue at that position. R and Y indicate purine and pyrimidine residues, respectively.

[c] Positions of the residues are with reference to the constant residues of yeast $tRNA^{Phe}$.

[d] Number of invariant residues out of 129 prokaryotic and eukaryotic tRNAs.

* These residues, R, Y, G/C, or A/C (>90% probability) can be identified as A, C, G, or U with 60% probability.

identified in each tRNA (*1, 3*). The data in Table I indicate 13 additional invariant residues occurring with about 90% probability: purine nucleotides in positions 9, 22, 26, 37, 52, and 73, pyrimidine nucleotides in positions 16, 25, and 62, G or C in positions 2, 30, and 40, and an A or C in position 38. With about 80% probability, five additional purine and pyrimidine nucleotides (R1, R7, R46, Y28, and Y66) and four specific nucleotides (C25, A37, C48, and G57) are identified as semiconserved residues.

Similarly, with at least 60% probability, all except 8 residues in the tRNA may be characterized as conserved or semiconserved, and 14 residues as specific nucleotides (compare residue numbers with asterisks in Table I).

The above invariant (23 earlier and 13 new) nucleotides therefore should be considered as parts of a "primordial" or "proto" tRNA from which at least these 129 tRNAs may have evolved (see Fig. 1).

B. Eukaryotic versus Prokaryotic tRNAs

In addition to the 36 invariant nucleotides described above, tRNAs from the eukaryotes contain 11 more in 57 sequences with >80% probability. The additional invariant residues, as shown in Fig. 2, in-

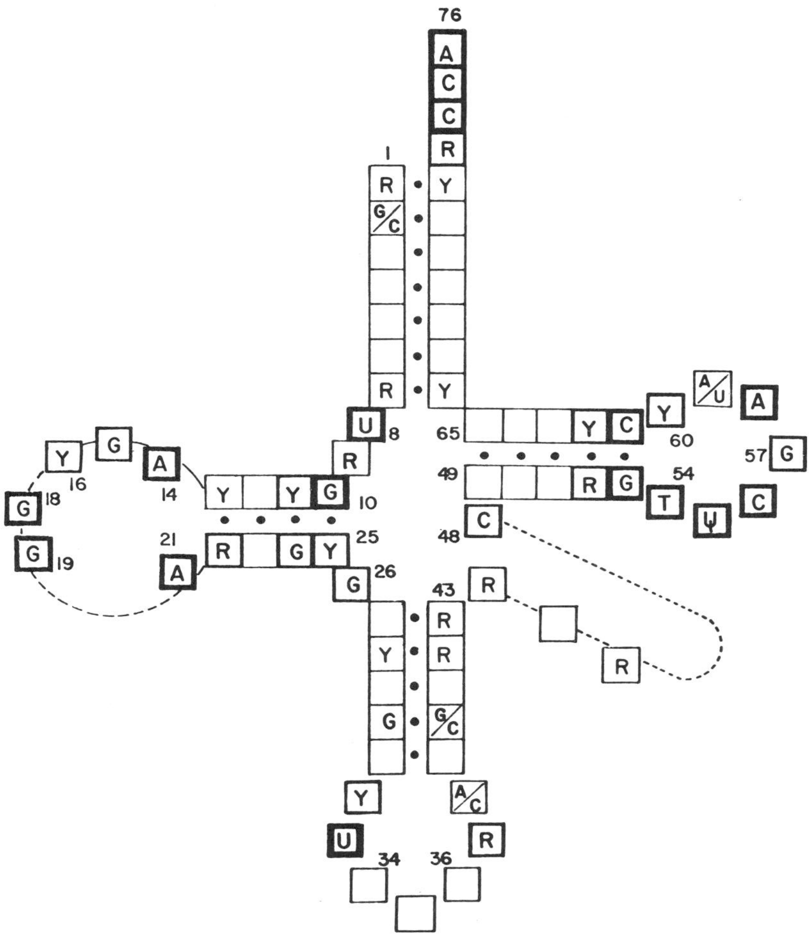

FIG. 2. A general structure of eukaryotic tRNAs based on 57 sequences with >80% probability, 22 invariant and 25 semiconserved nucleotides (see Fig. 1).

clude purine nucleotides in positions 1, 7, 42, 43, 44, and 46, pyrimidine nucleotides in positions 13, 28, 66, and 72, and either A or U in position 59. Also, several bases are identified with specific nucleotides in this "primordial" eukaryotic tRNA, for example, C in position 48 and G in positions 24, 26, 30, and 57.

From an analysis of the sequences of 63 prokaryotic tRNAs, a "primordial" tRNA sequence containing 48 "invariant" nucleotides may be derived (Fig. 3). The residues occurring with an average of

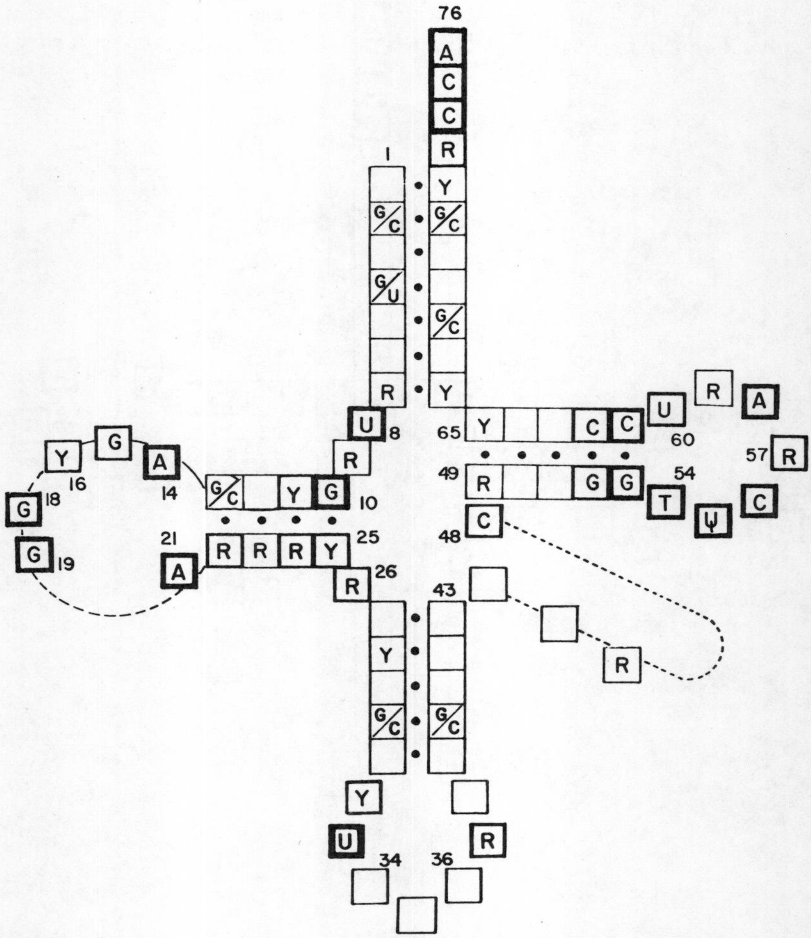

FIG. 3. A general structure of prokaryotic tRNAs based on 63 sequences with 85% probability; 21 invariant and 27 semiconserved nucleotide (see Fig. 1).

85% probability, in addition to the usual 36 residues, are: purine nucleotides in positions 7, 23, 46, 49, and 59; pyrimidine nucleotides in positions 28, 65, 66, and 72; G or C in positions 13, 68, and 71; G or U in position 4.

When the probability of a nucleotide occurring in a given position is lowered to 60%, about 20 additional "invariant" nucleotides may be identified, in addition to the 47 or 48 established for eukaryotic or prokaryotic tRNAs with >80% probability. The remaining eight or nine positions can be identified as variant nucleotide positions in 60% of the tRNAs in both eukaryotic and prokaryotic tRNAs; five positions are common to both: 17, 28, 34, 39, and 45. When invariant nucleotides with >60% probability are compared between the two kinds of tRNAs, 15 common and 11 different nucleotides are found in specific positions (see Table II).

TABLE II
NATURE OF ADDITIONAL INVARIANT RESIDUES IN PROKARYOTIC AND EUKARYOTIC tRNAs

Prokaryote tRNAs			Eukaryote tRNAs	
Invariant tRNAs[a]	Nature	Position	Nature	Invariant tRNAs[b]
47	G	1*	G	41
41	G/C	3^Δ	Y	42
50	G/C	5*	G/C	36
48	G/C	6		
42	G	7		
		9	G	37
45	C	11*	C	43
48	Y	12		
42	C	13		
42	C	16^Δ	hU	40
43	hU	20^Δ	Y	40
50	G	22*	G	34
39	A	23		
40	G	24		
49	C	25*	C	45
41	Y	27*	Y	45
44	C	28		
42	R	29*	R	38
39	G	30		
42	C	32*	C	42
		35	R	41
		36	Y	37
43	A	37*	A	40

(*Continued*)

TABLE II (*Continued*)

Prokaryote tRNAs			Eukaryote tRNAs	
Invariant tRNAs[a]	Nature	Position	Nature	Invariant tRNAs[b]
42	A	38*	A	39
39	C	40*	C	44
46	Y	41*	Y	38
39	G	42		
46	R	43		
41	R	44^Δ	A	44
39	G/m^7G	46*	G/m^7G	40
49	G	49^Δ	G/C	40
38	R	50^Δ	Y	38
40	R	51^Δ	G/C	38
		52	G	38
49	G	57		
		62	C	38
		63	G/C	43
38	Y	64^Δ	R	39
		65	G/C	37
42	C	66		
48	G/C	67^Δ	G/U	41
		68	G/C	39
40	Y	69^Δ	G/U	34
47	G/C	70^Δ	R	43
		71	G/C	39
48	C	72*	C	42
42	A	73*	A	45

[a] Invariate residues out of 63 prokaryotic tRNAs.

[b] Invariant residues out of 57 eukaryotic tRNAs.

*,Δ Indicate a common (*) or a different (Δ) residue between prokaryote and eukaryote tRNAs.

C. Evolution of tRNAs

An evolutionary theory should accommodate the interactions found *in vitro* between tRNAs and heterologous cognate synthetases. The probability that one organism can synthesize tRNAs capable of functioning in protein synthesis in another organism should be very low, but heterologous interactions do occur. Primordial sequences can be postulated on the assumption that the genetic code as we understand it today was preserved throughout the course of evolution. A "primordial" tRNA that existed billions of years ago and evolved into the tRNAs found in both simple and complex organisms can thus be derived. This phylogenic relationship can be traced from the similarities among invariant residues of various tRNAs.

Evolutionary trees have been constructed on the basis of phylogenic relationships. At the base of the tRNA evolutionary structure is a "proto-tRNA" or a primordial tRNA structure. The phylogenic distance between tRNAs (or tRNA families) is expressed as "accepted point mutations" (*10*) or the actual number of changes in nucleotide composition during evolution. The number of mutations may not equal the actual number of differences between sequences, since the nucleotide sequence present in an organism today may have undergone several changes before reaching the present status. Thus, the point mutations may be greater in number than the differences seen in a comparison of two sequences. A translation of the actual number of residue differences to accepted point mutation numbers are derived from the work of McLaughlin and Dayhoff (*10*).

Invariant residues in prokaryotic, eukaryotic, and the proposed primordial tRNA, calculated with >60% probability, have been compared. Twenty-two differences were found between the prokaryotic and eukaryotic tRNAs. Ten differences appear between the prokaryotes and the presumptive primordial tRNA, 16 between eukaryote and primordial tRNAs. According to McLaughlin and Dayhoff, 22, 10, and 16 residue differences are expressed as 27.2, 11, and 18.4 point mutations, respectively. These units of mutation define branch lengths on the evolutionary tree, thus illustrating the relationship among the evolutionary tRNAs derived here with >60% probability (Fig. 4).

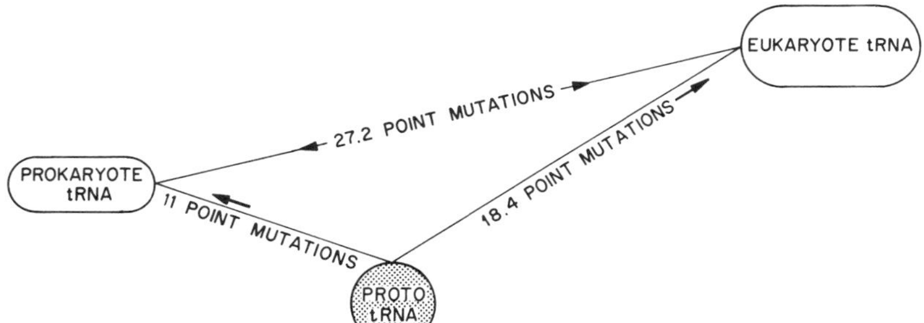

FIG. 4. Evolution of prokaryotic and eukaryotic tRNAs from a common "proto tRNA" type structure. Point mutations are derived from the variant residues found in the tRNAs from the three sources (see Figs. 1–3).

III. Three-Dimensional Structure of tRNAs

A. "Tertiary" Base-Pairs

The large number of invariant nucleotides in the tRNAs, and the fact that all the sequences can be written in the "cloverleaf" form indicate that the tRNAs may have similar "tertiary" hydrogen bonds, and therefore three-dimensional structures similar to that of yeast $tRNA^{Phe}$. Such hydrogen bonds in 129 tRNAs, based upon the $tRNA^{Phe}$ structure, are listed in Table III. The results are based on the probability of a given nucleotide occurring at positions involved in "tertiary" bonding in the $tRNA^{Phe}$ structure.

Several "tertiary" hydrogen bonds of $tRNA^{Phe}$ appear to be preserved in virtually all the tRNAs analyzed. For example, five base-pairs namely, 8·14 (U·A), 15·48 (G·C or A·U), 18·55 (G·Ψ or G·U), 19·56 (G·C), and 54·58 (T·m^1A or U·A) can occur in each tRNA. However, the remaining four can occur in only 64 to 78% of the tRNAs analyzed.

Recently, Kim (2) identified the tertiary base-pairs in the D_4V_5 tRNAs. The data are reproduced in Table IV. Most such pairs in yeast $tRNA^{Phe}$ can be replaced with equivalent ones in other tRNAs without distorting their backbone structures significantly. The three-

TABLE III
POSSIBLE TERTIARY BASE-PAIRS IN tRNAs

Residue numbers (bases in $tRNA^{Phe}$)	Type of bonding	Probability (%)	Bonding preserved (%)
(1) 8 · 14 (U · A)	U · A	68	—
	^{4}S · A	32	100
(2) 9 · 23 · 12 (A · A · U)	A · A · U	40	—
	mG · G · C	25	65
(3) 10 · 45 (G · G)	G · G	64	64
(4) 15 · 48 (G · C)	G · C/m^5C	79	—
	A · U	17	96
(5) 18 · 55 (G · Ψ)	G · Ψ	93	—
	G · U	7	100
(6) 19 · 56 (G · C)	G · C	100	100
(7) 13 · 22 · 46 (C · G · G)	C · G · C	60	—
	U · A · A	18	78
(8) 26 · 44 (G · A)	G · A	71	71
(9) 54 · 58 (T · m^1A)	T · m^1A	35	—
	U · A	65	100

Average frequence of tertiary base-pairs in tRNAs: 86%

TABLE IV
REPLACEABLE TERTIARY BASE-PAIRS IN D_4V_5 tRNAs

In yeast $tRNA^{Phe}$ residue No. (bases)	In all D_4V_5 tRNAs (51 sequences examined)	
(1) 8 · 14 (U · A)	51 U · A	
(2) 9 · 23 (A · A)	33 A · A	
	10 G · C	
	3 G · G	
	3 A · C	t
	1 A · G	
	1 G · A	
(3) 10 · 45 (G · G)	38 G · G	
	2 G · U	
	1 G · A	*
(4) 15 · 48 (G · C)	42 G · C	
	7 A · U	
	1 G · A	t
	1 A · C	
(5) 18 · 55 (G · Ψ)	51 G · Ψ	
(6) 19 · 56 (G · C)	51 G · C	
(7) 22 · 46 (G · G)	46 G · G	
	3 A · A	
	1 C · A	t
	1 G · U	
(8) 26 · 44 (G · A)	29 G · A	
	13 A · G	
	3 A · C	
	2 A · U	
	1 C · U	
	1 G · G	t
	1 A · A	
	1 Ψ · G	
(9) 54 · 58 (T · A)	51 T · A	

*: cannot make approximately equivalent H-bonds; t: requires considerable twisting around the H-bond.

dimensional structure of yeast $tRNA^{Phe}$, therefore, can be regarded as representative of all tRNAs with the possible exception of those with long variable loops.

B. Stable Stem Regions

The energy associated with the G·C base-pair (three hydrogen bonds) is higher than that in the A·U base-pair (two hydrogen bonds). Thus, a helix in which >50% of the base-pairs are G·C should be more

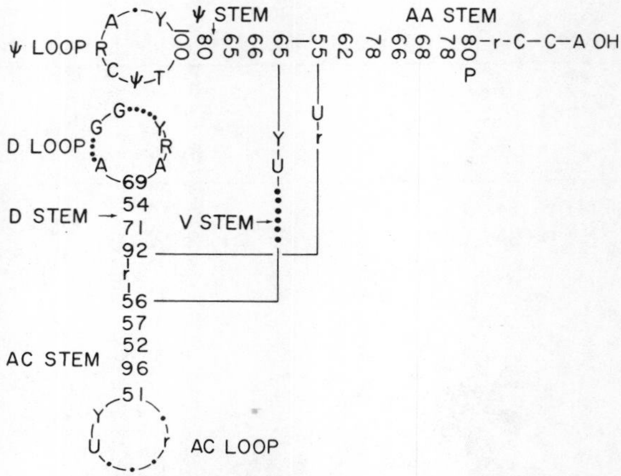

FIG. 5. Occurrence of the G·C base pairs in percent tRNAs at various positions of the two helices (S.-H. Kim, personal communication).

stable than one in which >50% are A·U. The probabilities of occurrence of G·C pairs in each position of the two helices of the tRNAs, as calculated by Kim (2), are shown in Fig. 5.

The G·C base-pair of the long helix formed by the AA and Ψ stems occurs with an average probability of 77%, the highest on the two ends of this helix. Similarily, the G·C base-pair of the short helix formed by D and AC stems occurs with an average probability of 66%, and with the lowest frequency at the base-pair located next to the AC loop. It is postulated that this base-pair may break easily during tRNA interaction with the ribosome and thus provide needed flexibility to the AC loop.

IV. Minor, Modified Residues

The presence of "modified" nucleotides[4] seems not to be responsible for the integrity of the tertiary structure of yeast $tRNA^{Phe}$. In the crystal structure,[2] modifying groups increase the accessible surface area of the bases by about 20%. The modifications may serve as specific and nonspecific recognition sites for various enzymes and protein factors.

A. Eukaryote-Specific Modifications

In the eukaryotic tRNAs, nucleotide modification is confined to twenty specific positions (Fig. 6). The residues with high incidences of

[4] See the article by Singer and Kröger on p. 151.

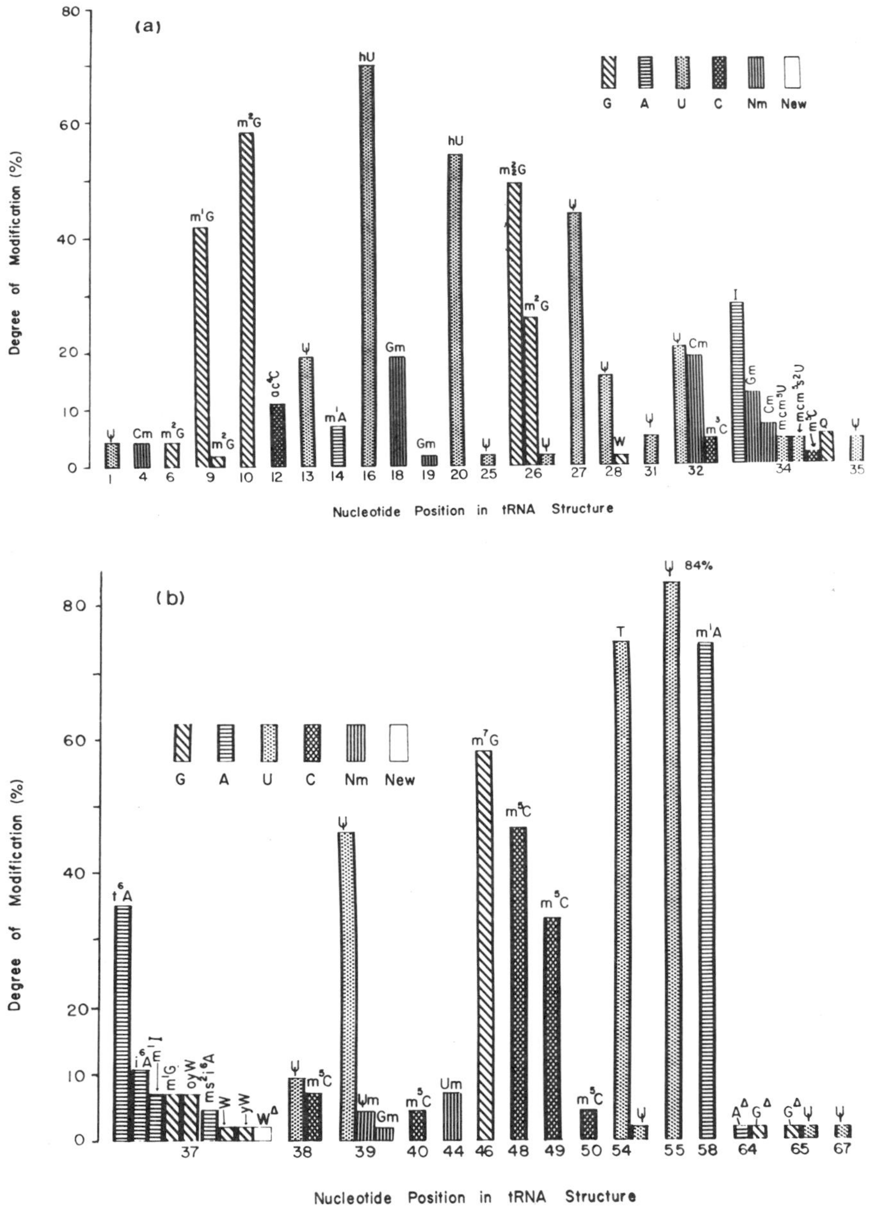

FIG. 6. (a) Modification of the nucleotides in eukaryotic tRNAs, residues 1 to 35. (b) Modification of the nucleotides in eukaryotic tRNAs, residues 36 to 76.

modification are located mostly in the short helix formed by D and AC stems (residues 10, 26, 27, 28, 48, and 49 are modified in 58, 77, 44, 20, 47, and 33% of the tRNAs, respectively). Residues 10 and 26, involved in the tertiary base-pairs, are generally present as 2-methylguanosines (see Tables III and IV). Pseudouridines are found in the AC stem (residues 27 and 28). Residue 48, base-pairing with 15, and residue 49 located at the middle of the long helix are present as 5-methylcytidines in 47 and 33% of the tRNAs, respectively.

Other specific modifications in the eukaryotic tRNAs, although seldom found (2 to 11% of the tRNAs), occur mostly in the base-paired nucleotides. For example, the long helix contains six (residues: 1, 4, 6, 50, 64, and 67) and the short helix four such nucleotides (residues: 2, 25, 31, and 40); two residues, 14 and 19, involved in the tertiary base-pairs are modified similarly. Thus, residues 35 (in the anticodon of yeast $tRNA^{Tyr}$) and 44 (modified as Ψ and Um) are the only ones that do not participate in base-pairing. Methylation, acetylation (m^2G, m^1A, m^5C, ac^4C), pseudouridylation, and ribose methylation (Cm, Um, Gm) are the most common modifications found in the base-paired nucleotides.[4]

B. Prokaryote-Specific Modifications

There are only three such modifications. Residue 8, present as 4-thiouridine in 63% of the prokaryotic tRNAs,[5] should make a tertiary base-pair with residue 14. Residues 22 and 41, both located in the short helix, are modified as m^1A and Ψ in only a few tRNAs (Fig. 7).

C. Modifications Common to Both Types of Organism

Modification of thirteen specific nucleotides is found in both prokaryotic and eukaryotic tRNAs (not including residues 34 and 37, see below). Three nucleotides of the D loop are dihydrouridine in most tRNAs in positions 16 and 20, and 2′-*O*-methylribose in position 18. Residue 39 of the short helix is pseudouridine in 45% of tRNAs of both kinds, yet another residue (No. 13) of the same helix, also pseudouridine, appears as such in only 5% of prokaryotic tRNAs. The pyrimidine nucleotide 32 located at the 5′ side of the anticodon is modified in eukaryotic tRNAs more often than in the prokaryotic tRNAs. Except for ribose methylation and pseudouridine occurrence, modifications of this nucleotide 32 are different in the two kinds of

[5] The proportion of 63% 4-thiouridine in position 8 reflects the large number of *E. coli* tRNAs analyzed. However, this modification (U to s^4U) is much lower in or absent from a number of other prokaryotic tRNAs.

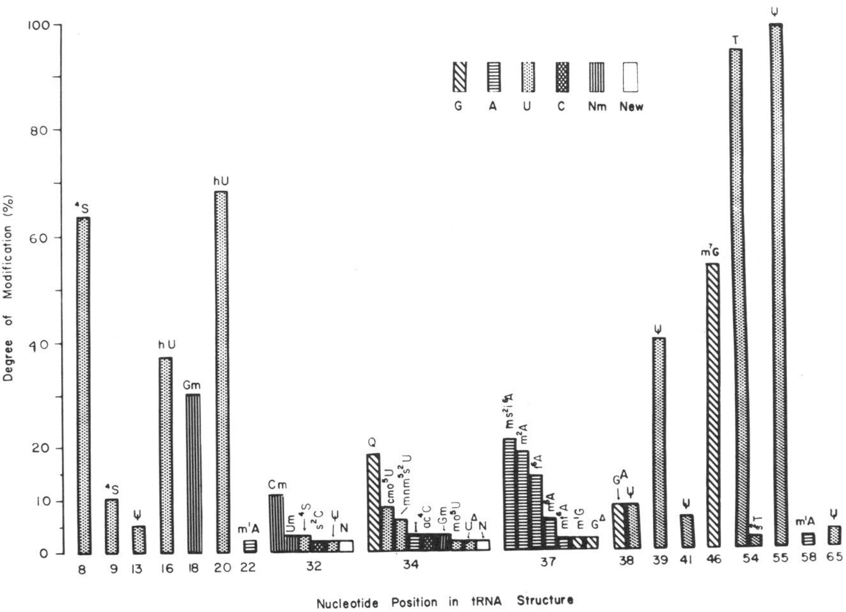

FIG. 7. Modification of the nucleotides in prokaryotic tRNAs, residues 1 to 76.

tRNAs. However, residue 38 located on the opposite side of the AC loop, is modified in an analogous manner in the two tRNAs. Similarily, residue 46 of the V loop and 54 and 55 of the Ψ loop bear common features in their modification groups and frequency. In contrast, two residues 9 and 58 involved in the tertiary base-pairs are modified rarely in prokaryotic, but very significantly (44 to 75%) in eukaryotic tRNAs.

D. Wobble Base Modifications[4]

The "wobble" base located in position 34 is modified in 61% of eukaryotic and 47% of prokaryotic tRNAs. (a) The modification A → I is very common in alanine, arginine, isoleucine, serine, threonine, and valine tRNAs of eukaryotes, but these tRNAs in prokaryotes contain a variety of modifications (I, mo^5U, cmo^5U or G). (b) The modification of G → Q is very common in prokaryotic asparagine, aspartate, histidine, and tyrosine tRNAs, but these tRNAs from yeasts contain a simple G. However, these same tRNAs from mammals contain either a simple Q ($tRNA^{Asn}$ and $tRNA^{His}$) or a modified Q (manQ in $tRNA^{Asp}$ and galQ in $tRNA^{Tyr}$). The absence of Q from yeast tRNAs probably indicates evolution in the eukaryotes independent of that of the prokaryotes.[6] (c) The modification of $U \rightarrow s^2U$ occurs in the prokaryotes (mnm^5s^2U in $tRNA^{Gln}$, $tRNA^{Glu}$, and $tRNA^{Lys}$) as well as in yeast tRNAs (mcm^5s^2U in $tRNA^{Glu}$ and $tRNA^{Lys}$). Although such mammalian tRNAs have not been sequenced yet, *in vivo* thiolation studies show the presence of 2-thiouridine derivatives in tRNAs from human lymphocytes (*11*) and 5-methyl-2-thiouridine occurs in mammalian glutamate and lysine tRNAs (*11a*).

In the wobble position, uridine is modified in many different ways and is present only as uridine in three tRNAs, *Staphylococcus* $tRNA_1^{Gly}$, baker's yeast $tRNA_1^{Leu}$, and *E. coli* psu^+ 7oc $tRNA^{Trp}$. With no exceptions, A is always modified to I. It has been suggested that the presence of simple U or A in this position may be lethal (*12*). Modification of the wobble nucleotide has been shown to enhance strict base-pairing with certain codons, while lowering or abolishing recognition with the other codons. For example, tRNAs with 2-thiouridine in position 34 recognize codons ending with A, but not with G (*12, 13*). Using the temperature-jump relaxation technique for the kinetic study of anticodon–anticodon association between two different tRNAs, Grosjean *et al.* have shown that $G \cdot s^2U$ base-pairing is forbidden for a steric reason, whereas a pyrimidine base (U or C)$\cdot s^2U$ interaction is allowed. The base-pair Q·C is less stable than G·C. However, Q·U is significantly more stable than G·U. Therefore, to achieve stable base-pairing between the codons ending with U and the wobble base, and to discriminate against binding of the tRNAs with codons ending with C, a modification of $G \rightarrow Q$ in the wobble position is necessary (*14*). The relationship between the modification of the wobble nucleotide and differences in codon recognition are shown in Table V.

In the bacteriophage MS2 RNA, the wobble codon is found more often in codons containing G or C, or both, in the first two positions, and less often in the codons containing A and U in these positions (*14a*). The degenerate code thus occurs only in the codes that can interact (hydrogen-bond) strongly with the tRNAs. The nonrandom distribution of the degenerate codes is therefore related to the stability of the codon–anticodon interaction in the prokaryotes.

[6] The absence of Q in tRNAs from livers of week-old rats and leukemic cells (R. P. Singhal, R. A. Kopper, and S. Nishimura, unpublished) and in animal tRNAs during development (Kasai *et al. NARes* **2**, 1931, 1975) has been reported. However, this deficiency of Q in animal tRNAs appears to be the result of incomplete post-transcriptional modification of the residues in the actively dividing cells. Bean chloroplast tRNA has little or no queuine modification (D. B. Dunn and M. D. M. Trigg, unpublished).

TABLE V
WOBBLE-BASE MODIFICATION AND CODON RECOGNITION

tRNAs encoding for:	Wobble-base modification	Prefers codons	Does not prefer codons
1. (A) In eukaryotes and procaryotes Arg$_2$ Ile, Thr; (B) In eukaryotes only Ala, Ser,[a] Val	A → I	NNA, NNC, NNU	NNG
(C) In yeasts Arg$_3$	U → mcm^{5}U	NNA	NNG[b]
(D) In prokaryotes only Ala, Ser, Val	U → cmo^{5}U	NNA, NNG, NNU	NNC
2. In prokaryotes and mammals[b] Asn, Asp, His, Tyr	G → Q or Q$^{\Delta}$	NNU	NNC[c]
3. (A) In prokaryotes Glu, Gln, Lys; (B) In yeasts Glu, Gln, Lys	U → mnm^{5}s^{2}U; U → mcm^{5}s^{2}U	NNA	NNG
(C) In mammals Glu, Lys	U → m^{5}s^{2}U		

[a] Instead of the usual modification of A → I in eukaryotic serine tRNAs, a different modification, U → mcm^{5}U in serine tRNA$_{2B}$, has been found in mammalian liver by D. B. Dunn and M. D. M. Trigg (*John Innes Institute Report* **67**, 121, 1976). This modified base also exhibits "wobble" characteristics (M. Staehelin, *Experentia* **27**, 1, 1971).

[b] Codon preference for this change is uncertain.

[c] Codon preference for only Q-tRNAs from *E. coli*[45] and not for Q- or Q$^{\Delta}$-tRNAs from mammals has been determined.

E. Purine Base in Position 37

Eukaryotic tRNAs are modified more than are the prokaryotic tRNAs (75% versus 62%) in this position. In the eukaryotes, all G residues in position 37, present in 19% of the tRNAs, are modified. In the prokaryotes, however, only one-fourth of the G's in this position are modified. Unmodified A is present to the same level (25 to 30%) in the two kinds of tRNAs.

An important correlation exists between the nature of the first base of the codon and modification of the purine base 37 located next to the anticodon (3′ side). The results are summarized in Table VI. As initially indicated by Nishimura (*13*), codons beginning with U or A

TABLE VI
RELATIONSHIP BETWEEN THE NATURE OF THE CODON AND THE MODIFIED BASE LOCATED ADJACENT TO THE ANTICODON

	Nature of the modified base 37 in tRNAs	
Codons	Prokaryotes	Eukaryotes
UNN	ms^2i^6A	i^6A
	m^1G (exception[a])	W ($tRNA^{Phe}$)
		m^1G and A (exceptions[b])
ANN	tc^6A	tc^6A
	A and m^1A (exceptions[c])	
CNN	m^1G or m^2A	m^1G
GNN	A, m^2A or m^6A	A or m^1I

[a] Phenylalanine tRNA of *Mycoplasma* contains m^1G.

[b] Leucine tRNAs from yeasts and tryptophan tRNAs from chicken cells and liver contain m^1G; tryptophan tRNA of yeast contains unmodified A.

[c] All prokaryote initiator tRNAs so far sequenced contain unmodified A, but initiator tRNA of *N. crassa* (mitochondria only) contains m^1G.

contain i^6A or t^6A. The base-pair A·U is weakly bonded. It has been suggested that the presence of a hydrophobic chain (isopentenyl chain on N^6) or a hydrophilic group (carbamoylthreonine on N^6) stabilizes the A·U bond, and thus overcomes the weak interaction associated with this bond. Since G·C is a strong base-pair, such stabilizing forces are not necessary. Accordingly, tRNAs that recognize the codons beginning with C or G contain no or little modification in this position (one methyl group on A, I, or G; see Table VI). The nucleotides in both positions 34 and 37 appear to be less modified as far as the extent of the modification is concerned (e.g., Q cf. manQ or galQ; m^1G cf. W) in prokaryotic than in eukaryotic tRNAs. The modification of the purine base adjacent to the anticodon appears to enhance interaction between the ribosome and the tRNAs. Such interaction can be achieved by introduction of the hydrophobic or hydrophilic modifying groups in A37 of tRNAs, which interact with the codons beginning with A or U. Thus, these modifications of A37 can enhance base-pairing of the A·U type between residue 36 of the tRNAs and the first letter of the codon.

V. Changes in the T-Ψ-C-G Sequence

The sequence located between residues 54 and 57 is a constant feature of most tRNAs. Changes in that sequence (see Table VII) appear to occur in the following four ways: (*a*) T (m^5U) is further mod-

TABLE VII
CHANGES IN T-Ψ-C SEQUENCE OF Ψ LOOP

Changes	tRNA	Source
(*a*) T → U[Δ]		
(1) Tm-Ψ-C	Lys	Liver; *Drosophila melanogastor*[a]
(2) s^2T-Ψ-C	Initiator	*Thermus thermophilus*[b]
(3) U-Ψ-C[c]	Phe	*Mycoplasma;* yeast mitochondria; rabbit liver
	Initiator	*Mycoplasma*
	Lys	Rabbit liver
	Val_1	Mouse myeloma and mammals
	Gly	Wheat germ
(4) Ψ-Ψ-C	Trp	Calf liver; chicken embryo cells; retro-transcriptase primer in one mouse sarcoma virus
	Pro	Mouse and chicken cells; retro-transcriptase primer in murine leukemia virus
(*b*) T → A		
A-Ψ-C	Initiator	Wheat germ
	Ala	*Bombyx mori*
(*c*) Ψ → U		
A-U-C	Initiator	Most eukaryotes
(*d*) TΨ → UG		
U-G-C	Gly (cell wall)	*Staphylococcus aureus* and *S. epidermidis*
	Initiator	*N. crassa* (mitochondria)

[a] Gross *et al., NARes* **1,** 35 (1974); Silverman *et al., NARes* **6,** 435 (1979).

[b] In *T. thermophilus,* besides the initiator tRNA, most other tRNAs are modified similarily, as reported by Watanabe *et al., FEBS Lett* **43,** 59 (1974). This modification (U → s^2T) appears to be necessary for the structural integrity of tRNAs at the high temperature at which the organism grows.

[c] Since both U and m^5U are found in position 54 of the same sequence, this sequence (UΨC) in several tRNAs could be the result of incomplete modification of U to m^5U.

ified (methylated or thiolated) or is replaced by U or Ψ; (*b*) T is replaced by a purine nucleotide, A; (*c*) Ψ is replaced by unmodified U; (*d*) both T and Ψ of the TΨCG sequence are replaced by U and a purine nucleotide, G, respectively. The first type of modification (*a*) is found in a variety of tRNAs, including "primer" tRNAs serving for the "reverse transcriptase" in DNA synthesis. While modification of types *b* and *c* are specific to eukaryotic initiator tRNAs, type *d*, which is most drastic, seems confined to the cell wall synthesizing $tRNA^{Gly}$ (exceptions: type *b,* tRNA of *B. mori;* type *c,* initiator tRNA of *N. crassa*).

VI. Structural Features of Initiator tRNAs

The structures of the initiator tRNAs of prokaryotes differ from those of eukaryotes. The prokaryotic initiator tRNAs maintain an excellent homology by having a unique feature in their structures: there is no hydrogen bonding between residue 1 (5′-terminal nucleoside) and residue 72 (fifth nucleoside from the 3′ terminus). Even initiator tRNAs from *Mycoplasma mycoides* sp. (*15*) and *Thermus thermophilus* (*16*) have this constant feature. Except for this lack of bonding between residues 1 and 72, the sequences of the prokaryotic initiator tRNAs are very similar to those tRNAs involved in chain elongation. However, the eukaryotic initiator tRNAs contain a unique sequence, A-U-C-G (or A-Ψ-C-G, as in wheat germ) in place of the invariant T-Ψ-C-G sequence in the Ψ loop of the noninitator tRNAs (exception: $tRNA^{Ala}$ from *Bombyx mori*). A base-pair between G 2 and C71 appears to be another unique feature of both eukaryotic and prokaryotic initiator tRNAs (exception, in wheat germ this pair is replaced by A2·U71). Among the eukaryote initiators, sequence homology ranges between 73 and 100%. Thus, the initiator tRNA sequences appear to be conserved among all vertebrates examined to date (*17*, *17a*).

VII. Mitochondrial tRNAs[7]

Mitochondria contain a partially autonomous protein-synthesizing machinery. Mitochondrial DNA transcribes no fewer than 19 different tRNAs (*18*). The absence of certain tRNAs in mitochondria [glutamine, asparagine, histidine, and proline tRNAs in HeLa cells (*19*) and asparagine tRNA in yeast (*20*)] has been explained by the limitations of current techniques (*21*) and modification of the amino acid after aminoacylation (e.g., Glu → Gln) (*22*).

These tRNAs contain more A and U residues and very few modified nucleotides. For example, yeast mitochondrial $tRNA^{Phe}$ contains 15% C and 20% G and lacks the usual modified residues, such as yW and T (*23*). Another unusual feature of yeast and *Neurospora* mitochondrial tRNAs is the complete absence of 7-methylguanylate; however, this modification does occur in mammalian mitochondrial tRNAs. If one accepts the hypothesis that mitochondria, like chloroplasts, evolved from prokaryotes by endosymbiosis, then the absence of wyebutosine is not surprising. Similarily, chloroplast $tRNA^{Phe}$ contains

[7] See article by Barnett *et al.* in Vol. 21 of this series.

6-isopentenyladenylate with a 2-methylthio group, a predominantly prokaryotic modification.

VIII. Interaction between tRNA and Cognate Synthetase

One or more isoacceptors of a given tRNA are recognized by the same cognate aminoacyl-tRNA synthetase, therefore variant residues or segments of the tRNAs rich in residues that vary from one isoacceptor to another cannot be considered as residues or segments involved in the interaction between tRNA and the cognate synthetase. On the other hand, the invariant (conserved) residues and *un*altered parts of the tRNA can be regarded as potential interaction sites, both specific and nonspecific, between the two molecules. We have compared the structures of isoaccepting tRNAs in all cases where the sequences are known and have related these comparisons to the results obtained from chemical-modification studies[4] in conjunction with the assay of their biological activity.

A comparison of the isoacceptor sequences recognized by the same cognate aminoacyl-tRNA synthetase together with studies on the biological effects resulting from chemical modifications of tRNAs suggests that three segments of the tRNA can be regarded as potential tRNA and synthetase interaction sites. The segments are located in the anticodon loop, the D stem, and both the AA stem and the D stem. We have assembled the tRNAs thus analyzed in the above three groups. Data from isoacceptors are discussed in all instances where sequences are known.

A. Anticodon Group

In several tRNAs, the nucleotides of the anticodon loop appear to be involved in the recognition process and include yeast $tRNA^{Val}$, yeast $tRNA^{Lys}$, *E. coli* $tRNA^{Glu}$, *E. coli* $tRNA^{Trp}$, *E. coli* $tRNA^{Gly}$, and *E. coli* $tRNA^{fMet}$.

1. Yeast $tRNA^{Val}$

Three yeast $tRNA^{Val}$ and two $tRNA^{Lys}$ isoacceptors are compared within each group, and the positions of nucleotide change are shown by arrows. Nucleotide changes (variant residues) among three isoacceptors are shown in Fig. 8. The changes are noted in 35 of the 78 residues. When common residues of eukaryotic tRNAs and the variant residues between these isoacceptors are disregarded, only a very few residues remain available as possible synthetase recognition sites. The residues of the 3′ side of the AC loop are the most conserved ones. The

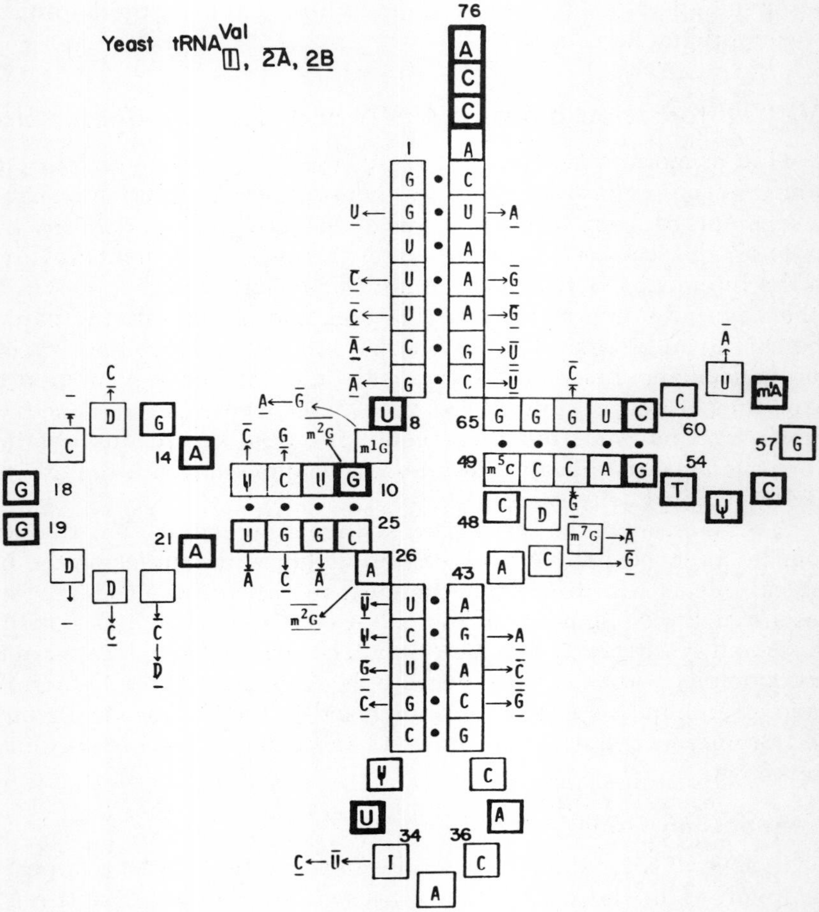

FIG. 8. Comparison of $tRNA^{Val}$ isoacceptor structures from yeast. Nucleotides of $tRNA_1^{Val}$ sequence are shown in the square boxes; changes from this structure are indicated outside the boxes with bars above and below the nucleotides for $tRNA_{2A}^{Val}$ and $tRNA_{2B}^{Val}$, respectively. A bar without nucleotide, outside the box, indicates deletion of that particular nucleotide in $tRNA_2^{Val}$ isoacceptors.

evidence from the chemical modification studies also indicates that the 3′ side of the AC loop is most likely the synthetase and tRNA interaction site for this tRNA. For example, Mirzabekov (*24*) used limited nuclease digestion on this tRNA and found that removal of a pentanucleotide, including three residues from the AC loop and the first G residue on the 3′ side of the AC stem, destroyed valine acceptor activity. This study was later modified to include only the two nucleotides on the 3′ end of the anticodon, eliminating the effect of the G residue and also the wobble nucleotide (*25*). Chambers *et al.* (*26, 27*) introduced a C → U change in the third position of the anticodon (residue 36) with the help of bisulfite. The result was a missense change (IAC → IAU); that is, the anticodon specific for the valine codon was thus abolished and the modified tRNA became specific for the isoleucine codon. At the same time, the chemical change also abolished its ability to esterify valine. Thus, position 36 was defined as an inactivation site. Since the sequences of the three isoacceptors showed changes in the wobble position (in one case I → U and in the other I → C), this particular residue cannot be involved in the interaction between tRNA and the enzyme, thus indicating the 3′ side of the anticodon as the potential interaction site.

2. *Escherichia coli* $tRNA^{Glu}$

A comparison of two isoacceptors of *E. coli* $tRNA^{Glu}$ showed only one nucleotide variation in the AA stem between the two sequences. Thus, little can be concluded from this comparison. However, experimental data based on chemical modification studies support the anticodon as a recognition site. Saneyoshi and Nishimura (*28*) modified the wobble nucleotide of *E. coli* $tRNA^{Glu}$ with cyanogen bromide, producing a thiocyanate derivative that had no biological activity. Singhal confirmed these results using cyanogen bromide and *p*-chloromercuribenzoate modifications (*29*). Furthermore, he converted cytidine residues located in the most exposed parts of the molecule into uridines using the bisulfite-mediated C → U conversion technique (*30*). A drop in aminoacylation ability accompanied changes in residues 32 and 36 (residues 33 and 37 according to the sequential numbering scheme used in refs. *29* and *30*). The two residues are located in the AC loop, position 32 in the first position on the 5′ side and position 36 in the third position of the anticodon. The oligonucleotide structure of the anticodon imposed limitations in distinguishing reactivities among the three cytidines of this fragment (*29*). (The pancreatic RNase digest produced pyrimidine mononucleotides and m^2A-Cp and RNase T1 produced a decanucleotide.) However, Hayatsu's group (*31*), using a deuterium-labeling technique, has found that the most reactive cytidines in this tRNA are located on the 3′ side of the anticodon.

3. Yeast $tRNA^{Lys}$

Two isoacceptors of yeast $tRNA^{Lys}$ show changes similar to those of the $tRNA^{Val}$ from yeast (Fig. 9). Although no chemical modification data are available for this tRNA, our results strongly suggest that the 3′

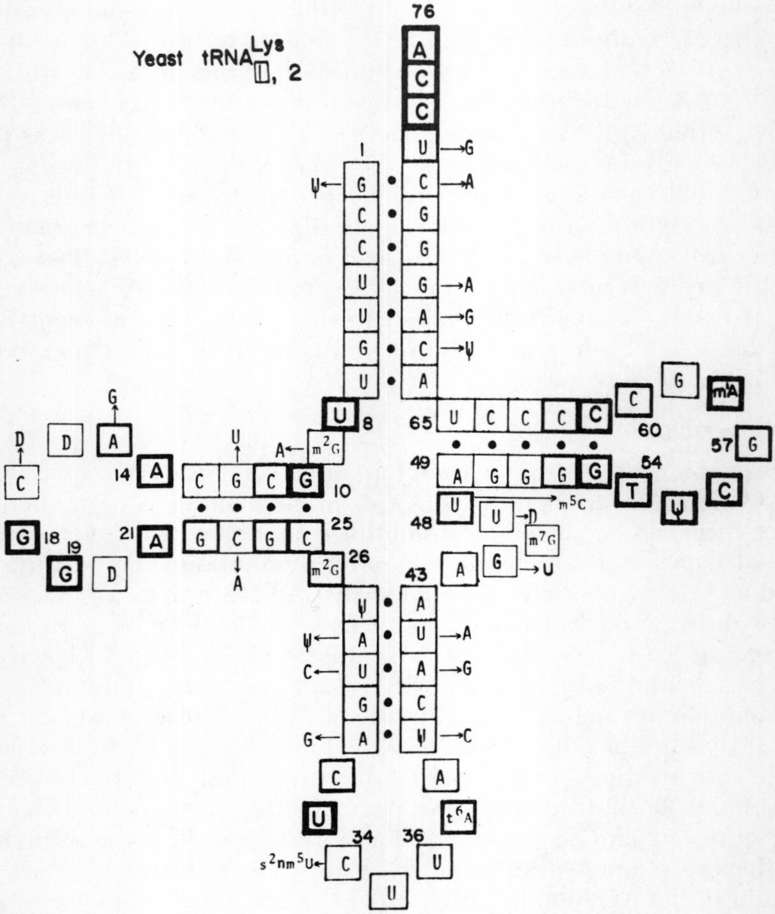

FIG. 9. Comparison of $tRNA^{Lys}$ isoacceptor structures from yeast. Nucleotides of $tRNA_1^{Lys}$ sequence are shown in the square boxes, and changes from this sequence in $tRNA_2^{Lys}$ structure are shown outside the boxes.

side of the anticodon nucleotides can be the interaction site, perhaps solely or in addition to one or two invariant base-pairs located in each AA stem and D stem.

4. *Escherichia coli* $tRNA^{Trp}$ AND $tRNA^{Gly}$

Anticodons of *E. coli* $tRNA^{Trp}$ and $tRNA^{Gly}$ have been suggested as the interaction site on the basis of gene mutation studies. Carbon and co-workers (*32–34*) obtained mutant missense suppressors of *E. coli* $tRNA^{Gly}$. When nucleotide changes in either of the two positions at the 3′ side of the AC loop were introduced, significant losses in the aminoacylation rate were observed. However, a nucleotide change in the wobble position had no effect on the interaction between tRNA and the synthetase. Missense mutation of *E. coli* $tRNA_3^{Gly}$ produced a C → A change at the 3′ side of the AC loop, and modification of the adjacent A into ms^2i^6A. These structural changes in the mutant tRNA reduced the rates of aminoacylation by the cognate synthetase (*33*). Similarly, a missense change in the third position of the anticodon of *E. coli* $tRNA_2^{Gly}$ caused a loss in the amino-acid acceptor activity (*34*).

The AC loop can also be regarded as a recognition site for *E. coli* $tRNA^{Trp}$. A mutational change of C → U in the first or second position (or both) of the AC loop of $tRNA^{Trp}$ resulted in a UAG or UAA suppressor codon accompanied by a total loss of tryptophan-accepting activity (*35*).

5. *Escherichia coli* Initiator tRNA

A comparison of *E. coli* initiator tRNAs indicates only one base change between their two isoacceptors. Two chemical alterations involving a photooxidation of G71 (*36*) and C → U changes in positions C35 and C75 (*37*) cause a loss of methionine acceptor activity for this tRNA. Thus, a modification of the wobble nucleotide or changes in the AA stem appear to abolish the interaction sites in the tRNA.

B. D-Stem Group

Transfer RNAs in which the D-stem can be considered as a recognition site are $tRNA^{Phe}$ from yeast and $tRNA^{Val}$, $tRNA^{Ile}$, $tRNA^{Met}$ from *E. coli.*

1. *Escherichia coli* $tRNA^{Val}$

The structures of three isoacceptors of *E. coli* $tRNA^{Val}$ are compared in Fig. 10. The AA stem, the Ψ stem and loop, the AC stem and loop, and the D loop contain base changes between the isoacceptors and, therefore, they cannot be the potential sites for synthetase and tRNA interaction. The nucleotides in the D stem, however, remain unaltered. Therefore, the D stem appears to be the recognition site for *E. coli* $tRNA^{Val}$. Wubbeler *et al.* (*38*), working with heterologous halves of yeast $tRNA^{Ala}$, and *E. coli* $tRNA_1^{Val}$, suggested that the 5′ (side) half-molecule, including the D region, are involved in the

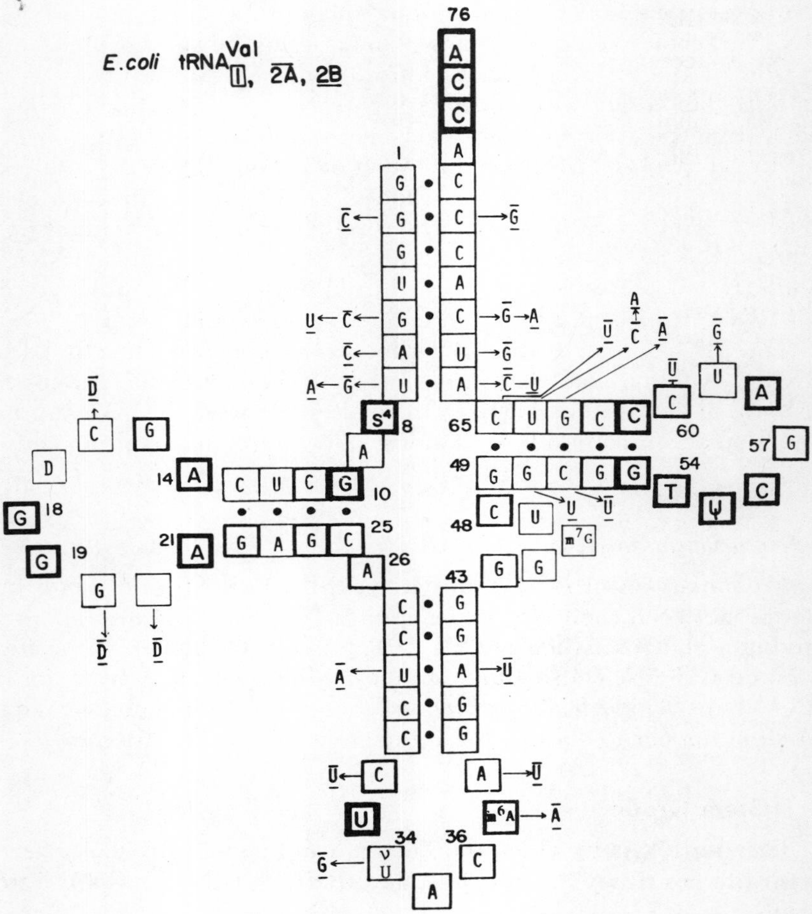

FIG. 10. Comparison of $tRNA^{Val}$ isoacceptor structures from *Escherichia coli*. Nucleotides of $tRNA_{1}^{Val}$ sequence are shown in the square boxes; changes from this structure are indicated outside the boxes with bars above and below the nucleotides for $tRNA_{2A}^{Val}$ and $tRNA_{2B}^{Val}$, respectively.

synthetase recognition. However, CCA-halves alone can be recognized by the synthetase, implying that the D region, although important, cannot be essential for recognition. Aminoacylation of eleven structurally similar tRNAs (including isoacceptors of *E. coli* valine tRNAs) with one synthetase showed that the D-stem region and the A73 in the fourth position from the 3′ end are necessary for the aminoacylation (*39*).

2. Yeast $tRNA^{Phe}$

No isoacceptors for this tRNA have been sequenced. However, the D stem has been implicated in recognition by others. Schoemaker *et al.* (*40*) photochemically cross-linked yeast $tRNA^{Phe}$ with yeast phenylalanyl-tRNA synthetase, *E. coli* isoleucyl-tRNA synthetase, and *E. coli* valyl-tRNA synthetase. Two parts of this $tRNA^{Phe}$ were found in each complex as common contact points between all three synthetases and the tRNA (residues 5–10 of the D stem and residues 46–51 of the V loop and Ψ stem); one unique binding region was found in each complex. These three enzymes appeared to recognize $tRNA^{Phe}$ by two distinct regions of the molecule. They were located close to each other and made physical contact with other parts of the molecule based upon three-dimensional structure. Other work also indicates that the D stem may be recognition site for this tRNA (*39*).

Contradictory evidence for the concept of D stem as the recognition site for these tRNAs arises from a consideration of the topography of the cognate synthetase-tRNA complex (*41*). Phenylalanyl-tRNA synthetase and $tRNA^{Phe}$ were allowed to complex, then digested with a nuclease. Transfer RNA segments in physical contact with the synthetase (i.e., protected from nuclease attack) were isolated and characterized. These segments came from the D loop and the anticodon region.

3. *Escherichia coli* $tRNA^{Ile}$

Isoacceptors of *E. coli* $tRNA^{Ile}$ were not available for comparison. However, photochemical cross-linking of *E. coli* $tRNA^{Ile}$ to *E. coli* isoleucyl-tRNA synthetase and to yeast valyl-tRNA synthetase suggested that the D stem and D loop regions of the molecule are involved in the recognition. The enzymes were cross-linked to two distinct areas of the tRNA: residues 11–15 and 20–23, located on either side of the D stem and D loop, respectively. A third area was also cross-linked to the tRNA molecule, specific to each synthetase (*42*).

4. *Escherichia coli* $tRNA^{Met}$

The only experimental evidence for D stem recognition in $tRNA^{Met}$ is found in the work on eleven tRNAs described earlier (*39*). A comparative study of isoacceptor structures shows only one base change in the D loop for this tRNA.

C. Acceptor Stem or D Stem or Both Stems

The tRNAs recognized by the AA stem or D stem (or both stems) include *E. coli* $tRNA^{Ser}$, rat liver $tRNA^{Ser}$, yeast $tRNA^{Ser}$, yeast $tRNA^{Leu}$, *E. coli* $tRNA^{Leu}$, yeast $tRNA^{Ala}$, and *E. coli* $tRNA^{Tyr}$.

1. *E. coli*, Rat Liver, and Yeast $tRNA^{Ser}$

Isoacceptors of $tRNA^{Ser}$ from *E. coli* and rat liver are compared in Figs. 11 and 12. *Escherichia coli* $tRNA^{Ser}$ contains 42 base differences spread over all parts of the molecule. When constant and variant residues are disregarded, the D stem and two base-pairs in the AA stem (G1·G72 and G2·C71) emerge as possible recognition sites (see Fig. 11). Serine tRNA from rat liver gives similar results though with fewer base changes (Fig. 12). The residues of Ψ stem, V loop, AC stem, and AC loop, and most of the AA stem are altered between the two isoacceptor sequences. However, no change is found in the D stem and D loop nucleotides. The fourth nucleotide from the 3′ end and the first, fourth, and seventh base-pairs of the acceptor stem can also be consid-

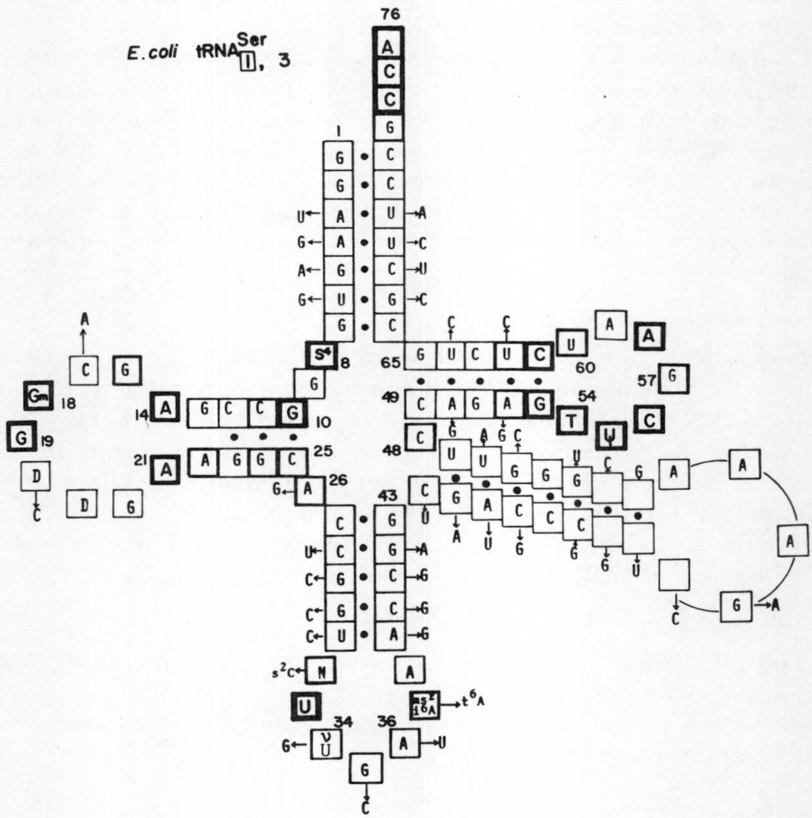

FIG. 11. Comparison of $tRNA^{Ser}$ isoacceptor structures from *Escherichia coli*. Nucleotides of $tRNA_1^{Ser}$ sequence are shown in the square boxes, and changes from this sequence in $tRNA_2^{Ser}$ structure are shown outside the boxes.

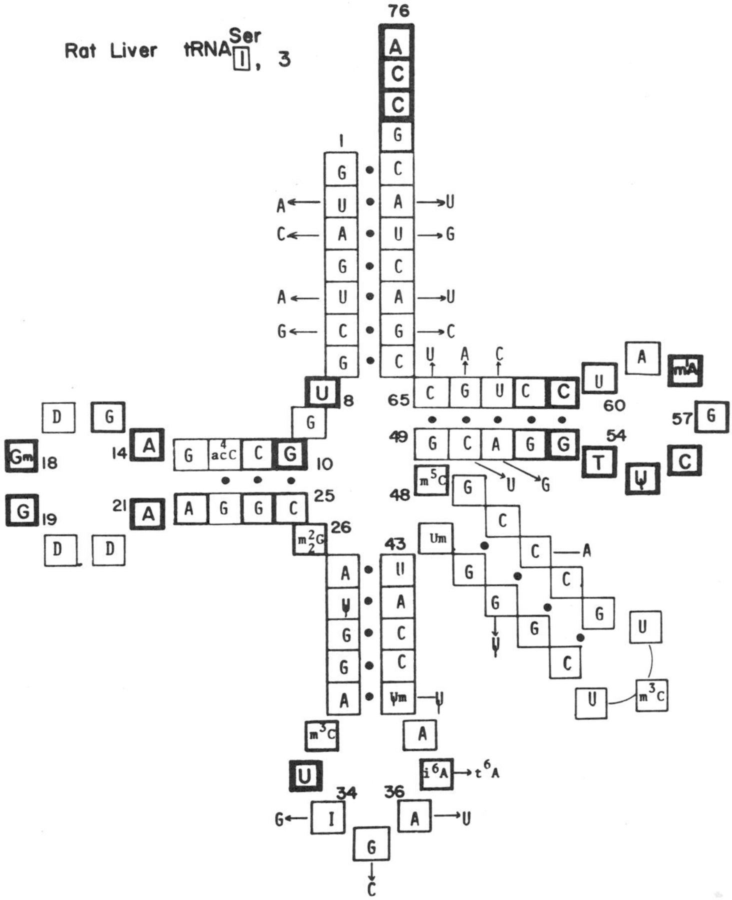

FIG. 12. Comparison oftRNA^{Ser} isoacceptor structures from rat liver. Nucleotides of tRNA$^{Ser}_{1}$ sequence are shown in the square boxes, and changes from this sequence in tRNA$^{Ser}_{3}$ structure are shown outside the boxes.

ered for the recognition. Two isoacceptors of yeast tRNA^{Ser} contain only three variable bases in the molcule; its placement in this group is suggested by experiments complexing this tRNA with the cognate synthetase followed by nuclease digestion (*41*). The CCA end was protected, but the anticodon loop was not.

2. YEAST AND *E. coli* tRNA^{Leu}

The isoacceptor sequences of baker's yeast tRNA$^{Leu}_{1,3}$ and *E. coli* K12 tRNA$^{Leu}_{1,2}$ were compared (figures not shown). Variable bases were found in all areas of the yeast $tRNA^{Leu}$ with the exception of the D stem. If invariant nucleotides are disregarded as specific recognition sites, the D stem and four base-pairs in the AA stem may then be considered. The *E. coli* $tRNA^{Leu}$ comparison reveals less obvious recognition sites. There are variable bases in the AA stem, D stem, wobble position, and V loop. Segments that can be recognized specifically by the synthetase are therefore three base-paired nucleotides in the AA stem, two base-pairs in the D stem, and the four *in*variable nucleotides of the AC loop. Presently, no experimental evidence is available to test the proposed recognition sites.

3. Yeast $tRNA^{Ala}$

Yeast $tRNA^{Ala}$ isoacceptors were not available for comparison; however, Chambers *et al.* (*43, 44*) have studied this tRNA by C → U chemical modification. When five C nucleotides in the tRNA were changed (72, 74, and 75 of the AA region, 36 of the anticodon, and 17 of the D loop), 47% of the acceptor activity was lost. A single conversion in the anticodon produced a missense change but did not prevent aminoacylation. Therefore, the remaining four positions in the AA stem and D loop may be the primary sites for this synthetase recognition.

4. *Escherichia coli* $tRNA^{Tyr}$

This has been studied by chemical modifications. Modification of the 3′-end cytidines with methoxyamine inhibits aminoacylation (*45*). Five mutants of *E. coli sup 3* tyrosine tRNA, each with a single base change in the AA stem (the fourth, fifth, and sixth positions from the 3′ end and the first and second positions from the 5′ end), caused mischarging of the $tRNA^{Tyr}$; that is, the tRNA was esterified with glutamine. It was concluded that the end position of the AA stem plays an important role in maintaining amino-acid specificity for both tyrosine and glutamine synthetases (*46*).

We have compared isoacceptors of $tRNA^{Tyr}$ sequences (*E. coli* su^- $tRNA^{Tyr}_{1,2}$, su^+ $tRNA^{Tyr}_3$, and *CA75* $tRNA^{Tyr}_{1,2}$); the results indicate that the base changes are in the V loop and wobble position. There are no changes in the AA stems. Therefore, the AA stem can be considered as one of the sites involved in the specific recognition by the synthetase.

D. Discussion of the Interaction Sites

For stable complex formation between *any* given tRNA and a particular enzyme, *general* interaction sites appear to be involved. A large number of the invariant nucleotides present in specific positions of both prokaryotic and eukaryotic tRNA can be considered as the binding sites for recognition of the *nonspecific* type. However, the *specific* site (or sites) responsible for unique interaction between a *specific* tRNA (isoacceptors) and its cognate synthetase should involve residues characterized as semiconserved and invariant positions of the tRNA.

The concept of interaction sites involving specific sequences is limited to consideration of the primary and secondary structures and does not include possible conformational requirements of the synthetase. A sequence change causes changes in the base-stacking with the adjacent nucleotides. This change, in turn, can alter conformation of neighboring or distal portions of the molecule. The sequence comparisons, chemical modification experiments and related studies, therefore, do not reveal such subtle conformational changes in defining the specific interaction sites.

The sequences of different tRNAs that can be aminoacylated by a single synthetase have been examined in the past in the hope that some common recognition sequence site would emerge. Thus, sets of isoacceptors for a given synthetase (*47, 48*), sets of heterologous cognate tRNAs (*49, 50*) and sets of heterologous noncognate (misacylated) tRNAs (*39, 51*) were compared. The heterologous misacylation studies (*39*) carried out in nonphysiological conditions indicated D stem and the fourth nucleotide from the 3′ end as determinants of recognition (K_m) for yeast Phe-tRNA synthetase, along with the size of the D loop and state of methylation of G10 influencing the V_{max} of the reaction. However, other workers (*51*) have interpreted their results to mean that the nucleotides near the 3′ end and the AC stem are involved in controlling the V_{max} of aminoacylation, but not the K_m (*5*).

Three interaction areas on the tRNA are suggested in this study as possible synthetase recognition sites. They are (a) the anticodon, (b) the D stem, and (c) the AA stem and/or D stem, or both stems. These sites are in agreement with the current experimental data and the comparisons of the isoacceptor structures. The nucleotides and parts of the molecule that differ among isoacceptors of a given tRNA cannot be considered as synthetase recognition sites. Therefore, the invariant residues or segments can be considered as specific sites involved in the recognition process. Our results do not contradict any conclusions drawn from experimental data (chemical modification, nuclease digestion, etc.). In the three-dimensional structure of $tRNA^{Phe}$, the anticodon, D stem, and AA stem are located along and around the diagonal of the molecule. This is the side of the tRNA suggested by Rich and

Schimmel as a possible recognition site (52). The synthetase may span the entire diagonal from the AC loop to the 3′ end, or it may interact only with certain segments of the molecule. That is, there may be recognition of the AC loop by one kind of synthetase, of the D stem by the other kind of synthetase, and the AA stem or the D stem or both stems by yet another kind of synthetase. A further subdivision of these three areas of the molecule may be necessary to encompass the recognition of all tRNAs.

An evolutionary relationship among tRNAs can also be involved in their recognition. Structurally similar tRNAs perhaps contain similar recognition sites. Dayhoff (53) has developed an evolutionary tree for *E. coli* tRNAs based on a matrix calculation of percent difference of bases among sequences. Branch lengths are measured that are proportional to the inferred number of point mutations per 100 residues. The tRNAs defined in the anticodon recognition group (glycine, tryptophan, and glutamine tRNAs) are all centrally located on this evolutionary tree. Thus, glutamine and glycine$_1$ tRNAs were derived directly from the hypothesized "proto-tRNA" organism. Glycine$_3$ and tryptophan tRNAs branch off the glycine tRNA$_1$. Valine, isoleucine, and methionine tRNAs (*E. coli*), which we considered in the D stem recognition group, branch out from a common ancestral point on one side of the evolutionary tree, from a common point, each closely related to the other. These tRNAs contain extra-large V loops (13–21 residues) in contrast to the other two groups of tRNAs, which have small V loops (4–5 residues). Yeast tRNA^{Phe}, which we included in the D stem recognition group, can overlap with the tRNAs of the anticodon group as supported by nuclease activity on the tRNA-synthetase complex (*41*). If the evolutionary trees for *E. coli* and yeast are similar in development, the tRNA^{Phe} of yeast may be involved in more specific interactions with the anticodon residues, since this tRNA is centrally located in the tree with other anticodon-recognizing tRNAs.

IX. Summary

A comparison of the sequences of tRNAs has provided new and supportive information concerning (a) secondary and tertiary structures; (b) minor modified nucleotides and their possible roles; and (c) binding sites for specific and nonspecific interactions between tRNA and the aminoacyl-tRNA synthetase. The statistical data reveal that invariant or conserved nucleotides are present in each prokaryotic and eukaryotic tRNA. These invariant positions, determined for >60 and >90% of the tRNAs, suggest evolutionary or "primordial" structures that may have differentiated into the highly specific structures required for the aminoacylation process. The study of variant residues indicate that the prokaryotic and eukaryotic tRNAs, although stemming from a common "proto" tRNA, have evolved in two directions. Interestingly, tRNA structure has not changed drastically during this evolution.

Our results indicate that the "tertiary" bonds (those that give or result from three-dimensional folding) in most tRNAs are analogous to those identified in yeast $tRNA^{Phe}$. Five such bonds occur in each tRNA analyzed, and another four are probably present in 64 to 78% of the tRNAs. Thus, each tRNA may exist in the L-shaped three-dimensional conformation exemplified by yeast $tRNA^{Phe}$.

The significance of modified nucleosides within the tRNA structure is as yet unknown, but they may serve as specific and nonspecific recognition sites during protein interaction. Eukaryotes and prokaryotes contain specific modifications as well as modifications common to both kinds of organism. Modifications in twenty different positions are specific to eukaryotic tRNAs, but modifications in only three positions are specific to prokaryotic tRNAs. A great variety of modification is found in two specific positions of both eukaryotic and prokaryotic tRNAs, namely, in the "wobble" position (34) and the hypermodified purine position (37). The wobble-base modification appears to promote binding with certain codons and to demote binding with other codons by giving specificity to the interaction between the modified wobble base and the third letter of the codon. The ambiguity in the coded message is thus suppressed by specific modification of the wobble base. The modification of the purine base adjacent to the anticodon (position 37) appears to enhance interaction between the ribosome and the tRNAs. Such interaction is achieved by the presence of a hydrophobic or hydrophilic side chain attached to A37 of tRNAs whose codons begin with A or U—cases where A·U base pairing occurs between residue 36 of the tRNAs and the first letter of the code. The sequence TΨC of the Ψ loop is generally modified in tRNAs that do not bind with the ribosome. Such tRNAs include eukaryotic initiator tRNAs (AUC, AΨC), glycine tRNAs involved in cell wall synthesis (UGC), and tRNAs serving the role of primer for reverse transcriptase (ΨΨC).

Examination of the variant residues of isoacceptors indicates up to three synthetase-tRNA (specific) interaction sites, located in (*a*) the anticodon loop, (*b*) the D stem, and (*c*) the AA stem and/or the D stem. One of these three sites is present in each tRNA analyzed. The results are in agreement with the conclusions drawn from chemical modification studies.

Although tRNAs appear to possess a common secondary structure and also perhaps similar tertiary structures, they can function with complete nonspecificity as well as with extreme specificity. This is due to discrete changes in their primary structures, on the one hand, and to specific modifications of certain residues after transcription, on the other. In this regard, tRNAs are very different from the proteins, which acquire functional specificity by virtue of their size and conformation. The tRNAs not only are smaller than most proteins, but also exhibit more functional features than most enzymes.

REFERENCES

1. S. H. Kim, This Series **17**, 181 (1976); *Adv. Enzymol.* **46**, 279 (1978).
2. S. H. Kim, *CSHSQB* (in press); personal communication.
3. A Rich and S. H. Kim, *Sci. Am.* **238**, 52 (1978).
4. B. F. C. Clark, This Series **20**, (1977).
5. J. Ofengand, *in* "Molecular Mechanisms of Protein Synthesis" (H. Weissback and S. Pestka, eds.), p. 7. Academic Press, New York, 1977.
6. A. Jack, J. E. Ladner, D. Rhodes, R. S. Brown, and A. Klug, *JMB* **111**, 315 (1977).
7. A. Rich and U. L. Rajbhandary, *ARB* **45**, 805 (1976).
8. G. Dirheimer, G. Keith, R. Martin, and J. Weissenbach, "Proceedings of the International Conference on tRNA, Polish Academy of Sciences, Poznan, Poland (1960)," p. 273 (1976).
9. M. Sundaralingam, *in* "Structure and Conformation of Nucleic Acids and Protein-Nucleic Acid Interactions" (M. Sundaralingam and S. T. Rao, eds.), p. 487. Univ. Park Press, Baltimore, Maryland, 1975.
10. M. O. Dayhoff and P. J. McLaughlin, *in* "Atlas of Protein Sequence and Structure" (M. O. Dayhoff, ed.), Vol. 5, p. 111. Natl. Biomed. Res. Found., Washington, D.C., 1972.
11. R. P. Singhal, B. L. Delmez, T. L. Street, G. W. Hiesterman, and T. J. Yeary, *Proc. Int. Symp. Biomol. Struct. Conform., Funct. Evol., 1978* (in press).
11a. F. Kimura-Harada, M. Saneyoshi, and S. Nishimura, *FEBS Lett.* **13**, 335 (1971).
12. J. A. McCloskey and S. Nishimura, *Acct. Chem. Res.* **10**, 403 (1977), and references cited therein.
13. S. Nishimura, This Series **12**, 49 (1972).
14. H. Grosjean, S. de Henau, and D. M. Crothers, *PNAS* **75**, 610 (1978).
14a. H. Grosjean, D. Sankoff, W. M. Jou, W. Fiers, and R. J. Cedergren, *J. Mol. Evol.* **12**, 113 (1978).
15. R. T. Walker and U. L. Rajbhandary, *NARes* **5**, 57 (1978).
16. K. Watanabe, Y. Kuchino, Z. Yamaizumi, M. Kato, T. Oshima, and S. Nishimura, *NARes, Suppl.* **5**, s473 (1978).
17. A. M. Gillum, N. Urquhart, M. Smith, and U. L. Rajbhandary, *Cell* **6**, 395 (1975).
17a. S. Silverman, J. Heckman, G. J. Cowling, A. D. Delaney, R. J. Dunn, I. C. Gilham, G. M. Tener, D. Soll, and U. L. Rajbhandary, *NARes* **6**, 421 (1979).
18. L. Angevev, N. Davidson, W. Murphy, D. Lynch, and G. Attardi, *Cell* **9**, 81 (1976).
19. D. Lynch and G. Attardi, *JMB* **102**, 125 (1976).
20. M. C. Martin and M. Rabinowitz, *Bchem* **17**, 1628 (1978).
21. L. Aujame and K. B. Freeman, *NARes* **6**, 455 (1979).
22. N. C. Martin, M. Rabinowitz, and H. Fukuhara, *Bchem* **16**, 4672 (1977).

23. R. P. Martin, A. P. Sibler, J. M. Schneller, G. Keith, A. J. C. Stahl, and G. Dirheimer, *NARes* **5,** 4579 (1978).
24. A. D. Mirzabekov, L. V. Kazarinove, D. Lastity, and A. A. Bayer, *FEBS Lett.* **3,** 268 (1969).
25. A. D. Mirzabekov, *Nature NB* **229,** 21 (1971).
26. R. W. Chambers, A. Aoyagi, Y. Furukawa, H. Zawadzka, and O. S. Bhanot, *JBC* **248,** 5549 (1973).
27. O. S. Bhanot, S. Aoyagi, and R. W. Chambers, *JBC* **252,** 2566 (1977).
28. M. Saneyoshi and S. Nishimura, *BBA* **246,** 123 (1971).
29. R. P. Singhal, *Bchem* **13,** 2924 (1974).
30. R. P. Singhal, *JBC* **246,** 5848 (1971).
31. Y. Wataya, S. Iida, I. Kudo, Z. Ohashi, S. Nishimura, K. Suga, H. Takagi, T. Yokoshima, and H. Hayatsu, *EJB* **64,** 27 (1976).
32. J. Squires and J. Carbon, *Nature NB* **233,** 274 (1971).
33. J. Carbon and E. W. Fleck, *JMB* **85,** 371 (1974).
34. J. W. Roberts and J. Carbon, *JBC* **250,** 5530 (1975).
35. M. Yaniv, W. R. Folk, P. Berg, and L. Soll, *JMB* **86,** 245 (1974).
36. L. H. Schulman, *PNAS* **69,** 3594 (1972).
37. L. H. Schulman and J. P. Goddard, *JBC* **248,** 1341 (1973).
38. W. Wübbeler, C. Lossow, F. Fittler, and H. G. Zachau, *EJB* **59,** 405 (1975).
39. B. Roe and B. Dudock, *BBRC* **49,** 399 (1972); *Nature NB* **246,** 135 (1973); *Bchem* **12,** 4146 (1973).
40. H. J. P. Schoemaker, G. P. Budzik, R. Giege, and P. R. Schimmel, *JBC* **250,** 4440 (1975).
41. W. Horz and H. G. Zachau, *EJB* **32,** 1 (1973).
42. G. P. Budzik, S. S. Lam, H. J. P. Schoemaker, and P. R. Schimmel, *JBC* **250,** 4433 (1975).
43. L. H. Schulman and R. W. Chambers, *Bchem* **61,** (1968).
44. O. S. Bhanot and R. W. Chambers, *JBC* **252,** 2551 (1977).
45. A. R. Cashmore, *FEBS Lett.* **12,** 90 (1970).
46. A. Ghysen and J. D. Celis, *JMB* **83,** 333 (1974).
47. D. Söll and P. R. Schimmel, *in* "The Enzymes" (P. D. Boyer, ed.), 3rd ed., Vol. 10, p. 489. Academic Press, New York, 1974.
48. P. W. Piper and B. F. C. Clark, *FEBS Lett.* **47,** 56 (1974).
49. V. D. Axel'rod, V. M. Kryukov, S. N. Isaenko, and A. A. Bayev, *FEBS Lett.* **45,** 333 (1974).
50. M. Simsek, U. L. Rajbhandary, M. Boisnard, and G. Petrissant, *Nature* **247,** 518 (1974).
51. R. Giege, D. Kern, J.-P. Ebel, H. Grosjean, S. de Henau, and H. Chantrenne, *EJB* **45,** 351 (1974).
52. A. Rich and P. R. Schimmel, *NARes* **4,** 1649 (1977).
53. M. O. Dayhoff, ed., "Atlas of Protein Sequence and Structure," Vol. 5, Suppl. 2, p. 280. Natl. Biomed. Res. Found., Washington, D. C., 1976.

APPENDIX A: Structures of 178 tRNA Sequences

167 TRANSFER RNA SEQUENCES FROM DIFFERENT SOURCES[a]

Origin	Total No.	Sequence Code Nos.[b]
1. Viruses	15	3R(2), 3Q(3), 4Q, 7G(3), 4L, 1P, 3S, 4S, 3T, W6
2. Bacteria		
(1) *Escherichia*	48	1A, 1R, 2R, 1N, 1D, 1C, 1Q, 2Q, 1E, 2E, 3E, 2G, 3G(2), 4G(3), 1H(2), 1I, 1L(2), 2L, 3L, 1K, liM(5), 1M, 2M, 1F, 1S, 2S, 1T, 1W, 2W, 3W, 2Y(3), 1V(2), 2V(2), 3V(2)
(2) *Salmonella*	7	1G(2), 1H(2), 1L(2), 5V
(3) *Staphylococcus*	6	5(3), 6G(3)
(4) *Bacillus subtilis*	4	2K, 3iM, 3F, 2T
(5) *B. stearothermophilus*	3	2F, 1Y, 4V
(6) *Therm. thermophilus*	1	2iM
(7) *Mycoplasma*	2	4iM, 4F
3. Yeasts	37	2A, 3A, 4A, R4, R5, R6, D2, C2, E4, G8, I12, L5, L6, L7, K3, K4(2), iM9, M3, F9, F10, F11, S5, S6, S7(2), S8, T4, W5, Y3(2), Y4, V6 to V10
4. Plants	15	A7, G9, iM5 to iM8, F5 to F8, F12, F13, Y5
5. Mammals	24	N2(2), G12, G13, iM10(8), M4(2), F14(4), S9, S10, W7, V11(2), V12
6. Other eukaryotes	5	A5, A6, G10, G11, iM11

[a] The definitions of the Sequence Code numbers and the symbols for the nucleoside residues are given in Appendix Sections A and C.

[b] Number in parentheses indicate number of species having common sequences (mammals) or similar sequences with few exceptions (mutants).

Abbreviations and Definitions

A. tRNA Code Numbers

Each sequence is numbered arbitrarily and carries in addition a single letter indicating the usual amino-acid acceptance (the letters are those recommended by IUPAC—IUB CBN in, e.g., *J. Biol. Chem.* **243,** 3557). A number *preceding* the letter indicates prokaryotic origin; a number *following* the letter indicates eukaryotic origin. Thus 1A represents an alanine tRNA from a prokaryote; R4 represents an arginine tRNA from a eukaryote. The one-letter code is as follows (*loc. cit.*):

A alanine
R arginine
N asparagine
D aspartic acid
C cysteine
Q glutamine
E glutamic acid
G glycine
H histidine
I isoleucine
L leucine
K lysine
M methionine
iM initiator methionine
F phenylalanine
P proline
S serine
T threonine
W tryptophan
Y tyrosine
V valine

B. Hydrogen Bonding

Non-hydrogen-bonded residues of the cloverleaf structure (see Appendix figure) are shown in the dark background, and the base-paired residues located in the stem regions are shown in the light background.

C. Abbreviations

SPECIAL NOTATIONS

N Unidentified residue
R,Y A purine, a pyrimidine nucleoside
$\overset{d}{U}$ Uridine or dihydrouridine
μ Change occurs at the position of the mutant tRNA (see note following the reference of this sequence).
$\overset{\Delta}{U}$ Unidentified modification of uridine.
$\overset{*}{N}$ Special comment for this residue; see the reference for that sequence.

The symbols in parentheses are those recommended for use in text; those outside the parentheses, with notations above or below the primary symbol, are in the form recommended for use in long sequences (the aim being to reduce overall length).

URIDINE

D (hU) 5,6-dihydrouridine
T (m^5U) ribothymidine or 5-methyluridine
$\overset{s}{T}$ (s^2T) 2-thioribothymidine (5-methyl-2-thiouridine)

$\overset{\alpha}{\mathrm{U}}$ (mo^5U) 5-methoxyuridine

$\overset{\nu}{\mathrm{U}}$ (cmo^5U) 5-carboxymethoxyuridine

$\overset{\beta}{\mathrm{U}}$ (mcm^5U) 5-(methoxycarbonylmethyl)uridine

$\overset{\mathrm{s}}{\mathrm{U}}$ (s^4U) 4-thiouridine

$\overset{\epsilon}{\mathrm{U}}$ (mcm^5s^2U) 5-(methoxycarbonylmethyl)-2-thiouridine

$\overset{\eta}{\mathrm{U}}$ (mnm^5s^2U) 5-methylaminomethyl-2-thiouridine

$\overset{\pi}{\mathrm{U}}$ (acp^3U) 3-(3-amino-3-carboxypropyl)uridine

$\underset{\mathrm{m}}{\mathrm{U}}$ (U^m) 2′-*O*-methyluridine

CYTIDINE

$\overset{3}{\mathrm{C}}$ (m^3C) 3-methylcytidine

$\overset{5}{\mathrm{C}}$ (m^5C) 5-methylcytidine

$\overset{\kappa}{\mathrm{C}}$ (ac^4C) N^4-acetylcytidine

$\overset{\mathrm{s}}{\mathrm{C}}$ (s^2C) 2-thiocytidine

$\underset{\mathrm{m}}{\mathrm{C}}$ (C^m) 2′-*O*-methylcytidine

ADENOSINE

$\overset{\mathrm{m}}{\mathrm{A}}$ (m^1A) 1-methyladenosine

$\overset{2}{\mathrm{A}}$ (m^2A) 2-methyladenosine

$\overset{6}{\mathrm{A}}$ (m^6A) N^6-methyladenosine

$\overset{\mathrm{i}}{\mathrm{A}}$ (i^6A) N^6-isopentenyladenosine

$\overset{\mathrm{s}}{\mathrm{A}}$ (ms^2i^6A) N^6-isopentenyl-2-methylthioadenosine

$\overset{\mathrm{t}}{\mathrm{A}}$ (tc^6A) *N*-[*N*-(9-β-D-ribofuranosylpurin-6-yl)carbamoyl] threonine; N^6-(threoninocarbonyl)adenosine

$\overset{\tau}{\mathrm{A}}$ (mt^6A) *N*-[*N*-(9-β-D-ribofuranosylpurin-6-yl)*N*-methylcarbamoyl]threonine; N^6-methyl-N^6-(glycinocarbonyl) adenosine

$\overset{\rho}{\mathrm{A}}$ (ρt^6A) *N*-[*N*-(9-β-D-ribofuranosylpurin-6-yl)carbamoyl] threoninetris(hydroxymethyl)methylamide; N^6-[tris(hydroxymethyl)methylamidothreoninocarbonyl]adenosine (this compound is an artifact, and is produced by enzymic amide bond formation between tris buffer and t^6A during isolation)

$\underset{\mathrm{m}}{\mathrm{A}}$ (A^m) 2′-*O*-methyladenosine

GUANOSINE

- $\overset{m}{G}$ (m^1G) 1-methylguanosine
- $\overset{2}{G}$ (m^2G) N^2-methylguanosine
- $\overset{\alpha}{G}$ (m^2_2G) N^2,N^2-dimethylguanosine
- $\overset{+}{G}$ (m^7G) 7-methylguanosine
- $\underset{m}{G}$ (G^m) 2′-*O*-methylguanosine

OTHERS

- $\overset{m}{I}$ (m^1I) 1-methylinosine
- W (W) wyosine; 4,9-dihydro-4,6-dimethyl-9-oxo-3-β-D-ribofuranosyl-1*H*-imidazo[1,2-α]purine; 3-β-D-ribofuranosylwye
- $\overset{y}{W}$ (yW) wybutosine; α-(carboxyamino)wyosine-7-butyric acid, dimethyl ester
- $\overset{o}{W}$ (oyW) peroxywybutosine; α-(carboxyamino)-β-hydroperoxywyosine-7-butyric acid, dimethyl ester (the natural product is probably hydroxybutosine rather than the peroxy compound; Kasai *et al.*, *NARes* **6**, 993, 1979)
- Q (Q) queuosine; 7-{[(*cis*-4,5-dihydroxy-2-cyclopenten-1-yl) amino]methyl}-7-deazaguanosine
- $\overset{n}{Q}$ (manQ) β-D-mannosylqueuosine, found in mammalian $tRNA^{Asp}$ (sequence location unknown)
- $\overset{g}{Q}$ (galQ) β-D-galactosylqueuosine, found in mammalian $tRNA^{Tyr}$ (sequence location unknown)
- $\underset{m}{N}$ (N^m) 2′-*O*-methylribo-purine or -pyrimidine nucleoside

D. References and Footnotes for Sequence List

ALANINE

1A. *Escherichia coli* B: R. J. Williams, W. Nagel, B. Roe, and B. Dudock, *BBRC* **60**, 1215 (1974) U8 is either U or s^4U.

A2. *Torulopsis utilis:* S. Takemura and K. Ogawa, *J. Biochem.* (*Tokyo*) **74**, 322 (1973).

A3. Baker's yeast: R. W. Holley, J. Apgar, G. A. Everett, J. T. Madison, M. Marquise, S. N. Merrill, J. R. Penswick and A. Zamir, *Science* **147**, 1462 (1965); C. R. Merril, *Biopolymers* **6**, 1727 (1968); R. H. Reeves, H. Imura, H. Schwam, G. B. Weiss, L. H. Schulman, and R. W. Chambers, *PNAS* **60**, 1450 (1968).

A4. Brewer's yeast: J. R. Penswick, R. Martin, and G. Dirheimer, *FEBS Lett.* **50**, 28 (1975).

A5. *Bombyx mori:* K. U. Sprague, O. Hagenbüchle, and M. C. Zuniga, *Cell* **11**, 561 (1977). U40 is either U or Ψ.

A6. *Bombyx mori:* K. U. Sprague, O. Hagenbüchle, and M. C. Zuniga, *Cell* **11**, 561 (1977).
A7. *Neurospora crassa,* mitochondrial: S. Yin and U. L. Rajbhandary, Massachusetts Institute of Technology, Cambridge (personal communication). C49 is either C or Cm.

ARGININE

1R. *Escherichia coli* B: K. Murao, T. Tanabe, F. Ishii, M. Namiki, and S. Nishimura, (1972) *BBRC* **47**, 1332 (1972).
2R. *Escherichia coli* B: K. Chakraburtty, *NARes* **2**, 1787 (1975).
3R. Bacteriophage T4 and mutant phage T4 psu^+ 4 UGA: G. P. Mazzara, J. G. Seidman, W. H. McClain, H. Yesian, J. Abelson, and C. Guthrie, *JBC* **252**, 8245 (1977); S. H. Kao and W. H. McClain, *ibid.* p. 8254. U36 indicates that the mutant contains U → A change.
R4. Brewer's yeast: J. Weissenbach, R. Martin, and G. Dirheimer, *EJB* **56**, 527 (1975).
R5. Brewer's yeast: B. Kuntzel, J. Weissenbach, and G. Dirheimer, *Biochimie* **56**, 1069 (1974).
R6. Brewer's yeast: B. Kuntzel, J. Weissenbach, and G. Dirheimer, *Biochimie* **56**, 1069 (1974).

ASPARAGINE

1N. *Escherichia coli:* K. Ohashi, F. Harada, Z. Ohashi, S. Nishimura, T. S. Stewart, G. Võgeli, T. McCutchan, and D. Söll, *NARes* **3**, 3369 (1976).
N2. Mammals—rat liver, human liver, and human placenta: E. Y. Chen and B. A. Roe, *BBRC* **82**, 235 (1978).

ASPARTIC ACID

1D. *Escherichia coli* B: F. Harada, K. Yamaizumi, and S. Nishimura, *BBRC* **49**, 1605 (1972).
D2. Brewer's yeast: J. Gangloff, G. Keith, J. P. Ebel, and G. Dirheimer, *BBA* **259**, 210 (1972).

CYSTEINE

1C. *Escherichia coli:* G. P. Mazzara and W. H. McClain, *JMB* **117**, 1061 (1977).
C2. Baker's yeast: N. J. Holness and G. Atfield, *BJ* **153**, 447 (1976).

GLUTAMINE

1Q. *Escherichia coli* K12: M. Yaniv and W. R. Folk, *JBC* **250**, 3243 (1975). U34 is probably a derivative of 2-thiouridine.
2Q. *Escherichia coli* K12: M. Yaniv and W. R. Folk, *JBC* **250**, 3243 (1975).
3Q. Bacteriophage T4, mutant phage T4 psu^+ 2 oc, and mutant phage T4 C34 psu^+ 2 am: J. G. Seidman, M. M. Comer, and W. H. McClain, *JMB* **90**, 677 (1974); M. M. Comer, K. Foss, and W. H. McClain, *ibid.* **99**, 283 (1975). In position 34, phage T4 and mutant psu^+ 2 oc contain an unknown derivative of uridine, and mutant psu^+ 2 am contains a cytosine. In position 36, phage T4 mutants contain G → A mutation.
4Q. Bacteriophage T4 from precursor: C. Guthrie, *JMB* **95**, 529 (1975).

GLUTAMIC ACID

1E. *Escherichia coli* B: M. Uziel and A. J. Weinberg, *NARes* **2**, 469 (1975).
2E. *Escherichia coli* B: R. P. Singhal, *JBC* **246**, 5848 (1971); Z. Ohashi, F. Harada, and S. Nishimura, *FEBS Lett.* **20**, 239 (1972).

3E. *Escherichia coli* K12: K. O. Munninger and S. H. Chang, *BBRC* **46,** 1837 (1972). Residues at positions 33 and 35 are derivatives of pyrimidines (probably a modified urindine) having fluorescent properties.

E4. Baker's yeast: T. Kobayashi, T. Irie, M. Yoshida, K. Takeishi, and T. Ukita, *BBA* **336,** 168 (1974).

GLYCINE

1G. *Salmonella typhimurium* and *Salmonella typhimurium* sufD: C. W. Hill, G. Combriato, W. Steinhart, D. L. Riddle, and J. Carbon, *JBC* **248,** 4252 (1973); D. L. Riddle and J. Carbon, *Nature NB* **242,** 230 (1973). G18 is either G or Gm; in mutant sufD, an additional C is present between residues C34 and C35.

2G. *Escherichia coli:* C. W. Hill, G. Combriato, W. Steinhart, D. L. Riddle, and J. Carbon, *JBC* **248,** 4252 (1973). In position 35, mutation C → U; C. W. Hill, G. Combriato, and W. Dolph, *J. Bact.* **117,** 351 (1974).

3G. *Escherichia coli* and mutant TsuA36: J. W. Roberts and J. Carbon, *JBC* **250,** 5530 (1975). In position 36, mutation C → U (mutant TsuA36).

4G. *Escherichia coli* and mutants su$^+$A78 and *E. coli* ins: C. Squires and J. Carbon, *Nature NB* **233,** 274 (1971); J. Carbon and E. W. Fleck, *JMB* **85,** 371 (1974). G34-mutation in *E. coli* ins has a G → U change; A37-mutation in su$^+$78 has an A → ms^2i^6A change.

5G and 6G. *Staphylococcus epidermidis* Texas 26, *Staphylococcus epidermidis* strain 66, and *Staphylococcus aureus* H: R. J. Roberts, *Nature NB* **237,** 44 (1972); R. J. Roberts, *JBC* **249,** 4787 (1974); B. G. Barrell and B. F. C. Clark, "Handbook of Nucleic Acid Sequences." Joynson-Bruvvers Ltd., Oxford, 1974. The differences (tabulated below) between two isoacceptors have been reported from three different species of *Staphylococcus.*

Strain	Isoacceptor	Residues									
		10	16	17	25	28	37	42	46	56	59
S. epidermidis Texas 26	1A	G	U		C	A	C	U	G	C	G
S. epidermidis Strain 66		U	C	U	U	G	C	C	A	C	G
S. aureus H		G	C	U	C	G	C	C	G	C	A
S. epidermidis Texas 26	1B	U	C	U	U	G	C	C	A	C	A
S. epidermidis Strain 66		U	C		U	G	C	C	A	U	G
S. aureus H		U	C		U	A	U	U	G	U	A

7G. Bacteriophage T2, T4, T6: S. Stahl, G. V. Paddock, and J. Abelson, *NARes* **1,** 1287 (1974); B. G. Barrell, A. R. Coulson, and W. H. McClain, *FEBS Lett.* **37,** 64 (1973). U34 is probably related to mam^5s^2U.

G8. *Saccharomyces cerevisiae:* M. Yoshida, *BBRC* **50,** 779 (1973).

G9. Wheat germ: K. B. Marcu, R. E. Mignery, and B. S. Dudock, *Bchem* **16,** 797 (1977).

G10. *Bombyx mori:* J. P. Garel and G. Keith, *Nature* **269,** 350 (1977); M. C. Zuniga and J. A. Steitz, *NARes* **4,** 4175 (1977).

G11. *Bombyx mori:* M. Kawakami, K. Nishio, and S. Takemura, *FEBS Lett.* **87,** 288 (1978).

G12 and G13. Human placenta (GCC) and (CCC): R. C. Gupta, B. A. Roe, and K. Randerath, personal communication (1979).

Histidine

1H. Bacteria, *Escherichia coli* B and K12, *Salmonella typhimurium*, and mutant HisT: C. E. Singer and G. R. Smith, *JBC* **247**, 2989 (1972) (*E. coli* and *S. typhimurium*). Mutant HisT contains $\Psi \rightarrow$ U change in positions 38 and 39; C. E. Singer, G. R. Smith, R. Cortese, and B. N. Ames, *Nature NB* **238**, 72 (1972) (*S. typhimurium* mutant).

Isoleucine

1I. *Escherichia coli:* M. Yarus and B. G. Barrell, *BBRC* **43**, 729 (1971). U47 is probably 3*N*-(3-amino-3-carboxypropyl)uridine, S. Friedman, H. J. Li, K. Nakanishi, and G. van Lear, *Bchem* **13**, 2932 (1974).

I2. *Torulopsis utilis:* S. Takemura, M. Murakami, and M. Miyazaki, *J. Biochem. (Tokyo)* **65**, 553 (1969).

Leucine

1L. Bacteria, *Escherichia coli* B and K12, *Salmonella typhimurium* LT2, *S. typhimurium* his T1504 mutant: H. U. Blank and D. Söll, *BBRC* **43**, 1192 (1971) (*E. coli* K12); S. K. Dube, K. A. Marcker, and A. Yudelevich, *FEBS Lett.* **9**, 168 (1970). (*E. coli* B). Mutant hisT contains $\Psi \rightarrow$ U change in positions 38 and 39; H. S. Allavdeen, S. K. Yang, and D. Söll, *FEBS Lett.* **28**, 205 (1972).

2L. *Escherichia coli* K12: H. U. Blank and D. Söll, *BBRC* **43**, 1192 (1971).

3L. *Escherichia coli:* Z. Yamaizumi, Y. Kuchino, F. Harada, S. Nishimura, and J. A. McCloskey, *CSHSQB* (Transfer RNAs, in press).

4L. Bacteriophage T4: T. C. Pinkerton, G. Paddock, and J. Abelson, *JBC* **248**, 6349 (1973).

L5. Baker's yeast: S. H. Chang, S. Kuo, E. Hawkins, and N. R. Miller, *BBRC* **51**, 951 (1973).

L6. Baker's yeast: K. Randerath, L. S. Y. Chia, R. C. Gupta, E. Randerath, E. R. Hawkins, C. K. Brum, and S. H. Chang, *BBRC* **63**, 157 (1975); *EJB* **93**, 79 (1979).

L7. *Torulopsis utilis:* A. Murasugi and S. Takemura, *J. Biochem. (Tokyo)* **83**, 1029 (1978).

Lysine

1K. *Escherichia coli* B: K. Chakraburtty, A. Steinschneider, R. V. Case, and A. H. Mehler, *NARes* **2**, 2069 (1975).

2K. *Bacillus subtilis:* Y. Yamada and H. Ishikura, *NARes* **4**, 4291 (1977). U34 is either U or a derivative of 2-thiouridine.

K3. Yeast (haploid): S. J. Smith, H. S. Teh, A. N. Ley, and P. D'Obrenan, *JBC* **248**, 4475 (1973). G9 is either G or m^1G.

K4. Yeast and *Saccharomyces cerevisiae* haploid 2: J. T. Madison and S. J. Boguslawski, *Bchem* **13**, 524 (1974); C. J. Smith, H. S. Teh, A. N. Ley, and P. D'Obrenan, *JBC* **248**, 4475 (1973).

Methionine-Initiator

1iM. *Escherichia coli* CA265 (1 + 2), *Escherichia coli* K12 MO7 (1a, 1b, and 3): S. K. Dube and K. A. Marcker, *EJB* **8**, 256 (1969); S. K. Dube, K. A. Marcker, B. F. C. Clark, and S. Cory, *Nature* **218**, 231 (1968); B. Z. Egan, J. F. Weiss, and A. D. Kelmers, *BBRC* **55**, 320 (1973). G46 is m^7G in *E. coli* CA265(1) and *E. coli* K12 MO(1a + 1b); m^7G46 A46 in *E. coli* CA265(2) and *E. coli* K12 MO(3).

2iM. *Thermus thermophilus:* K. Watanabe, T. Oshima, and S. Nishimura, *NARes* **3,** 1703 (1976).
3iM. *Bacillus subtilis:* Y. Yamada and H. Ishikura, *FEBS Lett.* **54,** 155 (1975).
4iM. *Mycoplasma* sp: R. T. Walker and U. L. Rajbhandary, *NARes* **5,** 57 (1978).
iM5. *Anacystis nidulans:* B. Ecarot-Charrier and R. J. Cedergren, *FEBS Lett.* **63,** 287 (1976).
iM6. *Neurospora crassa* (mitochondrial): J. E. Heckman, L. I. Hecker, S. D. Schwartzbach, W. E. Barnett, B. Baumstark, and U. L. Rajbhandary, *Cell* **13,** 83 (1978). U38 is probably pseudouridine.
iM7. *Neurospora crassa* (cytoplasmic): A. M. Gillum, L. I. Hecker, M. Silberklang, S. D. Schwartzbach, U. L. Rajbhandary, and W. E. Barnett, *NARes* **4,** 4109 (1977).
iM8. Wheat germ: H. P. Ghosh, K. Ghosh, M. Simsek, and U. L. Rajbhandary, *CSHSQB* (tRNA), Abstracts, p. 6 (1978).
iM9. Yeast: M. Simsek and U. L. Rajbhandary, *BBRC* **49,** 508 (1972).
iM10. Mammals—rabbit liver, sheep mammary glands, salmon testes, salmon liver, human placenta, mouse myeloma cells, oocytes, and somatic cells of *Xenopus laevis:* M. Simsek, U. L. Rajbhandary, M. Boisnard, and G. Petrissant, *Nature* **247,** 518 (1974); A. M. Gillum, N. Urguhart, M. Smith, and U. L. Rajbhandary, *Cell* **6,** 395 (1975); A. M. Gillum, B. A. Roe, M. P. J. S. Anandaraj, and U. L. Rajbhandary, *ibid.* p. 407; P. W. Piper and B. F. C. Clark, *EJB* **45,** 589 (1974); M. Wegnez, A. Mazabraud, H. Denis, G. Petrissant, and M. Boisnard, *ibid.* **60,** 295 (1975). G26 is present mostly as m^2G and rarely as m^2_2G.
iM11. *Scenedesmus obliquus:* D. S. Jones, University of Liverpool, Great Britain (personal communication, 1978).

METHIONINE

1M and 2M. *Escherichia coli* CA265: S. Cory, K. A. Marcker, S. K. Dube, and B. F. C. Clark, *Nature* **220,** 1039 (1968); S. Cory and K. A. Marcker, *EJB* **12,** 177 (1970); Z. Ohashi, K. Murao, T. Yahagi, D. L. von Minden, J. A. McCloskey, and S. Nishimura, *BBA* **262,** 209 (1972). G18 is either G or Gm.
M3. Baker's yeast: H. Gruhl and H. Feldmann, *EJB* **68,** 209 (1976); O. Koiwai and M. Miyazaki, *J. Biochem.* (*Tokyo*) **80,** 951 (1976). G26 is either G or m^2_2G, U27 is either Ψ or U.
M4. Mammals—mouse myeloma and rabbit liver: P. W. Piper, *EJB* **51,** 283 (1975); G. Petrissant and M. Boisnard, *Biochimie* **56,** 787 (1974).

PHENYLALANINE

1F. *Escherichia coli:* B. G. Barrell and F. Sanger, *FEBS Lett.* **3,** 275 (1969).
2F. *Bacillus stearothermophilus:* G. Keith, C. Guerrier-Takada, H. Grossjean, and G. Dirheimer, *FEBS Lett.* **84,** 241 (1977).
3F. *Bacillus subtilis:* H. Arnold and G. Keith, *NARes* **4,** 2821 (1977).
4F. *Mycoplasma* sp: M. E. Kimball, K. S. Szeto, and D. Söll, *NARes* **1,** 1721 (1974).
F5. Bean chloroplast: P. Guillemaut and G. Keith, *FEBS Lett.* **84,** 351 (1977).
F6. *Euglena gracilis* (chloroplast): S. H. Chang, L. Hecker, M. Siberklang, C. K. Brum, W. E. Barnett, and U. L. Rajbhandary, *Cell* **9,** 717 (1976).
F7. *Euglena gracilis* (cytoplasmic): S. H. Chang, C. K. Brum, J. J. Schnabel, J. E. Heckman, U. L. Rajbhandary, and W. E. Barnett, *FP* **37,** 1768 (1978). U47 is probably a derivative of uridine.
F8. Blue green algae: S. H. Chang, F. K. Lin, L. I. Hecker, J. E. Heckman, U. L. Rajbhandary, and W. E. Barnett, *CSHSQB* (Transfer RNAs, in press). U39 is prob-

ably a modified uridine. 5-Methylcytidine is commonly found in eukaryotic tRNAs; the presence of this modification in this tRNA is the only example of prokaryotic tRNAs containing m^5C. However, Dunn and Trigg failed to detect m^5C in unfractionated prokaryotic tRNAs in general, and in tRNAs from three Genera of these algae (D. B. Dunn and M. D. M. Trigg, *John Innes Institute Report* **68,** 137, 1977).

F9. Yeast: U. L. Rajbhandary and S. H. Chang, *JBC* **243,** 598 (1968).

F10. *Saccharomyces cerevisiae* (mitochondrial): R. P. Martin, A. P. Sibler, J. M. Schneller, G. Keith, A. J. C. Stahl, and G. Dirheimer, *NARes* **5,** 4579 (1978).

F11. *Schizosaccharomyces pombe:* T. McCutchan, S. Silverman, J. Kohli, and D. Söll, *Bchem* 17, 1622 (1978). G10 and G26 are probably m^2G.

F12. Wheat germ: B. S. Dudock and G. Katz, *JBC* **244,** 3069 (1969).

F13. *Lupinus luteus:* A. J. Rafalski, J. Barciszewski, K. Gulewicz, T. Twardowski, and G. Keith, *Acta Biochim. Pol.* **24,** 301 (1977).

F14. Mammals—rabbit liver, calf liver, bovine liver, and human placenta: G. Keith and G. Dirheimer, *BBA* **517,** 133 (1978); B. A. Roe, M. P. J. S. Anandaraj, L. S. Y. Chia, E. Randerath, R. C. Gupta, and K. Randerath, *BBRC* **66,** 1097 (1975).

Proline

1P. Bacteriophage T4: J. G. Seidman, B. G. Barrell, and W. H. McClain, *JMB* **99,** 733 (1975). U8 is either U or s^4U.

Serine

1S. *Escherichia coli* B: H. Ishikura, Y. Yamada, and S. Nishimura, *FEBS Lett.* **16,** 68 (1971); Y. Yamada and H. Ishikura, *BBA* **402,** 285 (1975).

2S. *Escherichia coli* B: Y. Yamada and H. Ishikura, *FEBS Lett.* **29,** 231 (1973). C32 is probably 2-thiocytidine.

3S. Bacteriophage T4: W. H. McClain, B. G. Barrell, and J. G. Seidman, *JMB* **99,** 717 (1975).

4S. Bacteriophage T4 mutant psu^+1 am: W. H. McClain, C. Guthrie, and B. G. Barrell, *JMB* **81,** 157 (1973).

S5. Brewer's yeast: H. G. Zachau, D. Dũtting, and H. Feldmann, *ZpChem.* **347,** 212 (1966).

S6. Brewer's yeast: H. G. Zachau, D. Dũtting, and H. Feldmann, *ZpChem* **347,** 212 (1966).

S7. Yeast (UCG) major and suppressor SUP-R11: P. W. Piper, *JMB* **122,** 217 (1978). G35 → U35 in SUP-R11 mutant.

S8. Yeast (UCG) minor: P. W. Piper, *JMB* **122,** 217 (1978). U44 is either U or Um.

S9. Rat liver: T. Ginsberg, H. Rogg, and M. Staehelin, *EJB* **21,** 249 (1971).

S10. Rat liver: H. Rogg, P. Muller, and M. Staehelin, *EJB* **53,** 115 (1975).

Threonine

1T. *Escherichia coli:* L. Clarke and J. Carbon, *JBC* **249,** 6874 (1974).

2T. *Bacillus subtilis:* T. Hasegawa and H. Ishikura, *NARes* **5,** 537 (1978).

3T. Bacteriophage T4: C. Guthrie, C. A. Scholla, H. Yesian, and J. Abelson, *NARes.* **5,** 1833 (1978).

T4. Yeast: J. Weissenbach, I. Kiraly, and G. Dirheimer, *Biochimie* **59,** 381 (1977). C48 is either C or m^5C; A49 is either A or G, and U65 is either U or C.

Tryptophan

1W. *Escherichia coli* CA244: D. Hirsch, *JMB* **58,** 439 (1971).

2W. *Escherichia coli* psu^+ UGA: D. Hirsch, *JMB* **58,** 439 (1971).

3W. *Escherichia coli* psu$^+$ 7am: M. Yaniv, W. R. Folk, P. Berg, and L. Soll, *JMB* **86**, 245 (1974).
4W. *Escherichia coli* psu$^+$ 7oc: M. Yaniv. W. R. Folk, P. Berg, and L. Soll, *JMB* **86**, 245 (1974).
W5. Brewer's yeast: G. Keith, A. Roy, J. P. Ebel, and G. Dirheimer, *Biochimie* **54**, 1405 (1972).
W6. Chicken cells (Rous sarcoma virus primer): F. Harada, R. C. Sawyer, and J. E. Dahlberg, *JBC* **250**, 3487 (1975). The sequence was derived from material using normal chicken cells rather than the virus-associated tRNA.
W7. Bovine liver: M. Fournier, J. Labouesse, G. Dirheimer, C. Fix, and G. Keith, *BBA* **521**, 198 (1978). G7 is either G or m^2G; D16 is either D or C; C34 is either C or Cm; G46 is either G or m^7G; D47 is either D or C; G57 is either G or A.

TYROSINE

1Y. *Bacillus stearothermophilus:* R. S. Brown, J. R. Rubin, D. Rhodes, H. Guilley, A. Simoncsits, and G. G. Brownlee, *NARes.* **5**, 23 (1978).
2Y. *Escherichia coli, Escherichia coli* psu$^+$ 3am, and *Escherichia coli* A2 psu$^+$ 3oc: H. M. Goodman, J. Abelson, A. Landy, S. Brenner, and J. D. Smith, *Nature* **217**, 1019 (1968); S. Altman, S. Brenner, and J. D. Smith, *JMB* **56**, 195 (1971). Q34 → C34 in mutant psu$^+$ 3am; Q34 → probably a modified U34 in mutant A2psu$^+$3oc, S. Altman, *NARes* **3**, 441 (1976); in *E. coli,* U47:2 is either U or C, C47:3 is either C or A.
Y3. Baker's yeast and yeast sup-5am: J. T. Madison and H. K. Kung, *JBC* **242**, 1324 (1967); P. W. Piper, M. Wasserstein, F. Engback, K. Kaltoft, J. E. Celis, J. Zeuthen, S. Liebman, and F. Sherman, *Nature* **262**, 757 (1976). G34 → C34 in sup-5am mutant.
Y4. *Torulopsis utilis:* S. Hashimoto, S. Takemura, and M. Miyazaki, *J. Biochem.* (*Tokyo*) **72**, 123 (1972).
Y5. *Neurospora crassa* (mitochondrial): J. Heckman, B. Alzner-Deweerd, and U. L. Rajbhandary, *PNAS* **76**, 717 (1979); personal communication.

VALINE

1V. *Escherichia coli* B and K12: M. Yaniv and B. G. Barrel, *Nature* **222**, 278 (1969); F. Kimura, F. Harada, and S. Nishimura, *Bchem* **10**, 3277 (1971).
2V. *Escherichia coli* B and K12: M. Yaniv and B. G. Barrel, *Nature NB* **233**, 113 (1971).
3V. *Escherichia coli* B and K12: M. Yaniv and B. G. Barrel, *Nature NB* **233**, 113 (1971).
4V. *Bacillus stearothermophilus:* C. Takada-Guerrier, H. Grosjean, G. Dirheimer, and G. Keith, *FEBS Lett.* **62**, (1976).
5V. *Salmonella typhimurium* LT2: M. Yaniv, unpublished results, in C. E. Singer and G. R. Smith, *JBC* **247**, 2989 (1972).
V6. Baker's yeast: J. Bonnet, J. P. Ebel, G. Dirheimer, L. P. Shershneva, A. I. Krutilina, T. V. Venkstern, and A. A. Bayer, *Biochimie* **56**, 1211 (1974).
V7. Baker's yeast: V. D. Axel'rod, V. M. Kryukov, S. N. Isaenko and A. A. Bayer, *FEBS Lett.* **45**, 333 (1974).
V8. Baker's yeast: V. G. Gorbulev, V. D. Axel'rod, and A. A. Bayer, *NARes* **4**, 3239 (1977).
V9. Brewer's yeast: J. Bonnet, J. P. Ebel, and G. Dirheimer, *EJB* **51**, 295 (1975).
V10. *Torulopsis utilis:* T. Mizutani, M. Miyazaki, and S. Takemura, *J. Biochem.* (*Tokyo*) **64**, 839 (1968).
V11. Mammals—mouse myeloma and rabbit liver: P. W. Piper, *EJB* **51**, 295 (1975); P. Jank, N. Shinda-Okada, S. Nishimura, and H. J. Gross, *NARes.* **4**, 1999 (1977).
V12. Human placenta: E. Y. Chen and B. A. Roe, *BBRC* **78**, 631 (1977).

ADDENDUM: Additional 11 tRNA Sequences

References and Footnotes for Additional Sequence List

ASPARAGINE

N3. Walker 256 mammary carcinosarcoma: B. A. Roe, A. F. Stankiewicz, H. L. Rizi, C. Weisz, M. N. DiLauro, D. Pike, C. Y. Chen, and E. Y. Chen, *NARes* **6**, 673 (1979). This sequence differs from that of the typical mammalian $tRNA^{Asn}$ in only the wobble position: Q34 → G34.

ASPARTIC ACID

D3. Beef liver: a tentative structure proposed by V. N. Vakharia, C. M. Garcia, and R. P. Singhal (unpublished results).

CYSTEINE

C3. *Saccharomyces cerevisiae* (mitochondrial). *This sequence was deduced from gene analysis of mt DNA of $tRNA^{Cys}$: J. L. Bos, K. A. Osinga, G. Van der Horst, and P. Borst, *NARes* **6**, 3255 (1979).

GLUTAMIC ACID

E5. *Schizosaccharomyces pombe:* T. W. Wong, T. McCutchan, J. Kohli, and D. Soll, *NARes* **6**, 2057 (1979).

HISTIDINE

H2. *Saccharomyces cerevisiae* (mitochondrial). *This sequence was deduced from gene analysis of mt DNA of $tRNA^{His}$: J. L. Bos, K. A. Osinga, G. Van der Horst, and P. Borst, *NARes* **6**, 3255 (1979).

LYSINE

K5. *Drosophila melanogaster* (2): S. Silverman, I. C. Gillam, G. M. Tener, and D. Soll, *NARes* **6**, 435 (1979). N54 is probably Tm.

METHIONINE-INITIATOR

iM12. *Drosophila melanogaster:* S. Silverman, J. Heckman, G. J. Cowling, A. D. Delaney, R. M. Dunn, I. C. Gillam, G. M. Tener, D. Soll, and U. L. Raj-Bhandary, *NARes* **6**, 421 (1979).

iM13. *Asterina amurensis* (starfish ovary): Y. Kuchino, M. Kato, H. Sugisaki, and S. Nishimura, *NARes* **6**, 3459 (1979). This sequence differs from that of the mammalian initiator tRNA in nine positions: 16, 26, 49–51, 55, 63–65.

PHENYLALANINE

F15. Barley embryo: A. Janowicz, J. M. Wower, and J. Augustyniak, *Plant. Sci. Lett.* **14**, 177 (1979). This sequence is identical with that of $tRNA^{Phe}$ from wheat germ (see sequence code no. F12).

TRYPTOPHAN

8W. *Escherichia coli* temperature-sensitive due to a lesion in the gene for $tRNA^{Trp}$: S. P. Eisenberg, L. Soll, and M. Yarus, *JBC* **254**, 5562 (1979). This mutant varies in only one base change, that is G7 → A7.

SERINE

S11. *Schizosaccharomyces pombe* suppressor 3-e: A. Rafalski, J. Kohli, P. Agris, and D. Soll, *NARes* **6**, 2683 (1979). U34 is a mixture of s^2T and mcm^5U.

No.	Iso-acceptor	Organism	Aminoacyl Stem / D Stem	D Loop	D Stem / Anticodon Stem / Anticodon Loop
			1 2 3 4 5 6 7 8 9 10 11 12 13 14 15 16 17:1	2 18 19 20:1	2 3 21 22 23 24 25 26 27 28 29 30 31 32 33 34 35 36 37 38
ASPARTIC ACID					
D3	2	Beef liver	pG-G-C-C-A-C-G-U-A-G-U-G-$\overset{5}{C}$-A-A-D	G-G-D	A-G-C-A-C-G-G-G-G-A-U-U-Q-U-C-$\overset{\Delta}{N}$-A
CYSTEINE					
C3		*S. cerevisiae* (mto)*	pG-G-A-G-A-U-G-U-U-G-U-U-U-U-A-A	G-G-U-U	A-A-A-C-U-A-U-U-A-G-A-U-U-G-C-A-A-A
GLUTAMIC ACID					
E5		*S. pombe*	pU-C-C-G-U-U-G-U-$\overset{m}{G}$-G-U-C-C-A-A-C	G-G-C-D	A-G-G-A-U-U-C-G-U-C-G-C-U-$\overset{\eta}{U}$-U-C-A-C
HISTIDINE					
H2		*S. cerevisiae* (mto)*	pG-G-U-G-A-A-U-A-U-A-U-U-U-C-A-A-U	G-G-U	A-G-A-A-A-A-U-A-C-G-C-U-U-G-U-G-G-U
LYSINE					
K5	2	*D. melanogaster*	pG-C-C-C-G-G-C-U-A-$\overset{2}{G}$-C-U-C-A-G-D-C	G-G-D	A-G-A-G-C-A-ψ-G-A-G-A-C-U-C-U-U-$\overset{t}{A}$-A
METHIONINE-INITIATOR					
iM12		*D. melanogaster*	pA-G-C-A-G-A-G-U-$\overset{m}{G}$-$\overset{2}{G}$-C-G-C-A-G-U	G-G-A	A-G-C-G-U-$\overset{2}{G}$-C-U-G-G-G-C-C-C-A-U-$\overset{t}{A}$-A
iM13		*A. amurensis*	pA-G-C-A-G-A-G-U-$\overset{m}{G}$-$\overset{2}{G}$-C-G-C-A-G-U	G-G-A	A-G-C-G-U-G-C-U-G-G-G-C-C-C-A-U-$\overset{t}{A}$-A
SERINE					
S11		*S. pombe*	pG-U-C-A-C-U-A-U-G-U-C-$\overset{\kappa}{C}$-G-A-G-D	G-G-D-D	A-A-G-G-A-$\overset{\alpha}{G}$-ψ-U-A-G-A-$\overset{3}{C}$-U-$\overset{\bullet}{U}$-C-A-$\overset{i}{A}$-A

No.	Iso-acceptor	Anticodon Stem / Variable Loop	TψC Stem / TψC Loop / TψC Stem / Aminoacyl Stem	No.
		39 40 41 42 43 44 45 46 47:1 2 3 4 5 6 7 8 9 10 11 12 13 14 15 16	48 49 50 51 52 53 54 55 56 57 58 59 60 61 62 63 64 65 66 67 68 69 70 71 72 73 74 75 76	
ASPARTIC ACID				
D3	2	-U-C-C-C-C-A-A-G-A	U-C-C-A-A-G-T-ψ-C-G-A-U-U-C-U-U-G-G-C-G-U-G-G-C-C-G-C-C-A	D3
CYSTEINE				
C3		-U-C-U-A-C-U-U-A-U	U-A-A-G-A-G-U-U-C-G-A-U-U-C-U-C-U-U-C-A-U-C-U-C-U-U-C-C-A	C3
GLUTAMIC ACID				
E5		-C-G-A-C-G-G-G-A	G-$\overset{5}{C}$-G-G-G-G-T-ψ-C-G-A-C-U-C-C-C-C-G-C-A-A-C-G-G-A-G-C-C-A	E5
HISTIDINE				
H2		-G-C-G-U-U-A-A-A	U-C-U-G-A-G-U-U-C-G-A-U-U-C-U-C-A-G-U-A-U-U-C-A-C-C-C-C-A	H2
LYSINE				
K5	2	-ψ-C-U-C-A-G-G-$\overset{+}{G}$-D	C-G-U-G-G-G-$\overset{\bullet}{N}$-$\overset{\Delta}{U}$-C-G-$\overset{m}{A}$-G-C-C-C-C-A-C-G-U-U-G-G-G-C-G-C-C-A	K5
METHIONINE-INITIATOR				
iM12		-C-C-C-A-G-A-G-$\overset{+}{G}$-D	$\overset{5}{C}$-C-G-A-G-G-A-U-C-G-$\overset{m}{A}$-A-A-C-C-U-U-G-C-U-C-U-G-C-U-A-C-C-A	iM12
iM13		-C-C-C-A-G-A-G-$\overset{+}{G}$-D	$\overset{5}{C}$-C-G-A-G-G-A-ψ-C-G-$\overset{m}{A}$-A-A-C-C-U-C-G-C-U-C-U-G-C-U-A-C-C-A	iM13
SERINE				
S11		-ψ-C-U-A-A-U-G-G-G-C-U-U-U-G-C-C-C-G	$\overset{5}{C}$-C-C-A-C-G-T-ψ-C-A-$\overset{m}{A}$-A-U-C-C-U-G-C-U-G-G-U-G-A-C-G-C-C-A	S11

[a] Boldface nucleotides are not hydrogen-bonded and are located in the loop regions of the tRNA.

No.	Iso-acceptor	Organism	Aminoacyl Stem 1–7 · D Stem 8–13 · D Loop 14–21 · D Stem 22–25 · 26 · Anticodon Stem 27–31 · Anticodon Loop 32–38
ALANINE			
1A	1a	*E. coli B*	pG-G-G-G-G-C-A-U-A-G-C-U-C-A-G-C-D——G-G-G——A-G-A-G-C-G-C-C-U-G-C-U-U-U-G-C-A-C
A2	1	*T. utilis*	pG-G-G-C-G-U-G-U-G-G-C-G-U-A-G-C-C——G-G-D——A-G-C-G-C-G-ψ-U-C-G-C-U-U-I-G-C-I-ψ
A3	1	Yeast (Baker's)	pG-G-G-C-G-U-G-U-G-G-C-G-U-A-G-D-C——G-G-D——A-G-C-G-C-G-C-U-C-C-C-U-U-I-G-C-I-ψ
A4	2	Yeast (Brewer's)	pG-G-G-C-G-U-G-U-G-G-C-G-C-G-U-A-G-D-C-G-G-D——A-G-C-G-C-G-C-U-C-C-C-U-U-I-G-C-I-ψ
A5	1	*Bombyx mori*	pG-G-G-G-G-C-G-U-A-G-C-U-C-A-G-A-D——G-G-U——A-G-A-G-C-G-C-U-C-G-C-U-U-I-G-C-I-ψ
A6	2	*Bombyx mori*	pG-G-G-G-G-C-G-U-A-G-C-U-C-A-G-A-D——G-G-U——A-G-A-G-C-G-C-U-C-G-C-U-U-I-G-C-I-ψ
A7		*N. crassa* (mto)	pG-G-G-G-G-U-A-U-A-G-U-A-U-A-A-D-U——G-G-D——A-G-U-A-C-A-G-C-A-A-U-C-U-U-G-C-U-C
ARGININE			
1R	1	*E. coli B*	pG-C-A-U-C-C-G-U-A-G-C-U-C-A-G-C-D——G-G-U——A-G-A-G-U-A-C-U-C-G-G-C-U-I-C-G-A-A
2R	2	*E. coli B*	pG-C-A-U-C-C-G-U-A-G-C-U-C-A-G-C-D——G-G-A-U——A-G-A-G-U-A-C-U-C-G-G-C-U-I-C-G-A-A
3R		Phage T4	pG-U-C-C-C-G-C-U-G-G-U-G-U-A-A-U——G-G-A-D——A-G-C-A-U-A-C-G-A-U-C-C-U-U-C-U-A-A
R4	2	Yeast (Brewer's)	pψ-U-C-C-U-C-G-U-G-G-C-C-C-A-A-D——G-G-D-C——A-C-G-G-C-G-ψ-C-U-G-G-C-U-I-C-G-A-A
R5	3a	Yeast (Brewer's)	pG-C-G-C-U-C-G-U-G-G-C-G-U-A-A-D——G-G-C——A-A-C-G-C-G-ψ-C-U-G-A-C-U-U-C-U-A-A
R6	3b	Yeast (Brewer's)	pG-C-G-C-U-U-G-U-G-G-C-G-U-A-A-D——G-G-C——A-A-C-G-C-G-ψ-C-U-G-A-C-U-U-C-U-A-A
ASPARAGINE			
1N		*E. coli*	pU-C-C-U-C-U-G-U-A-G-U-U-C-A-G-D-C——G-G-D——A-G-A-A-C-G-G-C-G-G-A-C-U-Q-U-U-A-A
N2		Mammals	pG-U-C-U-C-U-G-U-G-G-C-G-C-A-A-D-C——G-G-D-U——A-G-C-G-C-G-ψ-ψ-C-G-G-C-U-Q-U-U-A-A

No.	Iso-acceptor	Anticodon Stem $39^{40}41^{42}43^{44}45^{46}47$	Variable Loop $47{:}1\ 2\ 3\ 4\ 5\ 6\ 7\ 8\ 9^{10}11^{12}13^{14}15^{16}$	TψC Stem / TψC Loop / TψC Stem / Aminoacyl Stem $48^{49}50^{51}52^{53}54^{55}56^{57}58^{59}60^{61}62^{63}64^{65}66^{67}68^{69}70^{71}72^{73}74^{75}76$	No.
ALANINE					
1A	1a	-G-C-A-G-G-A-G-$\overset{+}{G}$-U		C-U-G-C-G-G-T-ψ-C-G-A-U-C-C-C-G-C-G-C-G-C-U-C-C-C-A-C-C-A	1A
A2	1	-G-C-G-A-A-A-G-G-D		C-U-C-C-G-G-T-ψ-C-G-$\overset{m}{A}$-C-U-C-C-G-G-A-C-U-C-G-U-C-C-A-C-C-A	A2
A3	1	-G-G-G-A-G-A-G-G-$\overset{d}{U}$		C-U-$\overset{*}{C}$-C-G-G-T-ψ-C-G-A-U-U-C-C-G-G-A-C-U-C-G-U-C-C-A-C-C-A	A3
A4	2	-G-G-G-A-G-A-G-G-U		C-U-C-C-G-G-T-ψ-C-G-A-U-U-C-C-G-G-A-C-U-C-G-U-C-C-A-C-C-A	A4
A5	1	-$\underset{m}{G}$-$\overset{*}{U}$-G-A-G-A-G-$\overset{+}{G}$-U		A-$\overset{5}{C}$-C-G-G-G-A-ψ-C-G-$\overset{m}{A}$-U-A-C-C-C-G-G-C-G-C-C-U-C-C-A-C-C-A	A5
A6	2	-$\underset{m}{G}$-C-G-A-G-A-G-$\overset{+}{G}$-U		A-$\overset{5}{C}$-C-G-G-G-A-ψ-C-G-$\overset{m}{A}$-U-A-C-C-C-G-G-C-G-C-C-U-C-C-A-C-C-A	A6
A7		-A-$\overset{\Delta}{U}$-U-G-C-U-U-G		U-$\overset{5}{C}$-A-A-G-G-T-ψ-C-A-A-A-U-C-C-U-U-G-U-A-U-C-U-C-C-A-C-C-A	A7
ARGININE					
1R	1	-C-C-G-A-G-C-G-$\overset{+}{G}$-$\overset{\pi}{U}$		C-G-G-A-G-G-T-ψ-C-G-A-A-U-C-C-U-C-C-C-G-G-A-U-G-C-A-C-C-A	1R
2R	2	-C-C-G-A-G-C-G-$\overset{+}{G}$-$\overset{\pi}{U}$		C-G-G-A-G-G-T-ψ-C-G-A-A-U-C-C-U-C-C-C-G-G-A-U-G-C-A-C-C-A	2R
3R		-G-ψ-U-U-G-C-G-G		U-C-C-U-G-G-T-ψ-C-G-A-U-C-C-C-A-G-G-G-C-G-G-G-A-U-A-C-C-A	3R
R4	2	-C-C-A-G-A-A-G-A-D		U-$\overset{5}{C}$-C-A-G-G-T-ψ-C-A-$\overset{m}{A}$-G-U-C-C-U-G-G-C-G-G-G-G-A-A-G-C-C-A	R4
R5	3a	-ψ-C-A-G-A-A-G-A-D		U-A-U-G-G-G-T-ψ-C-G-$\overset{m}{A}$-C-C-C-C-C-A-U-C-G-A-G-U-G-C-G-C-C-A	R5
R6	3b	-ψ-C-A-G-A-A-G-A-D		U-A-U-G-G-G-T-ψ-C-G-$\overset{m}{A}$-C-C-C-C-C-A-U-C-G-A-G-U-G-C-U-C-C-A	R6
ASPARAGINE					
1N		-ψ-C-C-G-U-A-U-$\overset{+}{G}$-U		C-A-C-U-G-G-T-ψ-C-G-A-G-U-C-C-A-G-U-C-A-G-A-G-G-A-G-C-C-A	1N
N2		-C-C-G-A-A-A-G-$\overset{+}{G}$-D		U-G-G-U-G-G-N-ψ-C-G-$\overset{m}{A}$-G-C-C-C-A-C-C-C-A-G-G-G-A-C-G-C-C-A	N2

No.	Iso-acceptor	Organism	Aminoacyl Stem — D Stem — D Loop — D Stem — Anticodon Stem — Anticodon Loop (1 2 3 4 5 6 7 8 9 10 11 12 13 14 15 16 17:1 2 18 19 20:1 2 3 21 22 23 24 25 26 27 28 29 30 31 32 33 34 35 36 37 38)
ASPARTIC ACID			
1D	1	E. *coli B*	pG-G-A-G-C-G-G-U-A-G-U-U-C-A-G-D-C——G-G-D-D——A-G-A-A-U-A-C-C-U-G-C-C-U-Q-U-C-A-C
D2	1	Yeast (Brewer's)	pU-C-C-G-U-G-A-U-A-G-U-U-ψ-A-A-D——G-G-D-C——A-G-A-A-U-G-G-G-C-G-C-ψ-U-G-U-C-G-C
CYSTEINE			
1C		E. *coli*	pG-G-C-G-C-G-U-U-A-A-C-A-A-A-G-C——G-G-D——D-A-U-G-U-A-G-C-G-G-A-ψ-U-G-C-A-A-A
C2		Yeast (Baker's)	pG-C-U-C-G-U-A-U-G-G-C-G-C-A-G-D——G-G-D——A-G-C-G-C-A-G-C-A-G-A-ψ-U-G-C-A-A-A
GLUTAMINE			
1Q	1	E. *coli K12*	pU-G-G-G-G-U-A-U-C-G-C-C-A-A-G-C——G-G-D——A-A-G-G-C-A-C-C-G-G-U-U-U-U-U-G-A-ψ
2Q	2	E. *coli K12*	pU-G-G-G-G-U-A-U-C-G-C-C-A-A-G-C——G-G-D——A-A-G-G-C-A-C-C-G-G-A-U-U-C-U-G-A-ψ
3Q		Phage T4	pU-G-G-G-A-A-U-U-A-G-C-C-A-A-G-D-D——G-G-D——A-A-G-G-C-A-U-A-G-C-A-C-U-U-U-G-A-C
4Q		Phage T4 (precursor)	pU-G-G-G-A-A-U-U-A-G-C-C-C-A-A-G-D-D——G-G-D-A——A-G-G-G-C-A-U-A-G-C-A-C-U-U-U-G-A-C
GLUTAMIC ACID			
1E	1	E. *coli B*	pG-U-C-C-C-C-U-U-C-G-U-C-ψ-A-G-A——G-G-C-C-C——A-G-G-A-C-A-C-C-G-C-C-C-U-U-U-C-A-C
2E	2	E. *coli B*	pG-U-C-C-C-C-U-U-C-G-U-C-ψ-A-G-A——G-G-C-C-C——A-G-G-A-C-A-C-C-G-C-C-C-U-U-U-C-A-C
3E	1	E. *coli K12*	pG-U-C-C-C-C-U-U-C-G-U-C-ψ-A-G-A——G-G-C-C-C——A-G-G-A-C-A-C-C-G-C-C-C-U-U-U-C-A-C
E4	3	Yeast (Baker's)	pU-C-C-G-A-U-A-U-A-G-U-G-ψ-A-A-C——G-G-C-D——A-U-C-A-C-A-ψ-C-A-C-G-C-U-U-U-C-A-C
GLYCINE			
1G	1	S. *typhimurium*	pG-C-G-G-G-C-G-U-A-G-U-U-C-A-A-U——G-G-D——A-G-A-A-C-G-A-G-A-G-C-U-U-C-C-C-A-A
2G	1	E. *coli*	pG-C-G-G-G-C-G-U-A-G-U-U-C-A-A-U——G-G-D——A-G-A-A-C-G-A-G-A-G-C-U-U-C-C-C-A-A
3G	2	E. *coli*	pG-C-G-G-G-C-A-U-C-G-U-A-U-A-A-U——G-G-C-U——A-U-U-A-C-C-U-C-A-G-C-C-U-U-C-C-A-A

No.	Iso-acceptor	Anticodon Stem / Variable Loop 39 40 41 42 43 44 45 46 47:1 2 3 4 5 6 7 8 9 10 11 12 13 14 15 16	TψC Stem 48 49 50	TψC Stem 51 52 53	TψC Loop 54 55 56 57 58 59 60	TψC Stem 61 62 63 64 65	Aminoacyl Stem 66 67 68 69 70 71 72 73 74 75 76	No.
ASPARTIC ACID								
1D	1	-G-C-A-G-G-G-G-$\overset{+}{G}$-U————	C-G-C	-G-G-G	-T-ψ-C-G-A-G-U	-C-C-C-G-ψ	-C-C-G-U-U-C-C-G-C-C-A	1D
D2	1	-G-U-G-C-C-A-G-A————	U-$\overset{5}{C}$-G	-G-G-G	-T-ψ-C-A-A-U-U	-C-C-C-C-G	-U-C-G-G-G-G-A-G-C-C-A	D2
CYSTEINE								
1C		-ψ-C-C-G-U-C-U-A————	G-U-C	-C-G-G	-T-ψ-C-G-A-C-U	-C-C-G-G-A	-A-C-G-C-G-C-C-U-C-C-A	1C
C2		-ψ-C-U-G-U-U-G-$\overset{+}{G}$-D————	$\overset{5}{C}$-C-U	-U-A-G	-T-ψ-C-G-$\overset{m}{A}$-U-C	-C-U-G-A-G	-U-G-C-G-A-G-C-U-C-C-A	C2
GLUTAMINE								
1Q	1	-A-C-C-G-G-C-A-U-U————	C-C-C	-U-G-G	-T-ψ-C-G-A-A-U	-C-C-A-G-G	-U-A-C-C-C-C-A-G-C-C-A	1Q
2Q	2	-ψ-C-C-G-G-C-A-U-U————	C-C-G	-A-G-G	-T-ψ-C-G-A-A-U	-C-C-U-C-G	-U-A-C-C-C-C-A-G-C-C-A	2Q
3Q		-ψ-G-C-U-A-G-A-U-G————	C-A-A	-A-G-G	-T-ψ-C-G-A-G-U	-C-C-U-U-U	-A-U-U-C-C-C-A-G-C-C-A	3Q
4Q		-ψ-G-C-U-A-G-A-U-G————	C-A-A	-A-G-G	-T-ψ-C-G-A-G-U	-C-C-U-U-U	-A-U-U-C-C-C-A-G-C-C-A	4Q
GLUTAMIC ACID								
1E	1	-G-G-C-G-G-U-A-A————	C-A-G	-G-G-G	-T-ψ-C-G-A-A-U	-C-C-C-C-U	-G-G-G-G-A-C-G-C-C-A	1E
2E	2	-G-G-C-G-G-U-A-A————	C-A-G	-G-G-G	-T-ψ-C-G-A-A-U	-C-C-C-C-U	-A-G-G-G-G-A-C-G-C-C-A	2E
3E	1	-G-G-C-G-G-U-A-A————	C-A-G	-G-G-G	-T-ψ-C-G-A-A-U	-C-C-C-C-U	-G-G-G-G-A-C-G-C-C-A	3E
E4	3	-C-G-U-G-G-A-G-A————	C-$\overset{5}{C}$-G	-G-G-G	-T-ψ-C-G-A-C-U	-C-C-C-C-G	-U-A-U-C-G-G-A-G-C-C-A	E4
GLYCINE								
1G	1	-G-C-U-C-U-A-U-A————	C-G-A	-G-G-G	-T-ψ-C-G-A-U-U	-C-C-C-U-U	-C-G-C-C-C-G-C-U-C-C-A	1G
2G	1	-G-C-U-C-U-A-U-A————	C-G-A	-G-G-G	-T-ψ-C-G-A-U-U	-C-C-C-U-U	-C-G-C-C-C-G-C-U-C-C-A	2G
3G	2	-G-C-U-G-A-U-G-A————	U-G-C	-G-G-G	-T-ψ-C-G-A-U-U	-C-C-C-G-C	-U-G-C-C-C-G-C-U-C-C-A	3G

No.	Iso-acceptor	Organism	Aminoacyl Stem / D Stem / D Loop / D Stem / Anticodon Stem / Anticodon Loop (1 2 3 4 5 6 7 8 9 10 11 12 13 14 15 16 17:1 2 18 19 20:1 2 3 21 22 23 24 25 26 27 28 29 30 31 32 33 34 35 36 37 38)
4G	3	*E. coli*	pG-C-G-G-G-A-A-U-A-G-C-U-C-A-G-D-D——G-G-D——A-G-A-G-C-A-C-G-A-C-C-U-U-G-C-C-A-A
5G	1b	*Staph*	pG-C-G-G-G-A-G-U-A-U-U-U-C-A-A-C-U——U-U-D——A-G-A-A-U-A-C-G-U-U-C-C-U-U-C-C-C-G
6G	1a	*Staph*	pG-C-G-G-G-A-G-U-A-G-U-U-C-A-A-U——U-U-D——A-G-A-A-C-A-C-A-U-U-C-C-U-U-C-C-C-G
7G		Phage	pG-C-G-G-A-U-A-U-C-G-U-A-U-A-A-U——G-G-D——A-U-U-A-C-C-U-C-A-G-A-C-U-U-C-C-A-A
G8	1	*S. cerevisiae*	pG-C-G-C-A-A-G-U-G-G-U-U-ψ-A-G-D——G-G-D——A-A-A-A-U-C-C-A-A-C-G-ψ-U-G-C-C-A-ψ
G9	1	Wheat germ	pG-C-A-C-C-A-G-U-G-G-U-C-ψ-A-G-D——G-G-U——A-G-A-A-U-A-G-U-A-C-C-C-U-G-C-C-A-C
G10	1	*Bombyx mori*	pG-C-A-U-C-G-G-U-G-G-U-U-C-A-G-U——G-G-D——A-G-A-A-U-G-C-U-C-G-C-C-U-G-C-C-A-C
G11	2	*Bombyx mori*	pG-C-G-U-U-G-G-U-G-G-U-G-ψ-A-A-D——G-G-D-C——A-G-C-A-U-A-G-ψ-U-G-C-C-U-U-C-C-A-A
G12		Human placenta (GCC)	pG-C-A-U-U-G-G-U-G-G-U-U-C-A-G-U——G-G-D——A-G-A-A-U-U-C-U-C-G-C-C-U-G-C-C-A-C
G13		Human placenta (CCC)	pG-C-G-C-C-G-C-U-G-G-U-G-ψ-A-G-U——G-G-D——A-U-C-A-U-G-C-A-A-G-A-U-U-C-C-C-A-U
HISTIDINE			
1H	1	Bacteria	pG-G-U-G-G-C-U-A-U-A-G-C-U-C-A-G-D-D——G-G-D——A-G-A-G-C-C-C-U-G-G-A-U-U-Q-U-G-A-ψ
ISOLEUCINE			
1I	1	*E. coli*	pA-G-G-C-U-U-G-U-A-G-C-U-C-A-G-G-U——G-G-D-D——A-G-A-G-C-G-C-A-C-C-C-C-U-G-A-U-A-A
I2		*T. utilis*	pG-G-U-C-C-C-U-U-G-G-C-C-C-A-G-D-D——G-G-D-D——A-A-G-G-C-G-ψ-G-G-U-G-C-U-I-A-U-A-A
LEUCINE			
1L	1	Bacteria	pG-C-G-A-A-G-G-U-G-G-C-G-G-A-A-D-D——G-G-D-A——G-A-C-G-C-G-C-U-A-G-C-U-U-C-A-G-G-ψ
2L	2	*E. coli K12*	pG-C-C-G-A-G-G-U-G-G-U-G-G-A-A-D-D——G-G-D-A——G-A-C-A-C-G-C-U-A-C-C-U-U-G-A-G-G-ψ
3L	5	*E. coli*	pG-C-C-C-G-G-A-U-G-G-U-G-G-A-A-D-C——G-G-D-A——G-A-C-A-C-A-A-G-G-G-A-ψ-U-N-A-A-A-A

No.	Iso-acceptor	Anticodon Stem / Variable Loop / TψC Stem / TψC Loop / TψC Stem / Aminoacyl Stem ($39^{40}41^{42}43^{44}45^{46}47{:}1\ 2\ 3\ 4\ 5\ 6\ 7\ 8\ 9^{10}11^{12}13^{14}15^{16}48^{49}50^{51}52^{53}54^{55}56^{57}58^{59}60^{61}62^{63}64^{65}66^{67}68^{69}70^{71}72^{73}74^{75}76$)	No.
4G	3	-G-G-U-C-G-G-G-G-U——C-G-C-G-A-G-T-ψ-C-G-A-G-U-C-U-C-G-U-U-U-C-C-C-G-C-U-C-C-A	4G
5G	1b	-G-A-A-C-G-A-G-A——U-A-U-A-G-G-U-G-C-A-A-A-U-C-C-U-A-U-C-U-U-C-C-G-C-U-C-C-A	5G
6G	1a	-G-A-A-U-G-A-G-G——U-A-U-A-G-G-U-G-C-A-A-G-U-C-C-U-A-U-C-U-U-C-C-G-C-U-C-C-A	6G
7G		-ψ-C-U-G-A-U-G-A——U-G-U-G-A-G-T-ψ-C-G-A-U-U-C-U-C-A-U-U-A-U-C-C-G-C-U-C-C-A	7G
G8	1	-C-G-U-U-G-G-G——C-C-C-C-G-G-T-ψ-C-G-A-U-U-C-C-G-G-G-C-U-U-G-C-G-C-A-C-C-A	G8
G9	1	-G-G-U-A-C-A-G-A——C-C-C-G-G-G-U-ψ-C-G-A-U-U-C-C-C-G-G-C-U-G-G-U-G-C-A-C-C-A	G9
G10	1	-G-C-G-G-G-C-G-G——C-C-C-G-G-G-T-ψ-C-G-A-U-U-C-C-C-G-G-C-C-G-A-U-G-C-A-C-C-A	G10
G11	2	-G-C-A-G-U-U-G-A——U-C-C-G-G-G-T-ψ-C-G-A-U-U-C-C-C-G-G-C-C-A-A-C-G-C-A-C-C-A	G11
G12		-G-C-G-G-G-A-G-G——C-C-C-G-G-G-T-ψ-C-G-A-U-U-C-C-C-G-G-C-C-A-A-U-G-C-A-C-C-A	G12
G13		-U-C-U-U-G-C-G-A——C-C-C-G-G-G-T-ψ-C-G-A-U-U-C-C-C-G-G-G-C-G-G-C-G-C-A-C-C-A	G13
HISTIDINE			
1H	1	-ψ-C-C-A-G-U-U-G-U——C-G-U-G-G-G-T-ψ-C-G-A-A-U-C-C-C-A-U-U-A-G-C-C-A-C-C-C-C-A	1H
ISOLEUCINE			
1I	1	-G-G-G-U-G-A-G-G-U——C-G-G-U-G-G-T-ψ-C-A-A-G-U-C-C-A-C-ψ-C-A-G-G-C-C-U-A-C-C-A	1I
I2		-C-G-C-C-A-A-G-A-D——C-A-G-C-A-G-T-ψ-C-G-A-U-C-C-U-G-C-U-A-G-G-A-C-C-A-C-C-A	I2
LEUCINE			
1L	1	-G-ψ-U-A-G-U-G-U-C-C-U-U-A-C-G-G-A-C-G——U-G-G-G-G-G-T-ψ-C-A-A-G-U-C-C-C-C-C-C-C-C-U-C-G-C-A-C-C-A	1L
2L	2	-G-G-U-A-G-U-G-C-C-C-A-A-U-A-G-G-G-C-U——U-A-C-G-G-G-T-ψ-C-A-A-G-U-C-C-C-G-U-C-C-U-C-G-G-U-A-C-C-A	2L
3L	5	-ψ-C-C-C-U-C-G-G-C-G-U-U-C-G-C-G-C-U-G——U-G-C-G-G-G-T-ψ-C-A-A-G-U-C-C-C-G-C-U-C-C-G-G-G-U-A-C-C-A	3L

No.	Iso-acceptor	Organism	Aminoacyl Stem — D Stem — D Loop — D Stem — Anticodon Stem — Anticodon Loop (1 2 3 4 5 6 7 8 9 10 11 12 13 14 15 16 17:1 2 18 19 20:1 3 21 22 23 24 25 26 27 28 29 30 31 32 33 34 35 36 37 38)
4L		Phage T4	pG-C-G-A-G-A-A-U-G-G-U-C-A-A-A-D-D—G-G-D-A—A-A-G-G-C-A-C-A-G-C-A-C-U-U-A-A-A-A
L5	3	Yeast (Baker's)	pG-G-U-U-G-U-U-U-G-G-C-C-G-A-G-C—G-G-D-C-D-A-A-G-G-C-G-C-C-U-G-A-ψ-U-C-A-A-G-C
L6	1	Yeast (Baker's)	pG-G-G-A-G-U-U-U-G-G-C-C-G-A-G-D—G-G-D-D-D-A-A-G-G-C-G-ψ-C-A-G-A-ψ-U-U-A-G-G-C
L7		*T. utilis*	pG-G-A-U-C-U-U-U-G-G-C-C-G-A-G-C—G-G-D-D-U-A-A-G-G-C-G-C-U-C-G-A-C-U-C-A-A-G-A
LYSINE			
1K		*E. coli B*	pG-G-G-U-C-G-U-U-A-G-C-U-C-A-G-D-D—G-G-D—A-G-A-G-C-A-G-U-U-G-A-C-U-U-U-U-A-A
2K		*B. subtilis*	pG-A-G-C-C-A-U-U-A-G-C-U-C-A-G-U-D—G-G-D—A-G-A-G-C-A-U-C-U-G-A-C-U-U-U-U-G-A
K3	1	Yeast (haploid)	pG-C-C-U-U-G-U-U-G-G-C-G-C-A-A-D-C—G-G-D—A-G-C-G-C-G-ψ-A-U-G-A-C-U-C-U-U-A-A
K4	2	Yeast	pψ-C-C-U-U-G-U-U-A-G-C-U-C-A-G-D-D—G-G-D—A-G-A-G-C-G-ψ-ψ-C-G-G-C-U-U-U-U-A-A
METHIONINE-INITIATOR			
1iM		*E. coli*	pC-G-C-G-G-G-G-U-G-G-A-G-C-A-G-C-C-U—G-G-D—A-G-C-U-C-G-U-C-G-G-G-C-U-C-A-U-A-A
2iM		*T. thermophilus*	pC-G-C-G-G-G-G-U-G-G-A-G-C-A-G-C-C-U—G-G-D—A-G-C-U-C-G-U-C-G-G-G-C-U-C-A-U-A-A
3iM		*B. subtilis*	pC-G-C-G-G-G-G-U-G-G-A-G-C-A-G-U-U-C—G-G-D—A-G-C-U-C-G-U-C-G-G-G-C-U-C-A-U-A-A
4iM		*Mycoplasma*	pC-G-C-G-G-G-G-U-A-G-A-G-C-A-G-U-U—G-G-D—A-G-C-U-C-G-C-C-G-G-G-C-U-C-A-U-A-A
iM5		*A. nidulans*	pC-G-C-G-G-G-G-U-A-G-A-G-C-A-G-C-C-U—G-G-D—A-G-C-U-C-G-U-C-G-G-G-C-U-C-A-U-A-A
iM6		*N. crassa* (mto)	pU-G-C-G-G-A-U-U-A-U-U-G-U-A-A-D—A-G-D—A-A-C-A-U-A-U-U-U-G-G-C-U-C-A-U-G-U
iM7		*N. crassa* (cyt)	pA-G-C-U-G-C-A-U-G-G-C-G-C-A-G-C—G-G-A—A-G-C-G-C-G-C-Y-G-G-G-C-U-C-A-U-A-A
iM8		Wheat germ	pA-U-C-A-G-A-G-U-G-G-C-G-C-A-G-C—G-G-A—A-G-C-G-U-G-G-ψ-G-G-G-C-C-C-A-U-A-A
iM9		Yeast	pA-G-C-C-G-C-G-U-G-G-C-G-C-A-G-D—G-G-A—A-G-C-G-C-G-C-A-G-G-G-C-U-C-A-U-A-A
iM10		Mammals	pA-G-C-A-G-A-G-U-G-G-C-G-C-A-G-C—G-G-A—A-G-C-G-U-G-C-U-G-G-G-C-C-C-A-U-A-A

No.	Iso-acceptor	Anticodon Stem / Variable Loop (39–48)	TψC Stem / TψC Loop / TψC Stem / Aminoacyl Stem (49–76)	No.
4L		-ψ-G-C-U-G-C-G-G-A-A-U-G-A-U-U-U-C-C-U—	U-G-U-G-G-G-T-ψ-C-G-A-G-U-C-C-C-A-C-U-U-C-U-C-G-C-A-C-C-A	4L
L5	3	-ψ-C-A-G-G-U-A-U-C-G-U-A-A-G-A-U-G—	C-A-A-G-A-G-T-ψ-C-G-A-A-U-C-U-C-U-U-A-G-C-A-A-C-C-A-C-C-A	L5
L6	1	-ψ-C-U-G-A-U-A-U-C-U-U-C-G-G-A-U-G—	C-A-A-G-G-G-T-ψ-C-G-A-A-U-C-C-C-U-U-A-G-C-U-C-U-C-A-C-C-A	L6
L7		-ψ-C-G-A-G-U-A-U-C-G-U-A-A-G-A-U-G—	C-A-U-G-A-G-T-ψ-C-G-A-A-U-C-U-C-A-U-A-G-G-A-U-C-C-A-C-C-A	L7
LYSINE				
1K		-ψ-C-A-A-U-U-G-G-U—	C-G-C-A-G-G-T-ψ-C-G-A-A-U-C-C-U-G-C-A-C-G-A-C-C-C-A-C-C-A	1K
2K		-ψ-C-A-G-A-G-G-G-U—	C-G-A-A-G-G-T-ψ-C-G-A-G-U-C-C-U-U-C-A-U-G-G-C-U-C-A-C-C-A	2K
K3	1	-ψ-C-A-U-A-A-G-G-U—	U-A-G-G-G-G-T-ψ-C-G-A-G-C-C-C-C-C-U-A-C-A-G-G-G-C-U-C-C-A	K3
K4	2	-C-C-G-A-A-A-U-G-U—	C-A-G-G-G-G-T-ψ-C-G-A-G-C-C-C-C-C-U-A-ψ-G-A-G-G-A-G-C-C-A	K4
METHIONINE-INITIATOR				
1iM		-C-C-C-G-A-A-G-G-U—	C-G-U-C-G-G-T-ψ-C-A-A-A-U-C-C-G-G-C-C-C-C-C-G-C-A-A-C-C-A	1iM
2iM		-C-C-C-G-A-A-G-G-U—	C-G-C-G-G-G-T-ψ-C-A-A-A-U-C-C-C-G-C-C-C-C-C-G-C-A-A-C-C-A	2iM
3iM		-C-C-C-G-A-A-G-G-U—	C-G-C-A-G-G-T-ψ-C-A-A-A-U-C-C-U-G-C-C-C-C-C-G-C-A-A-C-C-A	3iM
4iM		-C-C-C-G-G-A-G-G-C—	C-G-C-A-G-G-U-ψ-C-G-A-G-U-C-C-U-G-C-C-C-C-C-G-C-A-A-C-C-A	4iM
iM5		-C-C-C-G-A-A-G-G-U—	C-A-G-A-G-G-T-ψ-C-A-A-A-U-C-C-U-C-U-C-C-C-C-G-C-C-A-C-C-A	iM5
iM6		-C-C-G-A-A-U-G-A—	C-A-U-A-G-G-U-G-C-A-A-A-U-C-C-U-G-U-A-U-C-C-G-C-A-U-C-C-A	iM6
iM7		-C-C-C-G-G-A-G-G-U—	C-A-C-U-C-G-A-U-C-G-A-A-A-C-G-A-G-U-U-G-C-A-G-C-U-A-C-C-A	iM7
iM8		-C-C-C-A-C-A-G-G-D—	C-C-C-A-G-G-A-ψ-C-G-A-A-A-C-C-U-G-G-C-U-C-U-G-A-U-A-C-C-A	iM8
iM9		-C-C-C-U-G-A-U-G-D—	C-C-U-C-G-G-A-U-C-G-A-A-A-C-C-G-A-G-C-G-C-G-G-C-U-A-C-C-A	iM9
iM10		-C-C-C-A-G-A-G-G-D—	C-G-A-U-G-G-A-U-C-G-A-A-A-C-C-A-U-C-C-U-C-U-G-C-U-A-C-C-A	iM10

No.	Iso-acceptor	Organism	Aminoacyl Stem / D Stem / D Loop / D Stem / Anticodon Stem / Anticodon Loop (1 2 3 4 5 6 7 8 9 10 11 12 13 14 15 16 17:1 2 18 19 20:1 2 3 21 22 23 24 25 26 27 28 29 30 31 32 33 34 35 36 37 38)
iM11		*S. obliquus*	pA-G-C-U-G-A-G-U-G-G-C-G-C-A-G-U———G-G-A———A-G-C-G-U-G-A-U-G-G-G-C-U-C-A-U-A-A
METHIONINE			
1M	1	*E. coli* CA265	pG-G-C-U-A-C-G-U-A-G-C-U-C-A-G-U-D———G-G-D-D———A-G-A-G-C-A-C-A-U-C-A-C-U-C-A-U-A-A
2M	2	*E. coli* CA265	pG-G-C-U-A-C-G-U-A-G-C-U-C-A-G-D-D———G-G-D-D———A-G-A-G-C-A-C-A-U-C-A-C-U-C-A-U-A-A
M3		Yeast (Baker's)	pG-C-U-U-C-A-G-U-A-G-C-U-C-A-G-D-A———G-G-A———A-G-A-G-C-G-U-C-A-G-ψ-C-U-C-A-U-A-A
M4		Mammals	pG-C-C-U-C-G-U-U-A-G-C-G-C-A-G-D-A———G-G-D———A-G-C-G-C-G-ψ-C-A-G-ψ-C-U-C-A-U-A-A
PHENYLALANINE			
1F		*E. coli*	pG-C-C-C-G-G-A-U-A-G-C-U-C-A-G-D-C———G-G-D———A-G-A-G-C-A-G-G-G-G-A-ψ-U-G-A-A-A-A
2F		*B. stearothermophilus*	pG-G-C-U-C-G-G-U-A-G-C-U-C-A-G-U-C———G-G-D———A-G-A-G-C-A-A-A-G-G-A-C-U-G-A-A-A-A
3F		*B. subtilis*	pG-G-C-U-C-G-G-U-A-G-C-U-C-A-G-U-D———G-G-D———A-G-A-G-C-A-A-C-G-G-A-C-U-G-A-A-A-A
4F		*Mycoplasma*	pG-G-U-C-G-U-G-U-A-G-C-U-C-A-G-U-C———G-G-D———A-G-A-G-C-A-G-C-A-G-A-C-U-G-A-A-G-C
F5		Bean (cpt)	pG-U-C-G-G-G-A-U-A-G-C-U-C-A-G-U-D———G-G-D———A-G-A-G-C-A-G-A-G-G-A-C-U-G-A-A-A-A
F6		*Euglena grac.* (cpt)	pG-C-U-G-G-G-A-U-A-G-C-U-C-A-G-D-U———G-G-U———A-G-A-G-C-G-G-A-G-G-A-C-U-G-A-A-A-A
F7		*Euglena grac.* (cyt)	pG-C-C-G-A-C-U-U-A-G-C-U-C-A-G-D-D———G-G-G———A-G-A-G-C-G-ψ-ψ-A-G-A-C-U-G-A-A-W-A
F8		Blue green algae	pG-C-C-A-G-G-A-U-A-G-C-U-C-A-G-U-U———G-G-D———A-G-A-G-C-A-G-A-G-G-A-C-U-G-A-A-N-A
F9		Yeast	pG-C-G-G-A-U-U-U-A-G-C-U-C-A-G-D-D———G-G-G———A-G-A-G-C-G-C-C-A-G-A-C-U-G-A-A-W-A
F10		*S. cerevisiae* (mto)	pG-C-U-U-U-U-A-U-A-G-C-U-U-A-G-D———G-G-D———A-A-A-G-C-G-A-U-A-A-A-ψ-U-G-A-A-G-A
F11		*S. pombe*	pG-U-C-G-C-A-A-U-G-G-U-G-ψ-A-G-D-D———G-G-G———A-G-C-A-ψ-G-A-C-A-G-A-C-U-G-A-A-W-A
F12		Wheat germ	pG-C-G-G-G-G-A-U-A-G-C-U-C-A-G-D-D———G-G-G———A-G-A-G-C-G-ψ-C-A-G-A-C-U-G-A-A-W-A
F13		*L. luteus*	pG-C-G-G-G-G-A-U-A-G-C-U-C-A-G-D-D———G-G-G———A-G-A-G-C-G-ψ-C-A-G-A-C-U-G-A-A-W-A

No.	Isoacceptor	Anticodon Stem (39–43)	44–47	Variable Loop (1–16)	TψC Stem (48–53)	TψC Loop (54–60)	TψC Stem (61–65)	Aminoacyl Stem (66–72)	73–76	No.
iM11		-C-C-C-A-U-	A-G-G-D	—	C-A-C-A-G-G-	A-U-C-G-A-A-A-	C-C-U-G-U-	C-U-C-A-G-C-U-	A-C-C-A	iM11
METHIONINE										
1M	1	-ψ-G-A-U-G-	G-G-G-U	—	C-A-C-A-G-G-	T-ψ-C-G-A-A-U-	C-C-C-G-U-	C-G-U-A-G-C-C-	A-C-C-A	1M
2M	2	-ψ-G-A-U-G-	G-G-G-U	—	C-A-C-A-G-G-	T-ψ-C-G-A-A-U-	C-C-C-G-U-	C-G-U-A-G-C-C-	A-C-C-A	2M
M3		-ψ-C-U-G-A-	A-G-G-U	—	C-G-A-G-A-G-	T-ψ-C-G-A-A-C-	C-U-C-U-C-	C-U-G-G-A-G-C-	A-C-C-A	M3
M4		-ψ-C-U-G-A-	A-G-G-D	—	C-G-U-G-A-G-	T-ψ-C-G-A-U-C-	C-U-C-A-C-	A-C-G-G-G-G-C-	A-C-C-A	M4
PHENYLALANINE										
1F		-ψ-C-C-C-C-	G-U-G-U	—	C-C-U-U-G-G-	T-ψ-C-G-A-U-U-	C-C-G-A-G-	U-C-C-G-G-G-C-	A-C-C-A	1F
2F		-ψ-C-C-U-U-	G-U-G-U	—	C-G-G-C-G-G-	T-ψ-C-G-A-U-U-	C-C-G-U-C-	C-C-G-A-G-C-C-	A-C-C-A	2F
3F		-ψ-C-C-G-U-	G-U-G-U	—	C-G-G-C-G-G-	T-ψ-C-G-A-U-U-	C-C-G-U-C-	C-C-G-A-G-C-C-	A-C-C-A	3F
4F		-ψ-C-U-G-C-	G-U-G-U	—	C-G-G-C-G-G-	U-ψ-C-A-A-U-U-	C-C-G-U-C-	C-A-C-G-A-C-C-	A-C-C-A	4F
F5		-ψ-C-C-U-C-	C-U-G-U	—	C-A-C-C-A-G-	T-ψ-C-A-A-A-U-	C-U-G-G-U-	U-C-C-U-G-G-C-	A-C-C-A	F5
F6		-ψ-C-C-U-U-	G-U-G-U	—	C-A-C-C-A-G-	T-ψ-C-A-A-A-U-	C-U-G-G-U-	U-C-C-U-A-G-C-	A-C-C-A	F6
F7		-ψ-C-U-A-A-	A-G-G-U	—	C-C-C-U-G-G-	T-ψ-C-G-A-U-C-	C-C-G-G-G-	A-G-ψ-C-G-G-C-	A-C-C-A	F7
F8		-U-C-C-U-C-	G-U-G-U	—	C-G-G-C-G-G-	T-ψ-C-A-A-U-U-	C-C-G-C-C-	U-C-C-C-G-G-C-	A-C-C-A	F8
F9		-ψ-C-U-G-G-	A-G-G-U	—	C-C-U-G-U-G-	T-ψ-C-G-A-U-C-	C-A-C-A-G-	A-A-U-U-C-G-C-	A-C-C-A	F9
F10		-ψ-U-U-A-U-	U-U-A-C	—	A-U-G-U-A-G-	U-ψ-C-G-A-U-U-	C-U-C-A-U-	U-A-A-G-G-G-C-	A-C-C-A	F10
F11		-ψ-C-U-G-U-	U-G-G-U	—	C-A-U-C-G-G-	T-ψ-C-G-A-U-C-	C-C-G-G-U-	U-U-G-U-G-A-C-	A-C-C-A	F11
F12		-ψ-C-U-G-A-	A-G-G-D	—	C-G-C-G-U-G-	T-ψ-C-G-A-U-C-	C-A-C-G-C-	U-C-A-C-C-G-C-	A-C-C-A	F12
F13		-ψ-C-U-G-A-	A-G-G-D	—	C-A-C-G-U-G-	T-ψ-C-G-A-U-C-	C-A-C-G-U-	U-C-A-C-C-G-C-	A-C-C-A	F13

No.	Iso-acceptor	Organism	Aminoacyl Stem 1 2 3 4 5 6 7 8 9 · D Stem 10 11 12 13 · D Loop 14 15 16 17:1 2 18 19 20:1 2 3 21 · D Stem 22 23 24 25 26 · Anticodon Stem 27 28 29 30 31 · Anticodon Loop 32 33 34 35 36 37 38
F14		Mammals	pG-C-C-G-A-A-A-U-A-G-C-U-C-A-G-D-D——G-G-G——A-G-A-G-C-G-ψ-ψ-A-G-A-C-U-G-A-A-W-A
PROLINE			
1P		Phage T4	pC-U-C-C-G-U-G-U-A-G-C-U-C-A-G-U-U-U——G-G-D——A-G-A-G-C-G-C-C-U-G-A-U-U-U-G-G-G-A
SERINE			
1S	1	*E. coli B*	pG-G-A-A-G-U-G-U-G-G-C-C-G-A-G-C——G-G-D-D-G—A-A-G-G-C-A-C-C-G-G-U-C-U-U-G-A-A-A
2S	3	*E. coli B*	pG-G-U-G-A-G-G-U-G-G-C-C-G-A-G-A——G-G-C-D-G—A-A-G-G-C-G-C-U-C-C-C-C-U-G-C-U-A-A
3S		Phage T4	pG-G-A-G-G-C-G-U-G-G-C-A-G-A-G-U——G-G-D-D-U—A-A-U-G-C-A-C-C-G-G-U-C-U-N-G-A-A-A
4S		Phage T4 (psu$^+$1 am)	pG-G-A-G-G-C-G-U-G-G-C-A-G-A-G-U——G-G-D-D-U—A-A-U-G-C-A-C-C-G-G-U-C-U-C-U-A-A-A
S5	1	Yeast (Brewer's)	pG-G-C-A-A-C-U-U-G-G-C-C-C-G-A-G-D——G-G-D-D——A-A-G-G-C-G-A-A-A-G-A-ψ-U-I-G-A-A-A
S6	2	Yeast (Brewer's)	pG-G-C-A-A-C-U-U-G-G-C-C-C-G-A-G-D——G-G-D-D——A-A-G-G-C-G-A-A-A-G-A-ψ-U-I-G-A-A-A
S7		Yeast (UCG) major	pG-G-C-U-A-C-A-U-G-G-C-C-C-G-A-G-D——G-G-D-D——A-A-G-G-C-G-A-C-A-G-A-N-U-C-G-A-A-A
S8		Yeast (UCG) minor	pG-G-C-U-A-C-A-U-G-G-C-C-C-G-A-G-D——G-G-D-D——A-A-G-G-C-G-A-G-A-G-A-N-U-C-G-A-A-A
S9	1	Rat liver	pG-U-A-G-U-C-G-U-G-G-C-C-C-G-A-G-D——G-G-D-D——A-A-G-G-C-G-A-ψ-G-G-A-C-U-I-G-A-A-A
S10	3	Rat liver	pG-A-C-G-A-G-G-U-G-G-C-C-C-G-A-G-D——G-G-D-D——A-A-G-G-C-G-A-ψ-G-G-A-C-U-G-C-U-A-A
THREONINE			
1T		*E. coli*	pG-C-U-G-A-U-A-U-A-G-C-U-C-A-G-D-D——G-G-D——A-G-A-G-C-G-C-A-C-C-C-U-U-G-G-U-A-A
2T		*B. subtilis*	pG-C-C-G-C-U-G-U-A-G-C-U-C-A-A-U-D——G-G-U——A-G-A-G-C-A-A-C-U-G-A-C-U-U-G-U-A-A
3T		Phage T4	pG-C-U-G-A-U-U-U-A-G-C-U-C-A-G-D-A——G-G-D——A-G-A-G-C-A-C-C-U-C-A-C-U-U-G-U-A-A
T4	1a,1b	Yeast	pG-C-U-U-C-U-A-U-G-G-C-C-A-A-G-D-D——G-G-D——A-A-G-G-C-G-C-C-A-C-A-C-U-I-G-U-A-A

No.	Iso-acceptor	Anticodon Stem — Variable Loop — TψC Stem — TψC Loop — TψC Stem — Aminoacyl Stem 39 40 41 42 43 44 45 46 47:1 2 3 4 5 6 7 8 9 10 11 12 13 14 15 16 48 49 50 51 52 53 54 55 56 57 58 59 60 61 62 63 64 65 66 67 68 69 70 71 72 73 74 75 76	No.
F14		-ψ-C-U-A-A-A-G-G(+)-D——————C-C(5)-U-G-G-T-ψ-C-G-A(m)-U-C-C-C-G-G-G-U-U-U-C-G-G-C-A-C-C-A	F14
PROLINE			
1P		-ψ-C-A-G-G-A-G-G(+)-U——————C-C-A-A-G-G-T-ψ-C-A-A-A-U-C-C-U-U-G-U-A-U-G-G-A-G-A-C-C-A	1P
SERINE			
1S	1	-A-C-C-G-G-C-G-A-C-C-C-G-A-A-A-G-G-G-U-U——————C-C-A-G-A-G-T-ψ-C-G-A-A-U-C-U-C-U-G-C-G-C-U-U-C-C-G-C-C-A	1S
2S	3	-G-G-G-A-G-U-A-U-G-C-G-G-U-C-A-A-A-A-G-C-U-G-C-A-U-C-C-G-G-G-G-T-ψ-C-G-A-A-U-C-C-C-C-G-C-C-U-C-A-C-C-G-C-C-A	2S
3S		-A-C-C-G-G-C-A-G-U-C-G-C-U-C-C-G-G-C-G-A-C-U——————C-A-U-A-G-G-T-ψ-C-A-A-A-U-C-C-U-A-U-C-G-C-C-U-C-C-G-C-C-A	3S
4S		-A-C-C-G-G-C-A-G-U-C-G-C-U-C-C-G-G-C-G-A-C-U——————C-A-U-A-G-G-T-ψ-C-A-A-A-U-C-C-U-A-U-C-G-C-C-U-C-C-G-C-C-A	4S
S5	1	-ψ-C-U-U-U-U(m)-G-G-G-C-U-C-U-G-C-C-C-G——————C(5)-G-C-A-G-G-T-ψ-C-A-A-A-U-C-C-U-G-C-A-G-U-U-G-U-C-G-C-C-A	S5
S6	2	-ψ-C-U-U-U-U(m)-G-G-G-C-U-U-U-G-C-C-C-G——————C(5)-G-C-A-G-G-T-ψ-C-G-A-G-U-C-C-U-G-C-A-G-U-U-G-U-C-G-C-C-A	S6
S7		-ψ-C-U-G-U-U(m)-G-G-G-C-U-C-U-G-C-C-C-G——————C(5)-G-C-U-G-G-T-ψ-C-A-A-A-U-C-C-U-G-C-U-G-U-G-G-U-C-G-C-C-A	S7
S8		-ψ-C-U-C-U-U(*)-G-G-G-C-U-C-U-G-C-C-C-G——————C(5)-G-C-U-G-G-T-ψ-C-A-A-A-U-C-C-U-G-C-U-G-U-G-G-U-C-G-C-C-A	S8
S9	1	-ψ(m)-C-C-A-U-U(m)-G-G-G-G-U-C(3)-U-C-C-C-C-G——————C(5)-G-C-A-G-G-T-ψ-C-G-A(m)-A-U-C-C-U-G-C-C-G-A-C-U-A-C-G-C-C-A	S9
S10	3	-ψ-C-C-A-U-U(m)-G-ψ-G-C-U-C(3)-U-G-C-A-C-G——————C(5)-G-U-G-G-G-T-ψ-C-G-A(m)-A-U-C-C-C-A-U-C-C-U-C-G-U-C-G-C-C-A	S10
THREONINE			
1T		-G-G-G-U-G-A-G-G(+)-U——————C-G-G-C-A-G-T-ψ-C-G-A-A-U-C-U-G-C-C-U-A-U-C-A-G-C-A-C-C-A	1T
2T		-ψ-C-A-G-U-A-G-G(+)-U——————U-G-G-G-G-G-T-ψ-C-A-A-G-U-C-C-U-C-U-U-G-C-C-G-G-C-A-C-C-A	2T
3T		-ψ-G-A-G-G-A-U-G(+)-U——————C-G-G-C-G-G-T-ψ-C-G-A-U-U-C-C-G-U-C-A-A-U-C-A-G-C-A-C-C-A	3T
T4	1a b	-ψ-G-U-G-G-A-G-A-D——————C(*)-A(*)-U-C-G-G-T-ψ-C-A-A(m)-A-U-C-C-G-A-U(*)-U-G-G-A-A-G-C-A-C-C-A	T4

No.	Iso-acceptor	Organism	Aminoacyl Stem / D Stem / D Loop / D Stem / Anticodon Stem / Anticodon Loop (1 2 3 4 5 6 7 8 9 10 11 12 13 14 15 16 17:1 2 18 19 20:1 2 3 21 22 23 24 25 26 27 28 29 30 31 32 33 34 35 36 37 38)
TRYPTOPHAN			
1W		E. *coli* CA244	pA-G-G-G-G-C-G-U-A-G-U-U-C-A-A-D-D—G-G-D—A-G-A-G-C-A-C-C-G-G-U-C-U-C-C-A-A-A
2W		E. *coli* (psu⁺ UGA)	pA-G-G-G-G-C-G-U-A-G-U-U-C-A-A-D-D—G-G-D—A-G-A-A-C-A-C-C-G-G-U-C-U-C-C-A-A-A
3W		E. *coli* (psu⁺ 7am)	pA-G-G-G-G-C-G-U-A-G-U-U-C-A-A-D-D—G-G-D—A-G-A-G-C-A-C-C-G-G-U-C-U-C-U-A-A-A
4W		E. *coli* (psu⁺ 7oc)	pA-G-G-G-G-C-G-U-A-G-U-U-C-A-A-D-D—G-G-D—A-G-A-G-C-A-C-C-G-G-U-C-U-U-U-A-A-A
W5		Yeast (Brewer's)	pG-A-A-G-C-G-G-U-G-G-C-U-C-A-A-D—G-G-D—A-G-A-G-C-ψ-ψ-ψ-C-G-A-C-U-C-C-A-A-A
W6		Rous Sarcoma virus (primer)	pG-A-C-C-U-C-G-U-G-G-C-G-C-A-A-C—G-G-D—A-G-C-G-C-G-ψ-C-U-G-A-C-U-C-C-A-G-A
W7		Bovine liver	pG-A-C-C-U-C-G-U-G-G-C-G-C-A-A-D—G-G-D—A-G-C-G-C-G-ψ-C-U-G-A-C-U-C-C-A-G-A
TYROSINE			
1Y		B. *stearothermophilus*	pG-G-A-G-G-G-G-U-A-G-C-G-A-A-G-U—G-G-C-U-A—A-A-C-G-C-G-G-C-G-G-A-C-U-Q-U-A-A-A
2Y		E. *coli*	pG-G-U-G-G-G-G-U-U-C-C-C-G-A-G-C—G-G-C-C-A—A-A-G-G-G-A-G-C-A-G-A-C-U-Q-U-A-A-A
Y3		Yeast (Baker's)	pC-U-C-U-C-G-G-U-A-G-C-C-A-A-G-D-D—G-G-D-D-D—A-A-G-G-C-G-C-A-A-G-A-C-U-G-ψ-A-A-A
Y4		T. *utilis*	pC-U-C-U-C-G-G-U-G-G-C-C-A-A-G-D-D—G-G-D-D-D—A-A-G-G-C-G-ψ-C-A-G-A-C-U-G-ψ-A-A-A
Y5		N. *crassa* (mto)	pA-G-G-A-G-G-G-U-U-C-C-G-U-U-U-G-U-U—G-G-A-U—G-A-C-G-G-G-ψ-U-A-A-G-C-U-G-U-A-A-A
VALINE			
1V	1	E. *coli*	pG-G-G-U-G-A-U-U-A-G-C-U-C-A-G-C-D—G-G-G—A-G-A-G-C-A-C-C-U-C-C-C-U-U-A-C-A-A
2V	2a	E. *coli*	pG-C-G-U-C-C-G-U-A-G-C-U-C-A-G-D-D—G-G-D-D—A-G-A-G-C-A-C-C-A-C-C-U-U-G-A-C-A-U
3V	2b	E. *coli*	pG-C-G-U-U-C-A-U-A-G-C-U-C-A-G-D-D—G-G-D-D—A-G-A-G-C-A-C-C-A-C-C-U-U-G-A-C-A-U
4V		B. *stearothermophilus*	pG-A-U-U-C-C-G-U-A-G-C-U-C-A-G-C-D—G-G-G—A-G-A-G-C-G-C-C-A-C-C-U-U-G-A-C-A-G

No.	Iso-acceptor	Anticodon Stem / Variable Loop / TψC Stem / TψC Loop / TψC Stem / Aminoacyl Stem 39 40 41 42 43 44 45 46 47:1 2 3 4 5 6 7 8 9 10 11 12 13 14 15 16 48 49 50 51 52 53 54 55 56 57 58 59 60 61 62 63 64 65 66 67 68 69 70 71 72 73 74 75 76	No.
TRYPTOPHAN			
1W		-A-C-C-G-G-G-U-$\overset{+}{G}$-U————U-G-G-A-G-T-ψ-C-G-A-G-U-C-U-C-U-C-C-G-C-C-C-C-U-G-C-C-A	1W
2W		--A-C-C-G-G-G-U-$\overset{+}{G}$-U————U-G-G-A-G-T-ψ-C-G-A-G-U-C-U-C-U-C-C-G-C-C-C-C-U-G-C-C-A	2W
3W		-A-C-C-G-G-G-U-$\overset{+}{G}$-U————U-G-G-A-G-T-ψ-C-G-A-G-U-C-U-C-U-C-C-G-C-C-C-C-U-G-C-C-A	3W
4W		-A-C-C-G-G-G-U-$\overset{+}{G}$-U————U-G-G-A-G-T-ψ-C-G-A-G-U-C-U-C-U-C-C-G-C-C-C-C-U-G-C-C-A	4W
W5		-ψ-C-G-A-A-G-G-$\overset{+}{G}$-D————U-G-C-A-G-T-ψ-C-A-$\overset{m}{A}$-U-U-C-C-U-G-$\overset{*}{\psi}$-C-C-G-U-U-U-C-A-C-C-A	W5
W6		-$\underset{m}{\psi}$-C-A-G-A-A-G-$\overset{+}{G}$-C————U-G-C-G-U-G-ψ-ψ-C-G-$\overset{m}{A}$-A-U-C-A-C-G-U-C-G-G-G-U-C-A-C-C-A	W6
W7		-$\underset{m}{\psi}$-C-A-G-A-A-G-$\overset{*}{G}$-$\overset{*}{D}$————U-G-C-G-U-G-ψ-ψ-C-$\overset{*}{G}$-$\overset{m}{A}$-A-U-C-A-C-G-U-C-G-G-G-U-C-A-C-C-A	W7
TYROSINE			
1Y		-ψ-C-C-G-C-U-C-C-C-U-U-U-G-G-G-U-U————C-G-G-C-G-G-T-ψ-C-G-A-A-U-C-C-G-U-C-C-C-C-C-U-C-C-A-C-C-A	1Y
2Y		-ψ-C-U-G-C-C-G-U-C-A-$\overset{*}{U}$-$\overset{*}{C}$-G-A-C-U-U————C-G-A-A-G-G-T-ψ-C-G-A-A-U-C-C-U-U-C-C-C-C-C-A-C-C-A-C-C-A	2Y
Y3		-ψ-C-U-U-G-A-G-A-D————$\overset{5}{C}$-G-G-G-C-G-T-ψ-C-G-$\overset{m}{A}$-C-U-C-G-C-C-C-C-C-G-G-A-G-A-C-C-A	Y3
Y4		-ψ-C-U-G-A-A-C-A-D————$\overset{5}{C}$-G-G-G-C-G-T-ψ-C-G-$\overset{m}{A}$-A-U-C-G-C-C-C-C-C-G-A-G-A-G-A-C-C-A	Y4
Y5		-C-U-U-A-A-U-G-A-C-U-A-A-U-U-A-G-U-$\overset{\Delta}{C}$-G-U————$\overset{\Delta}{C}$-G-A-A-G-G-T-ψ-C-A-A-U-U-C-C-U-U-U-C-U-C-U-C-C-U-A-C-C-A	Y5
VALINE			
1V	1	-G-G-A-G-G-G-G-$\overset{+}{G}$-U————C-G-G-C-G-G-T-ψ-C-G-A-U-C-C-C-G-U-C-A-U-C-A-C-C-C-A-C-C-A	1V
2V	2a	-G-G-U-G-G-G-G-$\overset{+}{G}$-$\overset{\pi}{U}$————C-G-G-U-G-G-T-ψ-C-G-A-G-U-C-C-A-C-U-C-G-G-A-C-G-C-A-C-C-A	2V
3V	2b	-G-G-U-G-G-G-G-$\overset{+}{G}$-$\overset{\pi}{U}$————C-G-U-U-G-G-T-ψ-C-G-A-G-U-C-C-A-A-U-U-G-A-A-C-G-C-A-C-C-A	3V
4V		-G-G-U-G-G-A-G-$\overset{+}{G}$-U————C-G-C-U-G-G-T-ψ-C-G-A-G-C-C-C-A-G-U-C-G-G-A-A-U-C-A-C-C-A	4V

No.	Iso-acceptor	Organism	Aminoacyl Stem 1–7	D Stem 8–13	D Loop 14–21	D Stem 22–25	26	Anticodon Stem 27–31	Anticodon Loop 32–38
5V	1	*S. typhimurium LT2*	pG-G-G-U-G-A-U-	U-A-G-C-U-C-	A-G-C-D——G-G-G——	A-G-A-G-C-	A-	C-C-U-C-C-	C-U-U-A-C-A-A
V6	1	Yeast (Baker's)	pG-G-U-U-U-C-G-	U-G-G-U-C-ψ-	A-G-D-C——G-G-D-D——	A-U-G-G-C-	A-	ψ-C-U-G-C-	ψ-U-I-A-C-A-C
V7	2a	Yeast (Baker's)	pG-G-U-C-C-A-A-	U-G-G-U-C-C-	A-G-D——G-G-D-D-C——	A-A-G-A-C-	G-	ψ-C-G-C-C-	ψ-U-U-A-C-A-C
V8	2b	Yeast (Baker's)	pG-U-U-C-C-A-A-	U-A-G-U-G-ψ-	A-G-C——G-G-C-D——	A-U-C-A-C-	G-	ψ-ψ-G-C-C-	ψ-U-C-A-C-A-C
V9	2	Yeast (Brewer's)	pG-G-U-U-U-C-G-	U-G-G-U-C-ψ-	A-G-D-C——G-G-D-D——	A-U-G-G-C-	A-	ψ-C-U-G-C-	ψ-U-I-A-C-A-C
V10		*T. utilis*	pG-G-U-U-U-C-G-	U-G-G-U-C-ψ-	A-G-D-D——G-G-D-C——	A-U-G-G-C-	A-	ψ-C-U-G-C-	ψ-U-I-A-C-A-C
V11		Mammals	pG-U-U-U-C-C-G-	U-A-G-U-G-ψ-	A-G-D——G-G-D-D——	A-U-C-A-C-	G-	ψ-U-C-G-C-	C-U-I-A-C-A-C
V12	1	Human placenta	pG-U-U-U-C-C-G-	U-A-G-U-G-ψ-	A-G-D——G-G-D-D——	A-U-C-A-C-	G-	ψ-U-C-G-C-	C-U-C-A-C-A-C

No.	Iso-acceptor	Anticodon Stem 39–43	Variable Loop 44–48	TψC Stem 49–53	TψC Loop 54–60	TψC Stem 61–65	Aminoacyl Stem 66–72	73–76	No.
5V	1	-G-G-A-G-G-	G-G-G-U———C-	G-G-C-G-G-	T-ψ-C-G-A-U-C-	C-C-G-U-C-	A-U-C-A-C-C-C-	A-C-C-A	5V
V6	1	-G-C-A-G-A-	A-C-G-D———C-	C-C-C-A-G-	T-ψ-C-G-A-U-C-	C-U-G-G-G-	C-G-A-A-A-U-C-	A-C-C-A	V6
V7	2a	-G-G-C-G-A-	A-G-A-D———C-	C-C-G-A-G-	T-ψ-C-G-A-A-C-	C-U-C-G-G-	U-U-G-G-A-U-C-	A-C-C-A	V7
V8	2b	-G-G-C-A-A-	A-G-G-D———C-	C-C-G-A-G-	T-ψ-C-G-A-U-C-	C-U-C-G-G-	U-U-G-G-A-U-C-	A-C-C-A	V8
V9	2	-G-C-A-G-A-	A-C-G-D———C-	C-C-C-A-G-	T-ψ-C-G-A-U-C-	C-U-G-G-G-	C-G-A-A-A-U-C-	A-C-C-A	V9
V10		-G-C-A-G-A-	A-C———C-	C-C-C-A-G-	T-ψ-C-G-A-U-C-	C-U-G-G-G-	C-G-A-A-A-U-C-	A-C-C-A	V10
V11		-G-C-G-A-A-	A-G-G-D———C-	C-C-C-G-G-	U-ψ-C-G-A-A-A-	C-C-G-G-G-	C-G-G-A-A-A-C-	A-C-C-A	V11
V12	1	-G-C-G-A-A-	A-G-G-D———C-	C-C-C-G-G-	U-ψ-C-G-A-A-A-	C-C-G-G-G-	C-G-G-A-A-A-C-	A-C-C-A	V12

Subject Index

Contents of Previous Volumes